Mammalian Cell Membranes VOLUME FOUR

Mammalian Cell Membranes

VOLUME FOUR

Membranes and Cellular Functions

Edited by

G. A. Jamieson Ph.D., D.Sc.
Research Director
American Red Cross Blood Research Laboratory
Bethesda, Maryland, USA
and
Adjunct Professor of Biochemistry
Georgetown University Schools of Medicine and Dentistry
Washington, DC, USA

and

D. M. Robinson Ph.D.
Professor of Biology, Georgetown University
and
Member, Vincent T. Lombardi Cancer Research Center
Georgetown University Schools of Medicine and Dentistry
Washington, DC, USA

BUTTERWORTHS
LONDON · BOSTON
Sydney · Wellington · Durban · Toronto

THE BUTTERWORTH GROUP

UK
Butterworth & Co (Publishers) Ltd
London: 88 Kingsway, WC2B 6AB

AUSTRALIA
Butterworths Pty Ltd
Sydney: 586 Pacific Highway, Chatswood, NSW 2067
Also at Melbourne, Brisbane, Adelaide and Perth

SOUTH AFRICA
Butterworth & Co (South Africa) (Pty) Ltd
Durban: 152–154 Gale Street

NEW ZEALAND
Butterworths of New Zealand Ltd
Wellington: 26–28 Waring Taylor Street, 1

CANADA
Butterworth & Co (Canada) Ltd
Toronto: 2265 Midland Avenue,
Scarborough, Ontario, M1P 4S1

USA
Butterworths (Publishers) Inc
Boston: 19 Cummings Park,
Woburn, Mass. 01801

First published 1977

© Butterworth & Co (Publishers) Ltd 1977

ISBN 0 408 70774 7

Library of Congress Cataloging in Publication Data (Revised)
Main entry under title:

Mammalian cell membranes.

  Includes bibliographical references and index.
    CONTENTS: v. 1. General concepts.   v. 2. The
diversity of membranes.   v. 3. Surface membranes of
specific cell types.   v. 4. Membranes and cellular functions.
v. 5. Responses of plasma membranes.
    1. Mammals—Cytology.   2. Cell membranes.
I. Jamieson, Graham A., 1929–   II. Robinson,
David Mason, 1932–   [DNLM: 1. Cell membrane.
2. Mammals. QH601 M265]
QL739.15.M35   599'.08'75   75-33317
ISBN 0-408-70774-7

Printed in Great Britain by
Butler & Tanner Ltd, Frome and London

Contents

Contributors

GILBERT ASHWELL
Laboratory of Biochemistry and Metabolism, National Institute of Arthritis, Metabolism and Digestive Diseases, National Institutes of Health, Bethesda, Maryland 20014, USA

S. L. BONTING
Department of Biochemistry, University of Nijmegen, Nijmegen, The Netherlands

DAVID O. CARPENTER
Neurobiology Department, Armed Forces Radiobiology Research Institute, Bethesda, Maryland 20014, USA

DAVID W. DEAMER
Department of Zoology, University of California, Davis, California 95616, USA

J. J. H. H. M. DE PONT
Department of Biochemistry, University of Nijmegen, Nijmegen, The Netherlands

J. G. EDWARDS
Department of Cell Biology, University of Glasgow, Glasgow G12 8QQ, Scotland

JOSEFINA EUGUI
Department of Biochemistry, University of Navarra, Pamplona, Spain

MILTON KERN
National Institute of Arthritis, Metabolism and Digestive Diseases, National Institutes of Health, Bethesda, Maryland 20014, USA

J. V. KLAVINS
Long Island Jewish–Hillside Medical Center, Queens Hospital Center Affiliation, Jamaica, NY 11432, USA

NATALIA LÓPEZ-MORATALLA
Department of Biochemistry, University of Navarra, Pamplona, Spain

CONTRIBUTORS

BERTRAM SACKTOR
Laboratory of Molecular Aging, Gerontology Research Center, National Institute on Aging, National Institutes of Health, Baltimore City Hospitals, Baltimore, Maryland 21224, USA

ESTEBAN SANTIAGO
Department of Biochemistry, University of Navarra, Pamplona, Spain

MICHAEL J. WEISS
Long Island Jewish–Hillside Medical Center, Queens Hospital Center Affiliation, Jamaica, NY 11432, USA

Preface

This series on 'MAMMALIAN CELL MEMBRANES' represents an attempt to bring together broadly based reviews of specific areas so as to provide as comprehensive a treatment of the subject as possible. We sought to avoid producing another collection of raw experimental data on membranes, rather have we encouraged authors to attempt interpretation, where possible, and to express freely their views on controversial topics. Again, we have suggested that authors should not pay too much attention to attempts to avoid all overlap with fellow contributors in the hope that different points of view will provide greater illumination of controversial topics. In these ways, we hope that the series will prove readable for specialists and generalists alike.

The first volume, entitled *General Concepts*, served to introduce the subject and covered the essential aspects of physical and chemical studies which have contributed to our present knowledge of membrane structure and function. The second volume, *The Diversity of Membranes*, was concerned with specific types of intra- and extracellular membranes, while the third volume, *Surface Membranes of Specific Cell Types*, as its title indicates, reviewed the knowledge that we have of the surface membranes of the various cell types which have been studied in any detail to this time. *Membranes and Cellular Functions* are covered in the present volume, which concerns ultrastructural, biochemical and physiological aspects. Since the cell surface represents the point of interaction with the cellular environment, Volume 5 is entitled *Responses of Plasma Membranes* and addresses itself to the way in which external influences are mediated by the plasma membrane.

As editors, our approach to our responsibilities has been rather permissive. With regard to nomenclature and useful abbreviations, we have used 'cell surfaces' and 'plasma membranes' where appropriate rather than 'cell membranes' since this last is nonspecific. Both British and American usage and spelling have been utilized depending upon personal preference of the authors and editors with, again, no attempt at rigid adherence to a particular style.

While the title of the series is 'MAMMALIAN CELL MEMBRANES', we have encouraged authors to introduce concepts and techniques from non-mammalian systems which may be useful in their application to eukaryotic cells. The aim of this series is to provide a background of information and, hopefully, a stimulation of interest to those investigators working in, or about to enter, this burgeoning field.

Finally, the editors would like to acknowledge the dedication and resourcefulness of their secretary and editorial assistant, Mrs Alice R. Scipio, in the coordination and preparation of these volumes.

G. A. JAMIESON
D. M. ROBINSON

1

The relation of membrane ultrastructure to membrane function

David W. Deamer

Department of Zoology, University of California, Davis, California

1.1   INTRODUCTION

The title of this chapter is very broad and could as well be applied to a book as to a short review. It was initially suggested by the editors, with the provision that changes would be acceptable. After some consideration, the title did appear to have certain advantages. It permits a wide-ranging introduction to the rest of this volume, while still providing a focus for material to be covered within the chapter. But what is membrane ultrastructure? The term implies that we may somehow visualize the fine structure and perhaps even the molecular organization of membranes, probably by electron-microscopic techniques. However, the classical techniques of electron microscopy have provided only glimpses of what might be termed membrane ultrastructure, for instance the trilaminar structures found in thin-sectioned membranes and the large particles protruding from negatively stained inner mitochondrial membranes. The problem, then, was to find a focus which could both serve as an introduction and still fulfill the theme of the chapter title. Fortunately, such a focus does exist.

The classical electron-microscopic techniques have been very useful in defining cell ultrastructure, but have been of remarkably little help in deducing the molecular organization of membranes. The contribution of fixation and staining artifacts to the resulting images becomes too high at the magnifications required for visualizing molecular structures in sectioned material. In 1957, Steere suggested a new method for preparing electron-microscopic specimens which promised to be relatively free of artifacts. In this method, now generally termed 'freeze-etching', a biological specimen is frozen, fractured under vacuum, and the fracture surface is replicated for viewing in the electron microscope. Often, but not necessarily, some sublimation or 'etching' of ice in the fracture surface is permitted and this may occasionally

Figure 1.1   Human diploid lung fibroblast grown in tissue culture (WI-38). The fracture face has revealed the nuclear membrane with characteristic pores (N), the endoplasmic reticulum (ER), mitochondria (M) and the plasma membrane (PM). In the freeze-etch micrographs shown in this chapter, the photographs have been printed so that the platinum shadow appears black. Thus, structures protruding above the fracture plane are darkened on one side by the platinum and have white shadows'. The arrow shows the direction of the platinum shadow and is 1 μm long. Ice (I) in the external medium has been etched, producing a rough surface characteristic of etched ice and leaving behind nonetchable glycerol/water mixtures (G). Glycerol is often added to freeze-etch preparations to protect against freezing damage caused by extensive ice crystals. Note the particulate structures on most of the membrane fracture faces. (Courtesy of L. Packer)

provide further information. The freeze-etch method was developed to the stage of being a reliable technique by Moor *et al.* (1961) and Moor and Muhlethaler (1963), but a greater potential became apparent when Branton (1966) provided evidence that membranes were split along hydrophobic planes by the fracture process, such that interior views of membrane structure were revealed. External features of membranes could also be observed under some conditions by allowing deep etching to expose the surfaces. Thus, freeze-etching is the only microscopic technique available which can provide information about unfixed internal membrane ultrastructure at a resolving power of 2–3 nm, as well as surface features of some membranes. For this reason it seems acceptable to use freeze-etch images of mammalian cell membranes as a focus for this chapter, particularly emphasizing the insight provided by recent findings. Since the role of freeze-etching in developing concepts of membrane structure has been reviewed (Branton and Deamer, 1972), it would not be useful to repeat this discussion. Instead we will briefly summarize the problems presented by freeze-etch images, then deal with several membrane systems which have been studied in considerable detail.

The power of freeze-etching, as well as some of its limitations, is shown in *Figure 1.1*. This is a micrograph of a human diploid lung cell grown in tissue culture, then carried through the freeze-etch process. The fracture face shows the cytoplasm from the nucleus to the plasma membrane. Clearly, freeze-etching provides beautifully detailed images of cytoplasmic structures in face view. However, since only membrane fracture surfaces are revealed, it is difficult to identify all of the structures. Some of the more obvious membranous organelles may be readily distinguished, including the nuclear membrane with its characteristic pores, the endoplasmic reticulum, and the plasma membrane. Mitochondria may be identified by their distinctive fracture faces, since these have been characterized in isolated mitochondrial fractions (Wrigglesworth, Packer and Branton, 1970). The rest of the spherical structures presumably represent lysosomes, microbodies and miscellaneous vacuoles. Nonmembranous structures such as microtubules, filaments and chromosomes are not clearly visualized.

The reader should note that many of the membrane fracture surfaces contain particulate structures, in some instances partially aggregated (mitochondria) and elsewhere randomly dispersed. Such particulate structures are ubiquitous features of freeze-etch images of metabolically active membranes, and represent an important problem for membrane biology. The primary theme of this chapter will be the nature of these particles. Secondarily, we will ask what the particles may tell us about development of membrane structures, and speculate on their general significance in membrane function.

1.2   FREEZE-ETCH IMAGES

In the discussion to follow, we will make the assumption that fracture planes in freeze-etch images pass along hydrophobic planes in the ice–membrane system, as originally suggested by Branton (1966). It will not be necessary to repeat the arguments here, since there is now general acceptance of this assumption and direct evidence in several membrane systems (Deamer and Branton, 1967; Pinto da Silva and Branton, 1970). In beginning a discussion of freeze-etch particles, we will first briefly review two tests of the freeze-etch method which tend to rule out contributions of artifacts to the particulate images of membranes.

1.2.1   Lipid systems

The first test was to view pure lipid systems by the freeze-etch technique. If the particles were artifacts of the freeze-etch method, one might expect to find them on simple lipid surfaces produced by fracturing lipid bilayers. To this end, we examined a number of different lipids, including fatty acids and their salts, cholesterol, and various phospholipids under different conditions of hydration and ionic environment. When care was taken to minimize contamination of the lipid fracture surfaces, the surfaces of all lamellar-phase lipids were uniformly smooth (*Figure 1.2a*). Furthermore, it was possible to

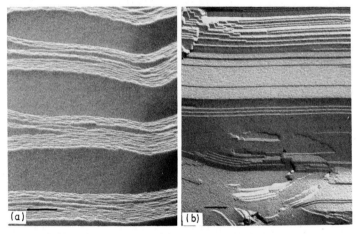

Figure 1.2   (a) Lamellar-phase lipid (L-α-dipalmitoylphosphatidylcholine). Fracture faces are smooth. The thinnest white lines have dimensions which suggest they represent the fracture plane breaking from one lipid bilayer to the next. The bar represents 0.1 μm. (b) Hexagonal-phase lipid (calcium cardiolipin). Fracture faces have a ribbed appearance. Each rib is produced by lipid molecules organized in rod-like structures with the lipid tails directed outward from the central axis of the rod. Repeat distance between the rods is 5.2 nm in the micrograph, which agrees with the 5.2 nm spacing found by X-ray diffraction analysis of the same lipid system. The bar represents 0.1 μm. (From Deamer et al., 1970, courtesy of Associated Scientific Publishers)

measure the repeat distance between lipid lamellae in some of the lipid crystals which happened to fracture across the lamellae, and the distances agreed very well with X-ray diffraction data from the same lipid system. We could conclude that the freeze-etch method did not disrupt the crystalline structures of the lipid, and that lipid bilayers presented smooth surfaces in freeze-etch images.

As a check on this, several hexagonal-phase lipids were also examined (*Figure 1.2b*). The freeze-etch images of these lipids showed a ribbed fracture surface, precisely that expected of hexagonally packed lipid rods. Again the X-ray diffraction data parameters agreed with measurements made directly from the micrographs, showing that the freeze-etch process could preserve fairly labile lipid phases.

More recent work by Costello and Gulik-Krzywicki (1976) generally confirmed these results for lipid systems. However, under some conditions, such as high water content and absence of an antifreeze agent (glycerol), extensive perturbations caused by freezing could be demonstrated. Thus, we should not accept freeze-fracture images as perfect reflections of biological ultrastructure, but instead should compare the results wherever possible with the same information provided by other techniques.

1.2.2 Lipoprotein crystals

A second test of the freeze-etch method was to determine whether it could visualize lipoprotein particles which were known to be present in a system.

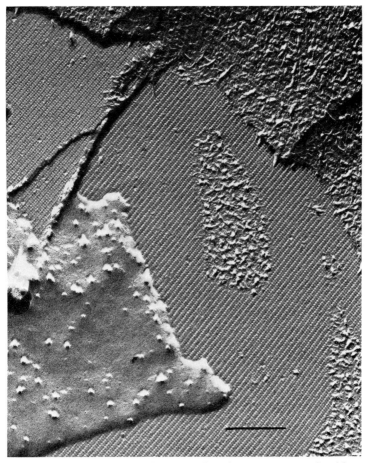

Figure 1.3 Amphibian (Xenopus) *yolk platelet. The spacings are produced by a crystalline array of lipoprotein subunits composed of lipovitellin and phosvitin. The subunits appear to be dimeric and have a long spacing of 17 nm and a short spacing of 8.5 nm. The bar represents 0.2 μm. (From Leonard, Deamer and Armstrong, 1972, courtesy of Academic Press)*

A convenient material for this was found in the form of the natural lipoprotein crystal of amphibian yolk platelets. Yolk platelets contain 20 percent lipid, composed of choline, ethanolamine and serine phospholipids, and two proteins, phosvitin and lipovitellin. The proteins are associated into complexes containing one vitellin and two phosvitin molecules, with the lipid associated primarily with the vitellin (Wallace and Dumont, 1968). The complexes are organized into a crystalline structure whose dimensions have been determined by X-ray diffraction (Ohlendorf *et al.*, 1975). Leonard, Deamer and Armstrong (1972) measured freeze-etch spacings in *Xenopus* yolk platelets and these may be compared with X-ray parameters (*Figure 1.3*). The freeze-etch method clearly resolved particulate structures within the yolk platelet crystal. The particles had the expected dimensions of the lipoprotein complex as characterized by biochemical studies, and crystal parameters measured directly from the freeze-etch micrographs are comparable to those derived by X-ray diffraction analysis of yolk platelets. We conclude that the freeze-etch process has the capacity to resolve lipoprotein particles, and that if such particles were present in biological membranes we might expect them to appear in freeze-etch images.

1.2.3   Sarcoplasmic reticulum

We can now turn to several membrane systems in which significant progress has been made toward resolving the nature of freeze-etch particles. The membranous structures chosen are the sarcoplasmic reticulum, rod outer segments, erythrocyte membranes, mitochondria and cell junctions.

The sarcoplasmic reticulum (SR) is a membranous elaboration of tubules and cisternae found in striated muscle cells. It is a highly specialized membrane, and has the specific property of releasing calcium into the contractile fibers of the muscle cell during contraction, then accumulating calcium during relaxation (*see* Martonosi, 1971; Inesi, 1972; *see also* W. G. Nayler, Vol. 3, Chapter 6, of this series). Because of the specialized nature of the SR, it is a good choice for analysis by the freeze-etch method since its function should be more readily related to ultrastructural elements visualized by freeze-etching. Although the property of calcium release does not seem to be preserved when SR is isolated as microsomes, calcium uptake is readily measured. Upon addition of calcium to isolated SR (*Figure 1.4*) the ATPase is activated and calcium accumulates within the vesicles of the microsomes. In this instance oxalate was present in order that calcium oxalate deposits within the vesicles could be visualized (*Figure 1.5*), and the filling of the microsomes with electron-dense calcium oxalate is readily apparent. These results demonstrate the usefulness of the SR system as a model for ion transport.

An important question in recent studies has been to define a molecular mechanism by which the SR performs its uptake function. Numerous biochemical investigations together suggest the following mechanism.

$$\text{ATP} + \text{E} \qquad \rightarrow \text{E} - \text{ATP} \qquad\qquad (1.1)$$

$$\text{E} - \text{ATP} + 2\text{CA}_{out}^{2+} \quad \rightarrow \text{E} - \text{P} - 2\,\text{Ca}_{out}^{2+} \qquad (1.2)$$

$$\text{E} - \text{P} - 2\,\text{Ca}_{out}^{2+} \quad \rightarrow \text{E} - \text{P} - 2\,\text{Ca}_{in}^{2+} \qquad (1.3)$$

$$\text{E} - \text{P} - 2\,\text{Ca}_{in}^{2+} \qquad \rightarrow \text{E} + \text{P}_i + 2\,\text{Ca}_{in}^{2+} \qquad (1.4)$$

In this model, ATP phosphorylates a carrier protein, E, as the initial step. Two Ca^{2+} ions are bound to the phosphorylated carrier and transported across the membrane. Calcium is released inside when the phosphate is

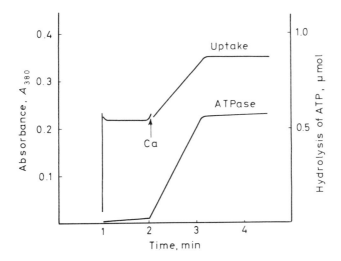

Figure 1.4   Calcium uptake coupled to ATP hydrolysis in sarcoplasmic reticulum derived from lobster abdominal muscle. At 1 min, an
aliquot of microsomes containing 0.24 mg protein was added, followed
by 1 μmol of Ca^{2+} at 2 min. The turbidity increase between 2 and
3 min reflects precipitation of calcium oxalate within the microsomal
vesicles. Note that ATP hydrolysis was activated upon addition of
Ca^{2+}, and ceased when all Ca^{2+} was accumulated. Conditions: 2 ml
of 1 mM Tricine buffer, pH 7.2, 5 mM $MgCl_2$, 5 mM ATP, 4 mM
potassium oxalate. ATPase activity was measured by the pH change
caused by hydrolysis of ATP

cleaved from the protein. This ion-transport mechanism is based on the following evidence:

1. The Ca:ATP ratio is close to 2 under a variety of conditions (Hasselbach, 1964).
2. ATP is bound to the protein prior to its phosphorylation (Inesi *et al.*, 1970).
3. The protein is phosphorylated during calcium transport (Makinose, 1969).
4. The mechanism is reversible, such that calcium gradients can drive the synthesis of ATP (Makinose and Hasselbach, 1971).
5. Ionophores release the calcium, indicating that calcium forms a gradient rather than binding to interior sites (Scarpa and Inesi, 1972).
6. Uncouplers which uncouple ion transport driven by chemiosmotic mechanisms have no effect on calcium uptake (Deamer and Baskin, 1972).

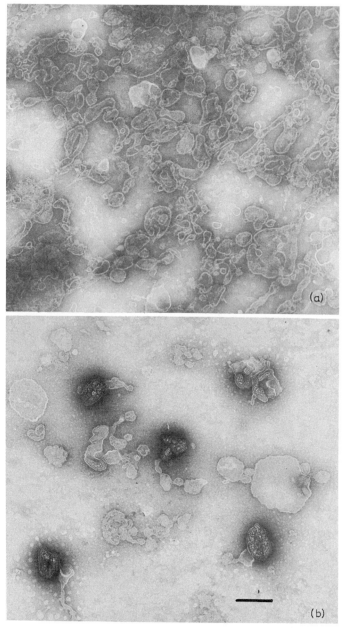

Figure 1.5   Negative stain preparation of sarcoplasmic reticulum isolated as microsomes from lobster abdominal muscle. (a) Prior to calcium uptake the vesicles appear empty. (b) After the preparation had accumulated 10 μmol calcium per milligram of protein, about half the vesicles contained electron-dense deposits of calcium oxalate. The white bubbles within the deposits are artifacts produced by the electron beam. Contaminating membranes do not have deposits. Conditions as in Figure 1.4. The bar represents 0.2 μm

The mechanism predicts that a carrier protein should be present in the membrane and that some portion of the protein must span the thickness of the membrane. Since this type of protein should be visible if the membrane is fractured along hydrophobic planes, we examined SR by the freeze-etch method (Deamer and Baskin, 1969). A freeze-etch micrograph of isolated SR (*Figure 1.6a*) shows that the membranes are highly particulate. The particulate structures have approximately the size and distribution expected of the Ca^{2+}-dependent ATPase, and it was suggested that they represent the ATPase of the SR.

An obvious next step in characterizing the particles was to isolate the ATPase and determine whether the purified material was similar in freeze-etch appearance to the original microsomal particles. MacLennan *et al.* (1971) reported a purification procedure based on detergent solubilization, and noted that particulate structures could be observed on preparations which were freeze-etched. However, the particles were not compared with those on the original microsomes. Deamer (1973) developed a different procedure and carried out freeze-etch analysis during the entire purification procedure. The method was to add controlled amounts of lysolecithin, a natural detergent, to suspensions of SR and use centrifugation to separate solubilized lipoprotein classes. Lysolecithin was chosen since freeze-etch microscopy had already been done on this lipid. It was found that at concentrations between 1 and 2 mg of lysolecithin per milligram of protein much of the protein and lipid of the microsomes was solubilized, but the ATPase formed vesicular structures which could be centrifuged to form a pellet. The ATPase activity of the pellet was approximately doubled by this procedure, and gel electrophoresis showed a single major band corresponding to 105 000 daltons. Other studies (MacLennan, 1970; Martonosi and Halpin, 1971; Meissner and Fleischer, 1971) using different isolation procedures have also demonstrated this band, and it is agreed that it represents the ATPase protein of the SR.

Figure 1.6a–d shows freeze-etch micrographs of isolated SR and the ATPase preparation. The numerous particulate structures of diameter 8 nm on the original vesicles remain throughout the isolation procedure, becoming even more clearly defined and concentrated within the fracture plane of the final ATPase preparation. Since the only polypeptide in the isolated material is the ATPase, we have suggested that the particles on the original membranes are correlated with the ATPase enzyme.

This is certainly insufficient evidence on which to base a detailed model of the SR membrane. However, if we assume that the fracture plane separates membranes in hydrophobic regions, and that the particles represent the ATPase as one or more globular polypeptide chains embedded in a lipid bilayer, it is possible to put together a general working model of the membrane. The ATPase represents approximately 70–80 percent of the membrane protein in purified SR, and the membrane is 38 percent phospholipid by weight with over 90 percent of the lipid being phospholipid. According to the calculations outlined by Branton and Deamer (1972), the lipid bilayer would comprise about 60 percent of the total membrane surface area with the rest being composed of penetrating protein. *Figure 1.7* shows how a globular protein of molecular weight 105 000 would be distributed in a membrane according to the specifications given above. The density and distribution of particles in

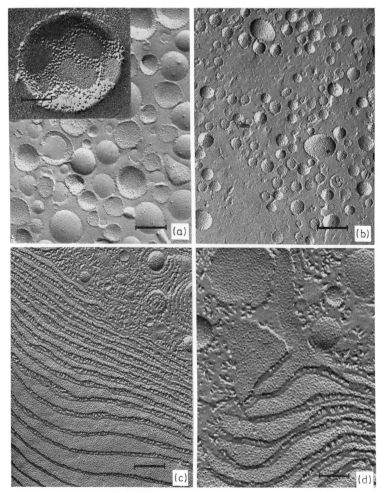

Figure 1.6   Freeze-etch micrographs of sarcoplasmic reticulum and isolated ATPase. (a) Microsomes prepared from lobster abdominal muscle are highly particulate and most of the 8-nm particles appear in the concave fracture face. The bar represents 0.2 μm. Inset: higher-magnification view of a vesicle following calcium uptake in the presence of oxalate as shown in Figure 1.5. Under these conditions many of the vesicles contain aggregates of the particles. The bar represents 0.1 μm. (b) Lysolecithin– ATPase complex isolated from the preparation shown in (a). The ATPase forms smaller vesicles, but the particles are similar in size to those on the original membrane and presumably represent the calcium-dependent ATPase. The bar represents 0.2 μm. (c) After partial drying the vesicular ATPase shown in (b) condensed into highly particulate lamellae. The bar represents 0.2 μm. (d) Higher-magnification view of (c). Note that all fracture faces are particulate. The bar represents 0.1 μm. (From Deamer, 1973, courtesy of the American Society of Biological Chemists)

the drawing resembles the image provided by freeze-etching for the SR microsomes (*Figure 1.6a*).

There are still uncertainties in the interpretation of the freeze-etch image:

1. The concave fracture face has nearly all the particles in the microsomes. It is still not clear what organization of lipid and protein would produce

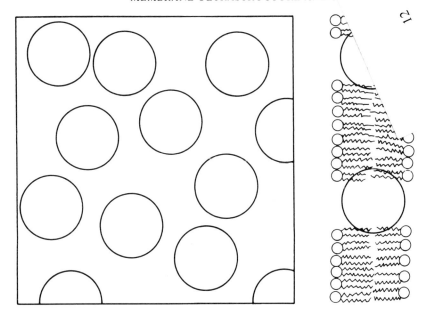

Figure 1.7   Frontal and cross-sectional views of ATPase distribution in a lipid bilayer. The circles represent a protein molecule of mol. wt 105 000 (6.5 nm in diameter) and show a random distribution of molecules calculated to comprise 40 percent of the membrane area. See text for details

such an image. Furthermore, there are no obvious depressions in the smooth face which one might expect to be left by removal of such large particles.

2. The measured diameter of the particles is 8–9 nm, whereas a globular protein of molecular weight 105 000 would be 6.5 nm in diameter. Since the layer of platinum–carbon shadow is about 1 nm thick, the final particle diameter of a 6.5 nm particle would be 8.5 nm, in agreement with the measurements from the micrographs. However, we are still uncertain whether we are visualizing a 'naked' protein embedded in a lipid bilayer, or whether the particle is a lipoprotein complex.

3. A simple calculation shows that the expected particle density would be 15 000–20 000 particles per μm^2 if every ATPase molecule in the membrane were visualized. However, the actual count is about 5000 particles per μm^{-2}. This suggests that the fracture plane may not reveal all ATPase sites, or alternatively, that the particles are in fact composed of 3–4 ATPase molecules. The latter suggestion would follow from the results of le Maire and Tanford (1976) which showed that detergent-solubilized SR ATPase must have 3–4 molecules for full activity.

These questions will be considered in detail in Section 1.3 of this chapter.

1.2.4   Erythrocyte membranes

Mammalian erythrocytes have been important in membrane research since they are a relatively pure cell population, and have the advantage that only a single type of membrane is present which may be readily isolated as erythrocyte 'ghosts'. Erythrocytes have the disadvantage that the cells have an age distribution, and that the membrane is relatively inactive metabolically. For instance, erythrocyte ATPase is one of the most thoroughly studied ion transport enzymes, yet there are only several hundred ATPase sites per cell, compared with several million polypeptide chains per cell involved in the structure of the plasma membrane.

Nonetheless, significant progress has been made in determining the molecular organization of lipid and protein in erythrocyte membranes, and freeze-etching has played an important role (*Figure 1.8*). Particulate structures are very clear, and there are approximately 500000 such structures per erythrocyte. There is, again, an asymmetric distribution of particles on

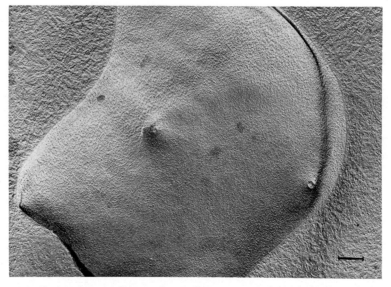

Figure 1.8 Beef erythrocyte ghost membrane. This micrograph shows the highly particulate PF face, which faces the cell interior. The outer EF fracture face (not shown) is relatively smooth. The bar represents 0.2 μm

the two fracture faces, with the majority of particles being on the cytoplasmic or protoplasmic fracture (PF) face; the PF face has 2800 particles per μm^2, and the extracellular fracture (EF) face, next to the circulating plasma, has 1400 particles per μm^2.

Since the erythrocyte membrane has relatively low enzyme activity, there is no obvious place to begin looking for a function of the particles. Certainly they cannot all represent sites of active ion transport, although a few may have this function. Another suggestion is that some of the particles may be sites for chloride–bicarbonate exchange (Bretscher, 1973). It is even possible that some of the particles have no specific function, but instead represent 'fossil' proteins remaining from the early development of the erthyrocyte.

Although it is not yet possible to assign function to the erythrocyte particles, considerable progress has been made toward defining their chemical nature. One of the simplest experiments has been to expose ghost membranes to proteolytic enzymes (Engstrom, 1970); the particles first aggregate, then disappear over a period of some hours. This result was consistent with a correlation between membrane protein and the particles, but did not permit a direct identification of the particles as proteins.

A more direct test of the chemical nature of the particles would be to label them with a specific marker which could be recognized in the freeze-etch image. Pinto da Silva and Branton (1970) were successful in using a ferritin-labeled antibody for this purpose, and found that sites on human erythrocyte

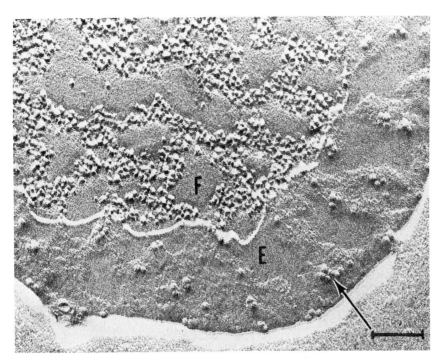

Figure 1.9   Human erythrocyte ghost following binding of ferritin-labeled anti-A antibody. The pattern of particles in the fracture face (F) may be seen continuing into the etch face (E) of this deep-etched preparation. Ferritin particles (arrow) appear exclusively within the zones of the pattern, suggesting that the particles in the fracture face carry the binding site of the anti-A antibody. In this instance the particles were experimentally aggregated to permit clearer interpretation of the relation between particles and ferritin binding. (Courtesy of P. Pinto da Silva)

ghosts for the A antigen could be marked by ferritin which had been conjugated to rabbit anti-human IgG. The ghosts were then freeze-etched, with deep etching carried out to expose membrane surfaces as well as the fracture surfaces. It is apparent (*Figure 1.9*) that ferritin appears only within the pattern of aggregated particles which continues from the fracture face to the exposed surface of the etched membranes. Thus, it could be concluded that the particles carry the sites which bind anti-A antibody. Similar experiments have subsequently been carried out by Marchesi *et al.* (1972) using phytohemagglutinin–ferritin as a surface marker, and again the bound ferritin followed the pattern of particles embedded in the fracture face. It is important to note that the particles can account for a major fraction of the total erythrocyte glycoprotein (Marchesi *et al.*, 1972). There are about 5×10^5 particles per erythrocyte membrane, and estimates of the phytohemagglutinin binding sites and antigen sites lie in the range $4–8 \times 10^5$. A proposed structure for the antigen-carrying major glycoprotein of erythrocyte membranes (Segrest *et al.*, 1972) has a portion of its chain specialized for interaction with the non-polar region of a lipid bilayer, and may account for some of the freeze-etch particles of the erythrocyte membrane (*see* Section 1.3).

1.2.5   Mitochondria

Mitochondria seem at first to be good candidates for freeze-etch studies. A great deal is understood of their biochemical, and related ultrastructural, aspects and they may be isolated in relatively pure preparations. However, freeze-etching has not been particularly useful in providing further under-

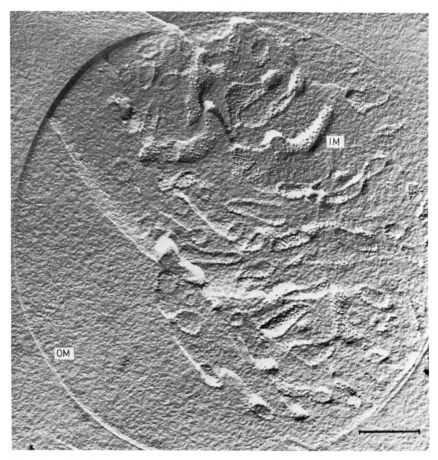

Figure 1.10   Beef heart mitochondrion. The outer membrane (OM) is seen as a circle surrounding the fracture faces of the complex inner membrane (IM). All fracture faces of the inner membrane are particulate

standing of mitochondrial membrane organization, largely because of the complexity of mitochondrial biochemical functions. In this sense, mitochondria are just the opposite of SR, in that they contain an array of metabolic and electron transport enzymes, coupling factors, and carrier proteins for a variety of metabolites and specific ions. It will obviously be a difficult task to specify function for the various particles found in freeze-etch images of mitochondrial membranes.

Although the intricate structure of the inner and outer membranes revealed

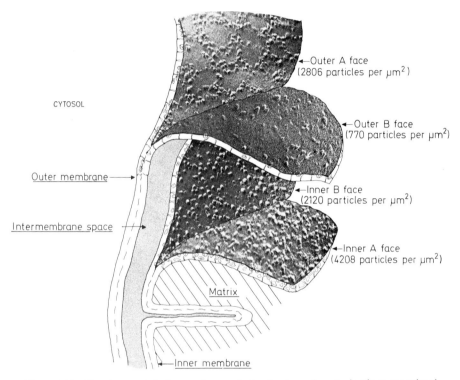

CYTOSOL

Outer A face
(2806 particles per μm²)

Outer B face
(770 particles per μm²)

Outer membrane

Inner B face
(2120 particles per μm²)

Intermembrane space

Inner A face
(4208 particles per μm²)

Matrix

Inner membrane

Figure 1.11   The characteristic fracture faces and ultrastructure of a mitochondrion are related in this composite view. (Courtesy of L. Packer)

in sectioned mitochondria is not clearly resolved by freeze-etching (*Figure 1.10*), particles are visible on all fracture faces. Wrigglesworth, Packer and Branton (1970) attempted to characterize the various fracture faces of the mitochondrion in regard to particle size and distribution. It was found that there are distinct patterns of particle distribution on the various fracture faces, and that the more metabolically active inner membrane has nearly twice as many particles as the outer membrane (*Figure 1.11*). The density of particles on the inner membrane, *ca.* 6000 particles per μm², is the highest recorded for any freeze-etched membrane.

Since the freeze-etch particles are the only molecular-sized structures which have been visualized in mitochondria (other than the ATPase seen in negatively stained preparations) there is naturally great interest in identifying their possible functions. One may imagine, for instance, that they represent coupling factors or protein components of the electron transport system; if coupling factors and cytochromes could be specifically removed from the membrane it might be possible to check this supposition.

The *petite* mutant of yeast lacks the enzyme activity and spectral components corresponding to cytochromes a, a_3, b and c_1 and partially or totally lacks mitochondrial DNA, depending on the yeast mutant strain.

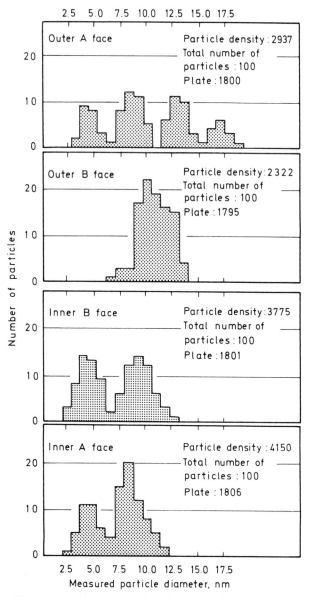

Figure 1.12a   *Membrane particles in mitochondria from wild-type yeast. Compare with Figure 1.12b. (See text for details)*

Oligomycin-sensitive ATPase is also absent. This offers the opportunity to compare loss of function in the mutants with possible structural alterations in the freeze-etch images (Packer, Williams and Criddle, 1973). The result is surprising (*Figure 1.12*); in spite of the very large deletions in functional proteins, the freeze-etch images of the membranes are quite similar in both the wild and mutant yeast strains.

A related study was carried out by Yamanaka and Deamer (1976), who

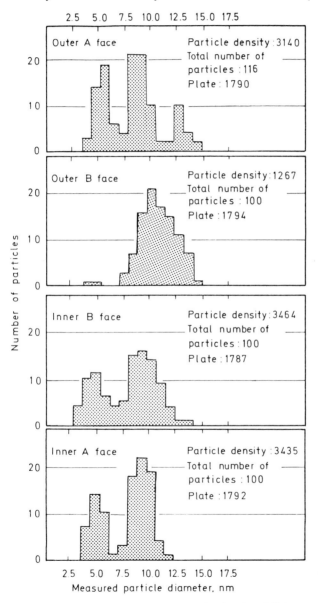

Figure 1.12b   Membrane particles in mitochondria from DNA⁻ petite mutant yeast. Compare with Figure 1.12a. (See text for details)

digested mitochrondrial, SR and erythrocyte membranes with nagarse, a bacterial protease. After extensive digestion of mitochondrial inner membrane vesicles, all the major polypeptides seen by gel electrophoresis were cleaved to low-molecular-weight fragments (10000–15000 daltons). Despite this, there was little alteration of the particle density or distribution in the fracture faces when digested membranes were compared with controls.

There are two possible explanations of this result. The first is that the particles are not related to the ATPase or electron transport enzymes, and are simply artifacts specific to the mitochondrial membrane. The second possibility is that the particles are hydrophobic apoproteins which are not affected by genetic alterations or enzymatic digestion. The problem is still to be resolved, but the results do illustrate both the potential power of freeze-etching in analyzing membrane structure, and the danger of assigning function to all particles which may appear in freeze-etch images.

1.2.6   Rod outer segments

When mammalian retinas are fractionated, one of the specific functional components which can be isolated is a structural element derived from the rod cells and termed 'rod outer segments'(ROS). ROS are stacks of membranous disks which play a role in the transduction of light energy into the electrochemical signal of the nerve impulse. When rod outer segments were first examined by freeze-etching (Clark and Branton, 1968), several different fracture faces were identified, and it was found that these faces were quite different from other freeze-etched membranes, in that they had a 'cobblestone' appearance. The complex freeze-etch image did not permit any straightforward conclusions about the organization of lipid and protein in this highly specialized membrane.

Approximately 85 percent of the protein of ROS is rhodopsin, a protein containing opsin as an apoprotein and retinal as a light-sensitive cofactor. The rhodopsin in isolated ROS still maintains the property of being chemically altered (bleached) by light, and therefore represents an important model system for the study of mechanisms by which light may produce a membrane change and associated nerve impulse. Chen and Hubbell (1973) and Hong and Hubbell (1973) have incorporated isolated rhodopsin into artificial lipid membranes in order to study such phenomena, and they utilized freeze-etch images as one of several parameters by which to compare the success of membrane reconstitution. In these experiments, rhodopsin preparations were placed in solutions of various lipids in the detergent dodecyltrimethylammonium bromide. The solution was then dialyzed, and as the detergent was removed lipid–protein complexes appeared as vesicular membranes. Some freeze-etch images of these are shown in *Figure 1.13*. The original ROS membranes can be prepared as vesicles by osmotic swelling followed by gentle homogenization. The vesicles have highly particulate surfaces (*Figure 1.13a*), the particles being 12 nm in diameter. Apparently the swelling process somehow spreads the 'cobblestone' membrane of ROS into vesicular membranes, which produce more typical freeze-etch images. In reconstituted vesicles containing various lipids, the particulate appearance was reproduced (*Figure 1.13b*) whereas lipid by itself produced only smooth lamellae (*Figure 1.13c*).

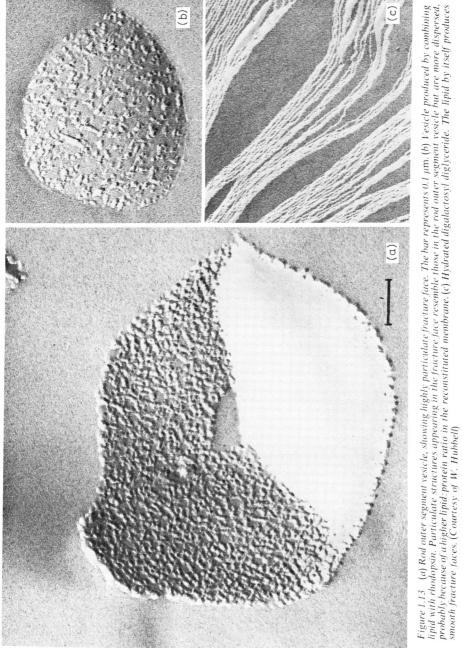

Figure 1.13 (a) Rod outer segment vesicle, showing highly particulate fracture face. The bar represents 0.1 μm. (b) Vesicle produced by combining lipid with rhodopsin. Particulate structures appearing in the fracture face resemble those in the rod outer segment vesicle but are more dispersed, probably because of a higher lipid:protein ratio in the reconstituted membrane. (c) Hydrated digalactosyl diglyceride. The lipid by itself produces smooth fracture faces. (Courtesy of W. Hubbell)

The particulate image was temperature-dependent; for instance, if the lipid used was d, l-(18:1)'phosphatidylcholine the particles aggregated at 5 °C and dispersed at 37 °C, suggesting that temperature-dependent phase changes in the lipid have the effect of aggregating the embedded protein particles. The image was also somewhat altered by light, in that exposure to light tended to disperse aggregates of particles under some conditions of temperature and lipid content.

It was concluded that the particles represent rhodopsin molecules embedded in lipid bilayers, and that the state of aggregation of the particles can be controlled by the physical state of the lipid in which they are embedded. By extension, the particles in the ROS membranes presumably also represent rhodopsin–lipid complexes. The success in reproducing the freeze-etch appearance of the original membranes by the reconstitution with rhodopsin and lipid shows that certain classes of protein can reassociate with lipid in such a way that they become embedded in the lipid bilayer. However, the conditions of detergent and dialysis under which the reconstitution occurs have no counterpart in cells, and membrane biogenesis still presents difficult conceptual problems.

1.2.7   Gap junctions

The last freeze-etch particles to be discussed are the structures associated with gap junctions. This species of intercellular junction is identified in lanthanum-stained sections of liver and other tissues as polygonal lattice structures within the junction. It is probable that such junctions are sites of intercellular transfer of ions and larger molecular species which do not normally cross membranes. The characteristic gap-junction structure may also be resolved by freeze-etching (Kreutziger, 1968; Goodenough and Revel, 1970). The junction is seen as an orderly array of particles (*Figure 1.14*). A second type of cell junction, the zonula occludens, is also visible in this image. If the reader will view the micrograph at an acute angle to the surface, the particles of the gap junction are seen to form three linear arrays at approximately 120° angles.

Three types of gap-junction particles have been identified in rat intestinal epithelial cells (Staehelin, 1972), of which two are shown in *Figure 1.15*. The type I particles are 8–9 nm in diameter and form extensive arrays similar to the liver-cell gap-junction particles while the type II particles are larger (10–11 nm in diameter), more dispersed (19–20 nm center to center spacing) and form highly organized hexagonal lattices. The type III particles (not shown) are smaller (6–8 nm center to center spacing) and form rectangular aggregates containing relatively few particles.

Goodenough and Stoekenius (1972) developed a method for isolating gap junctions which involves digestion of mouse liver plasma membranes with collagenase, followed by detergent solubilization. The gap junctions remain coherent through this procedure and may then be isolated by centrifugal techniques. When sodium dodecyl sulfate gel electrophoresis is applied to the protein of the gap junction, most of it is found to consist of a single major band of molecular weight 20000. The phospholipid associated with the protein is primarily phosphatidylcholine, together with smaller amounts of phosphatidylethanolamine and neutral lipid. X-ray diffraction of the wet pellet

Figure 1.14   Gap junction (GJ) in mouse liver cell plasma membrane. Zonula occludens (ZO) are also shown in this micrograph. (Courtesy of D. Goodenough)

gives repeat distances of 8.6 nm for the center to center spacing of the particles, a result consistent witth measurements taken from the freeze-etch images.

It was concluded that the distinctive particulate array in the liver cell gap junction is composed of a single lipoprotein species. This important result suggests that cells may synthesize classes of proteins which have the ability to merge with the membrane, then organize themselves into a functioning supramolecular structure. There remains the problem, however, that the particles are much larger than would be expected for a single protein of molecular weight 20 000. It is likely that the freeze-etch particles represent a complex of several polypeptides with lipid.

Figure 1.15   Gap junction in rat intestinal epithelial cells showing two types of gap-junction particles (I and II). The B fracture face on one cell composes most of the micrograph and the fracture has broken through to a neighboring A face, exposing the gap-junction particles. The bar represents 0.1 μm. (Courtesy of L. A. Staehelin)

1.3   MEMBRANE STRUCTURE

The direct evidence and indirect correlations discussed above strongly suggest that particulate freeze-etch images in membranes result from lipoprotein complexes embedded in a fluid matrix of the lipid bilayer. We may go on to ask the developmental question of how the protein becomes embedded in the membrane during membrane synthesis. This is an important question to resolve, since the timing and placement of specific proteins within membranes determines, in part, the development of differentiated character within cells.

There are several alternative mechanisms of protein incorporation into membranes. All are speculative at this juncture, and await further understanding of membrane structure before they may be critically judged. These alternatives are outlined below, and may conceptually be divided into mechanisms which form nonspecific, or specific, interactions between membrane protein and lipid. Possible electrostatic binding of proteins to membranes will not be discussed.

1.3.1   Nonspecific lipid–protein interactions

The simplest developmental scheme for binding of the hydrophobic proteins to membranes envisages that the protein is synthesized within the cell as usual, then finds its way to a membrane where a portion of its peptide chain becomes embedded in the nonpolar phase of the lipid bilayer, as shown in *Figure 1.16a*. As will be discussed, there is considerable evidence for such lipid–protein interaction from studies of protein binding to lipid monolayers.

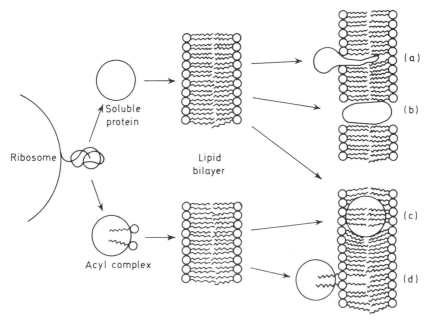

Figure 1.16   Models for the introduction of protein into preexisting membranes. (See text for details)

In this model there is no specific interaction between lipid chains and proteins. The lipid simply provides a nonpolar environment which stabilizes the binding of protein to the membrane.

A second possibility is that nearly all of the protein molecule may become embedded in a membrane, as shown in *Figure 1.16b*. This would imply unusual properties for the protein, since it would be required to be soluble both in aqueous and nonpolar phases. Singer (1971) noted this difficulty, and suggested that a bound ribosome might insert a protein directly into a membrane during protein synthesis (*see also* Dallner, Siekevitz and Palade, 1966).

1.3.2   Specific lipid–protein interaction

There is no doubt that lipid is required for many enzymatic processes (*see* Rothfield, 1971), but there are only a few initial studies which suggest that a specific lipid may be involved with a given protein in a membrane-catalyzed reaction. The problem is clouded by the fact that the physical properties of lipids make it difficult to combine them with isolated proteins. Lipid may be readily removed from a lipoprotein complex by detergent treatment or by extraction with organic solvents, but reconstitution of lipoproteins presents severe difficulties. Most phospholipids are very insoluble in aqueous phases, and may be added only as lipid suspensions or complexes with detergents. There is no guarantee that lipid added in this form will reproduce the original lipid environment required by an enzyme, and experimental differences between the effects of various lipids in reactivating enzymes may simply reflect differences in the physical properties of the lipids. The living cell, of course, forms phospholipid from soluble substrates and the problem does not arise.

The simplest mechanism by which specific lipid–protein interaction may occur is that protein embeds itself in the lipid layer as outlined previously, but that afterward lipid chains or headgroups interact with specific sites on the protein (*Figure 1.16c*). An important example of such specific lipid-binding sites on protein is the acyl-chain binding sites on serum albumin (Spector, John and Fletcher, 1969). Benson (1966) originally proposed specific lipid-binding sites on chloroplast membrane protein.

A variation of this mechanism is that a newly synthesized protein, specialized for incorporation into a membrane, first picks up monoacyl lipids (fatty acids, acyl CoA derivatives, or lysophosphatides from the cytosol) then carries the lipid to a membrane (*Figure 1.16c, d*). At the membrane interface a second acylation occurs to form a diacyl phospholipid (Lands, 1960) and the protein becomes firmly attached to the surface (Deamer, 1970) or the interior (Benson, 1966) of the membrane.

It must be emphasized that each of these models is speculative and that there is little evidence which permits a choice among them. Theyy may all be correct, in the sense that various proteins may interact with membranes by any one of the above mechanisms. In the discussion to follow, some of these models will be discussed in reference to the information provided by freeze-etch microscopy.

1.3.3   Origin of freeze-etch particles

The evidence which supports the contribution of protein penetration reactions to membrane development comes primarily from the fact that most proteins are capable of interacting with lipid monolayers. *Figure 1.17* shows a typical surface pressure change resulting from interaction of cytochrome *c* with a lipid monolayer. As the cytochrome *c* approaches and touches a lipid film, a portion of the polypeptide chain apparently penetrates the monolayer and becomes incorporated in the nonpolar lipid phase. Since this chain adds to the surface area of the monolayer, there is an increase in the surface pressure. Quinn and Dawson (1969) have shown that following the penetration reaction with lipid monolayers, cytochrome c is strongly bound, presumably by hydrophobic forces, and cannot be removed by conditions which would be expected to release electrostatic bonds between the lipid and protein. Morse and Deamer (1973) demonstrated that the reaction can be very rapid, and under the most favorable conditions is limited only by diffusion rates of the protein to the lipid interface. It does not require much extrapolation to suggest that similar reactions may occur within the cell.

In a recent study, Wehrli and Morse (1974) reacted erythrocyte apoprotein

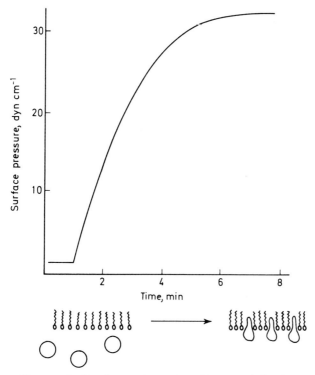

Figure 1.17 *Cytochrome* c *interaction with a stearyl phosphate monolayer. Cytochrome* c *was injected beneath the monolayer at 1 min. The increase in surface pressure probably results from penetration of the monolayer by hydrophobic portions of the protein, as shown in the lower drawing*

with lipid monolayers and examined the resulting films by the freeze-etch technique. *Figure 1.18* shows the freeze-etch images of erythrocyte ghost membrane, of lipid monolayers derived from the membranes and of the lipid monolayers following interaction with the apoprotein. The monolayer by itself is smooth (*Figure 1.18b*), but after penetration by the apoprotein the image (*Figure 1.18c*) resembles that of the original erythrocyte surface (*Figure 1.18a*). This striking result, although not providing direct evidence for the contribution of penetration reactions to membrane structure, certainly shows that freeze-etch particles may arise from such a mechanism. It should be added that Kimelberg and Papahadjopoulos (1971) and Vail. Papahadjopoulos and Muscarello (1974) found evidence for penetration of protein into

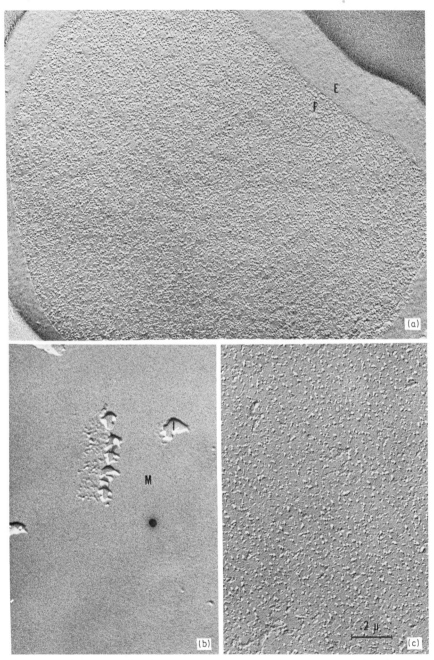

Figure 1.18   Interaction of erythrocyte membrane protein with erythrocyte lipid. (a) Erythrocyte ghost showing both fracture face (F) and etch face (E). (b) Erythrocyte lipid bilayer prepared and fractured according to the method of Deamer and Branton (1967). The lipid monolayer (M) facing the ice appears smooth, but occasional flaws in the monolayer permit etching of ice (I) beneath the monolayer. (c) Erythrocyte lipid prepared as in (b), but including interaction with erythrocyte membrane protein. The fracture face contains particles which resemble the original membrane particles. (Courtesy of E. Wehrli and P. D. Morse)

lipid bilayers of liposomes. However, this reaction is much less general and may occur under specialized conditions, in contrast with the ease of penetration of monolayers by most proteins. The difference may arise from the fact that monolayers at low surface pressures are loosely packed, whereas the molecules in lipid bilayers are relatively tightly packed and resist protein penetration. It is probable that some membrane proteins, for example cytochrome b_5 (Rogers and Strittmatter, 1973), are specialized to undergo penetration reactions with the lipid bilayers of membranes, and that most of the soluble proteins of the cell are unable to penetrate.

The apoprotein used by Wehrli and Morse in the experiment described above is a mixture of polypeptide chains derived from erythrocyte ghosts by solubilization in 2-chloroethanol. It is not known whether all of the polypeptides penetrate the film, or whether one or more is specialized to interact with lipid and produce the observed freeze-etch particles. In this regard, it is interesting that a partial structure for the major glycoprotein of erythrocyte membranes has been proposed (Morawiecki, 1964; Winzler, 1969; Segrest et al., 1972) which does suggest that this protein is specialized for involvement with membrane lipid. In the proposed structure, the central amino acid sequence of the protein is composed almost entirely of nonpolar amino acids. It was suggested that this region of the chain may be directed through the lipid bilayer of the membrane, thus anchoring the protein to the membrane structure. A protein spanning the membrane in this manner would presumably produce a particulate structure in the fracture face of a freeze-etch micrograph.

This correlation between protein structure and membrane properties, and the size and distribution of freeze-etch particles, is a very exciting finding. There remains, however, the problem of the disparity between the size of the particles as visualized by freeze-etching and the known sizes of the molecules which are proposed to produce the particulate image. The approximate molecular weight of the glycoprotein may be calculated to be about 90000 from its amino acid and carbohydrate composition and the chain proposed to span the membrane contains about 25 amino acid residues. The size of such a chain is much smaller than the diameter of the particles, which would be approximately correct only if the entire protein molecule were embedded in the bilayer.

One possible explanation is that more than a simple polypeptide chain is involved in the structure which produces the freeze-etch particles. Jost et al. (1973), in an investigation of the cytochrome oxidase system with spin labeling techniques, found that a fraction of the membrane lipid is not fluid but is, in fact, immobilized. It was proposed that this lipid forms a separate phase surrounding the cytochrome oxidase protein. This result, if applicable to other membrane proteins, suggests that some of the dimensions of the freeze-etch particle may arise from lipoprotein complexes, rather than simple polypeptide chains.

A second possibility (Bretscher, 1973) is that the particles represent component a, or complexes between this protein and the major glycoprotein. The latter possibility is attractive because of the correspondence between the particles and surface markers of the antigen, phytohemagglutinin and virus receptor sites, which are associated with the glycoprotein. It may be noted here that Segrest, Gulik-Krzywicki and Sardet (1974) demonstrated that the MN-glycoprotein membrane penetrating peptide could be isolated and incorporated into lecithin liposomes. When the liposomes were viewed by the freeze-fracture method, particles became visible only after a certain concentration of peptide in the lipid had been reached. It is clear from this result that single peptide fragments cannot be resolved by freeze-fracture, but that multimers of 10 or more peptide fragments are required to form visible particles. This finding supports the suggestion that at least some particles in natural membranes arise from polypeptide complexes, rather than single polypeptides.

1.3.4   Summary

We will summarize this section by schematically showing how freeze-etch particles may arise from the lipid–protein organization in erythrocyte membranes. *Table 1.1* presents some of the data available for the protein components of erythrocyte membranes. Three protein classes, the major glycoprotein, component *a* and tektin (Clarke, 1971), compose well over half the protein of the membrane. The evidence discussed here and reviewed elsewhere (Winzler, 1969; Bretscher, 1973; Marchesi *et al.*, 1973; Singer and Nicolson, 1972) suggests that the major glycoprotein has a portion of its peptide chain embedded in the membrane, that component *a* is representative of proteins which are almost entirely embedded within the lipid phase of the membrane, and that tektin forms long filamentous structures which are bound to the inner surface of the membrane. These three components are diagrammatically

Table 1.1 SUMMARY OF ERYTHROCYTE PROTEIN COMPONENTS
(Summary of data taken from Bretscher, 1973, Winzler, 1969 and Marchesi *et al.*, 1973)

Component	Percentage of total protein	Molecular weight	Composition	Number per cell	Possible functions
Component *a*	25	$\sim 100\,000$	10% carbohydrate	5×10^5	Ion channel
Major glycoprotein	10	$\sim 50\,000$	65% carbohydrate	7×10^5	Carries negative charge
Tektin	30	$\sim 200\,000$	Protein	2×10^5	Structural

shown in *Figure 1.19*. The possibility of specific lipid binding is not shown since this would esssentially result in the same image shown for component *a*.

Figure 1.19a shows the original membrane with the protein components interacting with a lipid bilayer. In *Figure 1.19b*, the membrane has been frozen, fractured and shadowed, and in *Figure 1.19c* the platinum–carbon replica has been cleaned of organic material. The platinum shadow is shown as heavy black lines delineating the original features of the fracture surface, but the thick support film of carbon is not included.

Note first that tektin cannot directly contribute to the freeze-etch image, since it is a surface protein. However, it is apparent that tektin could direct the arrangement of other membrane components and presumably has a role in lending structural support to what would otherwise be a seemingly fragile lipid bilayer. The glycoprotein, on the other hand, if embedded as visualized

(Marchesi *et al.*, 1972) could certainly produce a discontinuity in the fracture surface. Since the peptide chain by itself could not form a large enough particle to explain the freeze-etch image, a small amount of lipid is shown contributing to the particle dimensions. If a protein were partially embedded in the inner membrane surface this would have the advantage of explaining the asymmetric distribution of particles on fracture surfaces. One might expect that proteins embedded in a plasma membrane would tend to appear on the inner fracture surface, since the proteins synthesized in the cell would penetrate from the cytoplasmic side.

A totally embedded protein as shown for component *a* in *Figure 1.19* could

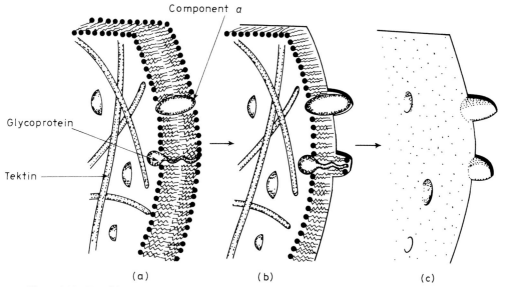

Component *a*

Glycoprotein

Tektin

(a)   (b)   (c)

Figure 1.19   Possible models for production of freeze-etch particles by lipoprotein structures in an erythrocyte membrane. (a) Original membrane. Component a, the major glycoprotein and tektin are shown interacting with a lipid bilayer. (b) The membrane has been frozen, fractured and shadowed with platinum. Component a and the major glycoprotein produce discontinuities in the replica. (c) The rest of the organic components of the membrane have been dissolved, leaving a platinum–carbon replica of the fracture face

account for the dimensions of the resulting particles visualized by freeze-etching. The ATPase of sR is probably the clearest example of this type of embedded protein. However, the model given for component *a* and the glycoprotein has the disadvantage that the particles should leave behind a series of depressions in the lipid layer which has pulled away. The opposite surfaces of most particulate fracture faces are remarkably smooth, and the fact that depressions can be seen in certain membrane structures, for instance the gap junction (Staehelin, 1972), suggests that the smooth surfaces do not result from some inadequacy of the freeze-etch method. It is interesting that a smooth fracture face opposite a particulate face is compatible with the notion that lipid chains are specifically binding to embedded membrane proteins, since a protein pulling away from such a lipid interaction during fracturing would leave behind a set of lipid molecules, rather than a depression in the surface (*see Figure 1.16c*).

In conclusion, a variety of different membrane proteins and lipoprotein complexes have been found to be associated with a particulate appearance in freeze-etch images. This supports the generalization that particulate subunits compose in part the structure of biological membranes. This concept was first suggested by Sjostrand (1963) and by Fernández-Morán et al. (1964), who independently proposed that particulate subunits best explain certain microscopic and functional data from mitochondrial membranes. Green et al. (1967) elaborated this suggestion into the generalization that all membranes are composed of such subunits and that membrane function is, in part, governed by highly specific ordering of, and by energy-dependent changes in order among, the subunits. However, the freeze-etch data reviewed here do not support the more general statement. It is clear that certain substructures within membranes, specifically the crystalline appearance of gap-junction particles, come very near the expected appearance of an ordered subunit membrane. The fact that freeze-etching is capable of demonstrating an ordered appearance in a few specialized membrane structures, but typically shows a random distribution of particles in most membranes, argues against a general crystalline subunit structure as proposed by Vanderkooi and Green (1970).

The concept that a major contribution to membrane structure is made by lipoprotein subunits embedded in fluid lipid bilayers was recently proposed by a number of independent investigators (Branton, 1969; Singer, 1971; Hendler, 1971; Branton and Deamer, 1972). The term 'fluid mosaic', suggested by Singer and Nicolson (1972), is receiving acceptance as best describing such a structure. However, the freeze-etch image also clearly shows that membrane organelles and even localized areas within single membranes have highly specialized structures, and there are still important questions to be answered about how freeze-etch images arise from the various suggested organizations of lipid and protein in membranes. Although the evidence supporting the general occurrence of embedded lipoprotein particles is strong, it is likely that a number of still unsuspected lipid–protein interactions in membranes await discovery.

REFERENCES

BENSON, A. (1966). *J. Am. Oil Chem. Soc.*, **43**:265.

BRANTON. D. (1966). *Proc. natn. Acad. Sci. U.S.A.*, **55**:1048.

BRANTON. D. (1969). *A. Rev. Pl. Physiol.*, **20**:209.

BRANTON, D. and DEAMER, D. (1972). *Membrane Structure*. Protoplasmatologia II, E, 1. Vienna; Springer-Verlag.

BRETSCHER, M. (1973). *Science, N.Y.*, **181**:622.

CHEN, Y. S. and HUBBELL, W. L. (1973). *Expl Eye Res.*, **17**:517.

CLARK. A. W. and BRANTON. D. (1968). *Z. Zellforsch. mikrosk. Anat.*, **91**:586.

CLARKE. M. (1971). *Biochem. biophys. Res. Commun.*, **45**:1063.

COSTELLO. M. J. and GULIK-KRZYWICKI. T. (1976). *Biophys. J.*, **16**:103a.

DALLNER, G.. SIEKEVITZ. P. and PALADE. G. E. (1966). *J. Cell Biol.*, **30**:97.

DEAMER. D. (1970). *J. Bioenergetics*, **1**:237.

DEAMER, D. (1973). *J. biol. Chem.*, **248**:5477.

DEAMER. D. and BASKIN. R. (1969). *J. Cell Biol.*, **42**:296.

DEAMER. D. and BASKIN, R. (1972). *Archs Biochem. Biophys.*, **153**:47.

DEAMER. D. and BRANTON. D. (1967). *Science, N.Y.*, **158**:655.

DEAMER, D., LEONARD, R., TARDIEU, A. and BRANTON, D. (1970). *Biochim. biophys. Acta*, **219**:47.

ENGSTROM. L. H. (1970). *Ph. D. Dissertation*, University of California, Berkeley.

FERNÁNDEZ-MORÁN. H.. ODA. T.. BLAIR. P. V. and GREEN. D. E. (1964). *J. Cell Biol.*, **22**:63.

GOODENOUGH, D. A. and REVEL. J. P., (1970). *J. Cell Biol.*, **45**:272.

GOODENOUGH. D. A. and STOEKENIUS, W. (1972). *J. Cell Biol.*, **54**:646.

GREEN, D. E.. ALLMAN. D. W.. BACHMANN. E.. BAUM. H.. KOPACZYK. K.. KORMAN. E. F.. LIPTON. S.. MACLENNAN, D. H., MCCONNELL, D. G., PERDUE, J. F., RIESKE, J. S. and TZAGOLOFF, A. (1967). *Archs Biochem. Biophys.*, **119**:312.

HASSELBACH, W.. (1964). *Prog. Biophys. biophys. Chem.*,**14**:169

HENDLER, R. W. (1971). *Physiol. Rev.*, **51**:66.

HONG, K. and HUBBELL, W. L. (1973). *Biochemistry*, **12**:4517.

INESI. G. (1972). *A. Rev. Biophys. Bioengng*, **1**:191.

INESI, G., MARING, E., MURPHY, A. J. and MCFARLAND, B. H. (1970). *Archs Biochem. Biophys.*,**138**:285.

JOST, P. C., GRIFFITH, O. H., CAPALDI, R. A. and VANDERKOOI, G. (1973). *Proc. natn. Acad. Sci. U.S.A.*, **70**:480.

KIMELBERG, H. K. and PAPAHADJOPOULOS, D. (1971). *Biochim. biophys. Acta.* **233**:805.

KREUTZIGER, G.. O. (1968). *Proc. Electron Microsc. Soc. Am.*, **26**:234.

LANDS. W. (1960). *J. biol. Chem.*, **235**:2233.

LE MAIRE, M. and TANFORD, C. (1976). *Biophys. J.*, **16**:82a.

LEONARD, R., DEAMER, D. and ARMSTRONG, P. (1972). *J. Ultrastruct. Res.*, **40**:1.

MACLENNAN. D. H. (1970). *J. biol. Chem.*, **245**:4508.

MACLENNAN, D. H., SEEMAN, P., ILES, G. H. and YIP, C. C. (1971). *J. biol. Chem.*, **246**:2702.

MAKINOSE. M. (1969). *Eur. J. Biochem.*, **10**:74.

MAKINOSE, M. and HASSELBACH. W. (1971). *FEBS Lett.*, **12**:271.

MARCHESI. V. T.. JACKSON. R. L., SEGREST. J. P. and KAHANE. I. (1973). *Fedn Proc. Fedn Am. Socs exp. Biol.*, **32**:1833.

MARCHESI, V. T., TILLACK, T. W., JACKSON, R. L., SEGREST, J. P. and SCOTT, R. E. (1972). *Proc. natn. Acad. Sci. U.S.A.*, **69**:1445.

MARTONOSI. A. (1971). *Biomembranes*, pp. 191–256. Ed. L. A. MANSON. New York; Plenum Press.

MARTONOSI. A. and HALPIN, R. A. (1971). *Archs Biochem. Biophys.*, **144**:66.

MEISSNER, G. and FLEISCHER, S. (1971). *Biochim. biophys. Acta*, **241**:356.

MOOR, H. and MUHLETHALER, K. (1963). *J. Cell Biol.*, **17**:609.

MOOR. H.. MUHLETHALER. K.. WALDNER. H. and FREY-WYSSLING. A. (1961). *J. biophys. biochem. Cytol.*, **10**:1.

MORAWIECKI, A. (1964). *Bioochim. biophys Acta*, **83**:339.

MORSE, P. and DEAMER, D. (1973). *Biochim. biophys. Acta*, **298**:769.

OHLENDORF. D. H.. COLLINS. M. L.. PURONEN, E. O., BANASZAK. L. J. and HARRISON. S. C. (1975). *J. molec. Biol.*, **99**:153.

PACKER. L.. WILLIAMS. M. and CRIDDLE. R. (1973). *Biochim. biophys. Acta*, **292**:92.

PINTO DA SILVA. P. G. and BRANTON, D. (1970). *J. Cell Biol.*, **45**:598.

QUINN, P. J. and DAWSON. R. M. C. (1969). *Biochem. J.*, **115**:65.

ROGERS. M. J. and STRITTMATTER. P. (1973). *J. biol. Chem.*, **248**:800.

ROTHFIELD. L. I. (1971). *Structure and Function of Biological Membranes*, pp. 251–282. Ed. L. I. ROTHFIELD. New York; Academic Press.

SCARPA, A. and INESI, G. (1972). *FEBS Lett.*, **22**:273.

SEGREST. J. P., JACKSON. R. L.. TERRY. W. and MARCHESI. V. T. (1972). *Fedn Proc. Fedn Am. Socs exp. Biol.*, **31**:736.

SEGREST, J. P., GULIK-KRZYWICKI, T. and SARDET, C. (1974). *Proc. natn. Acad. Sci. U.S.A.*, **71**:3294.

SINGER. S. J. (1971). *Structure and Function of Biological Membranes*, pp. 145–222. Ed. L. I. ROTHFIELD. New York; Academic Press.

SINGER. S. J. and NICOLSON. G. L. (1972). *Science, N. Y.*, **174**:720.

SJOSTRAND. F. S. (1963). *J. Ultrastruct. Res.*, **9**:340.

SPECTOR. A. A.. JOHN. K. and FLETCHER. J. E. (1969). *J. Lipid Res.*, **10**:56.

STAEHELIN. L. A. (1972). *Proc. natn. Acad. Sci. U.S.A.*, **69**:1318.

STEERE. R. L., (1957). *J. biophys. biochem. Cytol.*, **3**:45.

VAIL, W. J., PAPAHADJOPOULOS, D. and MUSCARELLO, M. A. (1974). *Biochim. biophys. Acta*, **345**:463.

VANDERKOOI, G. and GREEN, D. E. (1970). *Proc. natn. Acad. Sci. U.S.A.*, **66**:615.

WALLACE, R. A. and DUMONT, J. N. (1968). *J. cell. Physiol.*, Suppl. **72**:73.

WEHRLI, E. and MORSE, P. D. (1974). *J. supramolec. Struct.*, **2**:71.

WINZLER. R. J. (1969). *Red Cell Membrane*, pp. 157–171. Ed. G. A. JAMIESON and T. J. GREENWALT. Philadelphia; J. B. Lippincott.

WRIGGLESWORTH. J. M.. PACKER, L. and BRANTON, D. (1970). *Biochim. biophys. Acta*, **205**:125.

YAMANAKA, N. and DEAMER. D. (1976). *Biochim. biophys. Acta*, **426**:132.

2

Cell adhesion

J. G. Edwards
Department of Cell Biology, University of Glasgow

2.1   INTRODUCTION

An essential function of the plasma membrane of animal cells, perhaps in concert with extracellular substances, must be to effect the bonding together of cells into tissues, and to control the adhesion of cells to non-cellular surfaces. Cells can clearly be programmed to adopt widely differing life-styles in these respects. They may remain separate, suspended in fluid medium (red blood cells); individual, but capable of locomotion over surfaces (lymphocytes); or firmly bonded into a tissue in which the intercellular contacts must sustain tension (heart muscle). Transitions between these states are frequent even in adult life, for example the release of circulating blood cells from haemopoietic tissue, the aggregation of platelets and the wound-healing behaviour of fibroblasts. In embryonic development, the making and breaking of contacts between cells, and between cells and non-cellular surfaces according to what is apparently a detailed programme, seems to be a fundamental part of the mechanism by which simple arrays of cells give rise to more complex structures. The molecular events underlying this aspect of cell behaviour are at present largely obscure.

When one cell encounters another, or a non-cellular surface, whether long-lasting, transient or no adhesion results may often be decided by interactions which are less simple and direct than the bonding together of two surfaces as constituted at· the instant of encounter. (The phenomenon of *encounter* may be thought to be unimportant in those many tissues in which the cells remain in contact from the time of their origin. In Section 2.2 I shall suggest a contrary view.) This review will illustrate how different workers use the term 'cell adhesion' in different ways, some to describe the overall outcome of cell encounters, others with various more restricted meanings. It is clearly desirable that the various cellular activities which contribute to the outcome of these encounters be resolved experimentally, and it is important to determine what kind of event is being measured by an assay of 'cell adhesion' taking place outside the animal. The following speculative analysis of the stages of cell encounter is intended only to illustrate some possible interactions and to introduce some of the ideas considered later in more detail.

Stage 1: Approach of surfaces. As the surfaces of two cells approach within a few tens of nanometres of each other they are mutually attracted by an electrodynamic force, in the absence of more intimate molecular interactions, which acts continuously (Curtis, 1967; Weiss, 1967; Parsegian and Gingell, 1972). This attraction is opposed by electrostatic repulsion between the negatively charged surfaces. Even at this early stage, environmental information, relating to the presence of the other cell, could reach the cytoplasm of the cells, by way of an increase in negative electrostatic surface potential (Wolpert and Gingell, 1969). Transfer of information at any stage could have consequences for the future stability of the contact.

Stage 2: Initial molecular interaction. Enmeshing of surface macromolecules on apposed cells could depend on a transient stabilization of the contact established by the interactions of stage 1. The time course of these events is unknown, but it may be determined by the tangential diffusion of molecules in the two membranes (Frye and Edidin, 1970) and may be less than 1 s. Such interaction would depend on the close approach (less than 0.5 nm) of the cell surfaces and could be achieved by the protrusion of some molecular species a distance of 5–10 nm from the lipid of the plasma membrane (Weiss and Harlos, 1972). This interaction could constitute adhesive recognition by a mechanism of molecular complementarity (Weiss, 1946; Moscona, 1968; Roth, McGuire and Roseman, 1971b) and be transduced to the cell interior to regulate motile and biosynthetic phenomena (Abercrombie, 1970; Dulbecco and Stoker, 1970).

Stage 3: Junction formation. Further molecular interactions may be arbitrarily distinguished from stage 2 because they give rise to specializations of the membranes detectable by electron microscopy, or mechanistically distinguished because they depend on an active response from the cell, such as the extrusion of junction components from the cytoplasm (Sheffield, 1970). The time course here may extend indefinitely from a few seconds upwards (Heaysman and Pegrum, 1973a). In the longer term there may be synthesis and secretion of extracellular matrix materials. The decision to progress into this stage may depend on what the cell 'sees' in the earlier interactions. This may be simply because junctions can form only between cells already adherent, but could also depend on more complex regulatory consequences of stages 1 and 2. Moreover the formation of a gap junction (Gilula, Reeves and Steinbach, 1972), by permitting the exchange of small molecules between cells, could provide the basis for a further cellular decision on the future of the contact (Wolpert, 1971). The progress of contact through stages 2 and 3 may depend on the compliance of the cell surface or its active participation in permitting or generating appropriate geometries (Weiss, 1967) such as extrusion of processes (Bangham and Pethica, 1960) and expansion of the area of contact. The medium, which must drain from the space between the cells, may influence these processes as a consequence of its viscosity (Curtis, 1967).

Stage 4: Cell separation. Cell separation may ensue if any of the interactions described above fail to stabilize the adhesion against the prevailing distractive forces. One possibility is that the two plasma membranes continue to adhere but are disrupted (Weiss, 1967) or drawn out into retraction fibrils, perhaps because the adhesion of the membranes was not co-ordinated with an appropriate filament system within the cell. Cells could contribute to their separation by reversing some of the interaction which has already been established, for example by enzymatic modification of surface components (Roth, McGuire and Roseman, 1971b; Martz and Steinberg, 1973). For a cell to end a contact by its own pulling (Gustafson and Wolpert, 1967; Harris, 1973b) or by the action of interfacial forces (Steinberg, 1970), it must also establish a stronger adhesion elsewhere.

This review is not arranged in the sequence of the stages just described, but moves from experimental systems, which probably depend on all of them, towards what are, hopefully, simpler situations.

2.2   THE SIGNIFICANCE OF HISTOGENETIC REAGGREGATION

An extensive literature now illustrates that coherent tissues of developing animals can be experimentally disaggregated to yield suspensions of viable single cells, cells which under appropriate conditions can reassociate and reconstruct some structures and patterns characteristic of the tissue of their origin (Townes and Holtfreter, 1955; Trinkaus, 1969; Steinberg, 1970; Moscona, 1973). Among warm-blooded vertebrates, cells from a given tissue sometimes mingle with cells from the corresponding tissue of another species to form a chimeric tissue (Moscona, 1957). On the other hand, cells from different tissues segregate by tissue type—the remarkable phenomenon of 'sorting-out'. The same rule of tissue formation emerges from another experimental approach, the collecting-aggregate technique (Roth and Weston, 1967; *see* Section 2.7). Similarly, studies of the fusion of skeletal myoblasts in monolayer culture (Yaffe and Feldman, 1965) have shown that myoblasts of rat and horse will fuse to form chimeric muscle fibres, but will exclude the nuclei of cells of other tissues, including cardiac myocytes. The rule is not invariable, however, since mouse and chick embryonic myocardial cells sort out from each other in aggregates of heart ventricle cells (Burdick and Steinberg, 1969), nor has it been very widely explored to establish phylogenetic limits. The relevance, for real morphogenetic processes, of cell-sorting in reaggregates and tissue reconstruction *in vitro* has been called in question (Wolpert, 1971). One conclusion must surely be widely agreed upon, however, namely that reaggregation experiments show the instructions for assembly of tissues to be carried by the cells themselves, and that a further source of ordering information is not required. This follows directly from the existence of histogenetic reaggregation, whatever its mechanism. The criticism may still be made that the cells of many tissues shown to be capable of tissue reconstruction in aggregates are never separated *in vivo*, much less required by the developmental programme to sort out from other tissues which are anatomically remote. It is not a complete answer (Moscona, 1973) to point to the many instances, including presumptive heart cells, where dispersal and recombination of cells within the organism are actual features of embryonic development.

It can be argued, however, that cellular self-recognition of the kind seen in cell-sorting experiments may be necessary for the stability of embryonic tissues. First, if cells are to be free to exchange contacts and move about within tissues, they must somehow avoid mixing at the boundaries between differentiated tissues. That embryonic tissues do possess such 'fluid' properties is seen, for example, in the tissue-spreading experiments of Steinberg (1962a) discussed in Section 2.3. Secondly, when cells in developing tissues are growing and dividing, newly formed surface may be required to recognize contiguous cells *de novo*, even if previously formed surface remains in stable adhesion. In this way, for example, when cells in the germinal layer of an epithelium divide asymmetrically, one daughter cell may be given the freedom to make the transitory adhesions required to move away, whilst the other remains in place. There may therefore be dynamic aspects to the stability even of those tissues which do not undergo gross dispersal. So, although the sorting of liver from cartilage cells *in vitro* is quite unlike any process of normal development, the self-recognition it reveals may be of fundamental importance. This conclusion, like the one made earlier, does not require a knowledge of mechanism. Whether cell-sorting is adequately modelled by separation of immiscible liquids (Section 2.3) or whether it requires qualitatively specific adhesive recognition (Section 2.8) does not affect the issue.

The relevance of reaggregation *in vitro* to histogenesis *in vivo* is illustrated by a unique instance involving genetic analysis. The autosomal recessive mutation in 'reeler' mice affects the structures of the cerebellum and cerebrum. In homozygotes, grossly abnormal cellular arrangements are found in these specific regions of the brain. De Long and Sidman (1970) showed that cells which had been dissociated at the appropriate developmental stage from the cerebral isocortex of normal mouse embryos were able to reaggregate and establish cell alignments resembling those of normal tissue *in vivo*. Comparable cells from reeler embryos also reaggregated but became organized in a manner resembling that of the mutant *in vivo* showing, in particular, a lack of alignment of neurone cell bodies and processes.

Although reaggregation techniques are clearly of great value in the analysis of histogenesis at the cellular level, it remains to be seen whether the same procedures will provide a means for investigating the molecular basis of intercellular adhesion.

A series of experiments with inhibitors (Moscona and Moscona, 1963; Richard, Glaeser and Todd, 1968) clearly established that reaggregation requires macromolecular synthesis, in particular of protein. Together with the widely acknowledged efficacy of trypsin in dissociating cells (Moscona, 1963) this suggests that protein located outside the permeability barrier of the cells may be involved in interactions such as those described in stages 2 and 3 of Section 2.1. The suggestion that inhibition of reaggregation by periodate is evidence for the involvement of carbohydrate (Moscona, 1965) is, however, unacceptable. This reagent would react very rapidly with glucose in the medium used in these experiments and it may be that its effects were due to the reactive aldehydes so generated.

Another approach has been to use the progress of reaggregation as an assay for soluble aggregation-promoting factors derived in various ways from intact embryonic cells. For example, Lilien (1968) found that conditioned medium, taken from monolayer cultures of chick embryonic neural retina cells, enhanced the aggregation of freshly dissociated homologous cells, as assessed by the size of aggregates formed after 24 hours. Provided the conditioned medium was dialysed, it had no similar effect on liver, heart or limb-bud cells. Daday and Creaser (1970) reported that a preparation obtained from neural retina cells by washing with EDTA also promoted their aggregation. Particularly striking are the results described by Garber and Moscona (1972), in which a preparation from cerebral cells enhanced the aggregation of homologous cells only, being inactive even with cells from other regions of the brain. Moscona (1968) believed that these preparations contain discrete factors which act as cell ligands, but these have not yet been isolated or characterized. The investigation of the mode of action of such ligands would require short-term assays, since effects on aggregation observed over many hours may be a consequence of many different events.

2.3   THE DIFFERENTIAL ADHESION HYPOTHESIS

It is still widely supposed that the sorting of cells in reaggregates is evidence for the operation of specific mechanisms of intercellular adhesion. Both Curtis (1967) and Steinberg (1970) have argued that, on the contrary, such specificity is not required to account for segregation of cells in aggregates, neither does it account for the patterning of cells which accompanies segregation. Steinberg's studies of this problem led to a detailed hypothesis, the *differential adhesion hypothesis*, concerning the manner in which adhesive interactions between individual cells generate the configurations of embryonic cell populations (Steinberg, 1970). The configuration to which most attention has been given is the sphere within a sphere: the 'onion' configuration (Goel *et al.*, 1970), which commonly results when dissociated cells from two different embryonic tissues are mixed and reaggregated. In addition to sorting into two homogeneous tissues, cells from one of the tissues usually form one or more approximately spherical discrete islands within an enveloping sphere of the other tissue. Steinberg (1958) based an early model for the control of adhesion between like and unlike cells on a geometric lattice of sites capable of binding calcium ions. Although this model seems something of a numerological curiosity today, it led to the suggestion that, in an aggregate containing a mixture of two cell types X and Y, X cells would displace Y from already established contacts, if X cells adhere to each other more tightly than do Y cells, and this could result in the formation of one or more solid islands of X surrounded by Y. At that time, the main alternative explanation for the internalization of X cells was a proposal (Stefanelli, Zacchei and Ceccherini, 1961) that X cells might be attracted inwards by chemotactic response to a radial concentration gradient. To test this, Steinberg (1962b) chose a cell pair, namely chick embryonic heart and neural retina, in which the internally segregating component—heart—could be detected cytologically at low concentration. In aggregates containing only 1 percent of heart cells, these appeared to avoid the surface of the aggregates, but did not progress towards the centre. Thus central location of the internal component was a property of the heart cell population, but not of individual cells, and this was not consistent with the chemotactic mechanism. Steinberg drew an analogy between the sorting of cells in these mixed aggregates and the breaking of a dispersion of one liquid in another immiscible liquid of similar density. The basis of the analogy is that both systems are composed of two kinds of discrete units (molecules of the liquids corresponding to cells in the tissues) which are motile, and which cohere and adhere with different strengths. (In this context, cohesion refers to like–like interaction, adhesion to like–unlike.) Steinberg was then able to show that the time-course of sorting in the pair heart–neural retina was more as expected from this analogy than from a mechanism based on differential timing of surface changes proposed by Curtis (1961). To these early results Steinberg (1970) has added two series of observations which must surely be of great importance for the understanding of tissue organization. The first series showed that sorting of dissociated cells in aggregates is a change towards an equilibrium configuration which can be approached from another direction. A variety of pairs, which reaggregate and sort to yield the 'onion' configuration, yield the same configuration when juxtaposed as fragments of intact tissue (the externally segregating tissue engulfs the internal). Steinberg pointed out that in so doing, the covering tissue actually relinquishes like–like adhesions in favour of like–unlike. The second series of observations revealed the existence of a hierarchy of embryonic cell types with respect to their tendency to take up the inner position, either in reaggregation or fragment-spreading experiments. The hierarchy of chick embryonic tissues reported by Steinberg (1970), listed so that every member tends to envelop any of those which precede it, is the germinal layer of epidermis, limb-bud pre-cartilage, pigmented epithelium of retina, myocardium of heart ventricle, neural tube and liver. The elegance of these findings is that they are for the most part independent of details of technical procedures, in particular of tissue dissociation, so that it is reasonable to suppose they reflect the operation of rules of cell behaviour which operate *in vivo*, although then doubtless subject to complex structural restraints. One substantial exception is that the behaviour of heart is sensitive to certain details of procedure (Wiseman, Steinberg and Phillips, 1972; Armstrong and Niederman, 1972). A detailed formalism based on the immiscible liquid analogy (Steinberg, 1962a; Phillips and Steinberg, 1969) has been developed to systematize both the observations that binary cell systems approach configurational equilibria, and that a hierarchy exists. The movement of cells towards the onion configuration is seen as an approach to a 'thermodynamic' equilibrium, and the ranking of cells in the hierarchy as the sequence of relative strengths with which the cells cohere, defined as specific 'interfacial free energies' of the cell populations. This analogy may well prove self-consistent in codifying many phenomena at the level of cell populations where there is evidence for reversibility and the existence of equilibria. However, this is true only of tissues at those developmental stages before irreversibility results, perhaps from the formation of certain kinds of junctions and the synthesis of extracellular matrix and fibre networks. It also seems quite feasible that the equilibrium flattening of spherical aggregates in a centrifugal field could measure the same kind of cohesiveness which decides position in the onion configuration (Phillips and Steinberg, 1969). Whether this is really so depends on the demonstration that, for each tissue studied, the shape change of the aggregate results wholly from redistribution of cells within the aggregates and that there is no contribution from shape changes of individual cells.

The differential adhesion hypothesis does not require that sorting in reaggregates should result from the operation of specific adhesive mechanisms. A more complex question is whether the success of the hypothesis actually argues against the existence of such specificity; the answer seems to be that it does not. Strict application of the liquid drop analogy has shown (Steinberg, 1962a) that the required condition for cells of type X to form a sphere which is totally enclosed by cells of type Y is that the XY adhesions must be intermediate in strength (expressed as the value of the 'reversible work of adhesion') between strong XX and weaker YY adhesions. They must also be weaker than the mean strength of XX and YY. Simple models in which specific homotypic bonding for XX and YY, acting additively with a non-specific mechanism for XX, XY and YY, evidently suggest that XY adhesion will be weaker than XX or YY, a situation which should lead to incomplete coverage of X by Y, according to the liquid analogy. Indeed, the required relationship can be obtained simply on the assumption that there is a single adhesive mechanism common to X and Y, effected by sites which are more frequent on cells of type X than on Y—the site frequency model (Steinberg, 1963). It is, therefore, important to ask what configuration is attained by cell pairs for which the collecting aggregate technique points to the existence of adhesive specificity (Section 2.7), particularly for the pair 7-day-old chick embryonic neural retina and heart. Neural retina cells totally enveloped heart cells in aggregate-fusion experiments by Moyer and Steinberg (1972), and these authors explain the apparent conflict regarding specificity by the principle that the rate of a reaction (the collecting-aggregate measurement) is not determined by its free-energy change (from the equilibrium configuration). It seems likely that the differences between these two operational measurements of the strengths of cell adhesion are greater than this suggests. The collecting-aggregate technique compares the net rate of formation of adhesions over short times, in the face of relatively simple shearing forces tending to separate cells from aggregates. Steinberg's configurational measurement compares their stability in the face of complex distractive forces dependent on both competing attachments and cellular motility, and exerted over much longer times.

Indeed, nothing in the observations of Steinberg and his colleagues yet justifies proceeding from the level of cells within populations to any particular explanation of the phenomena in terms of the behaviour of individual cells and its molecular basis. In particular it is essential to realize that when Phillips and Steinberg (1969) referred to interfacial free energies, this applies not to the cell surface, but to the surface of a tissue generated by an array of independently motile cells. In terms of the introduction to this chapter, Steinberg's hypothesis deals with adhesion in the overall sense (stages 1–4) so that the actual aspects of cell behaviour which generate the population phenomena remain to be determined. As Abercrombie (1964) has pointed out, 'any cellular behavior or combination of kinds of behavior which sufficiently reduces the probability of separation of cells once collision has occurred will, in theory, serve instead of adhesion'.

2.3.1   **Differential adhesion and the movement of cells on solid surfaces**

For cell-sorting to occur in aggregates, cells must move. There has been debate whether the movements are passive, driven by the tendency of cells to equilibrate their adhesive interactions, or whether the cells supply meta-bolic energy for some kind of motility. In several instances, cell-sorting has been found to be inhibited by concentrations of cytochalasins similar to those required to inhibit the motility of cells in monolayer culture (Steinberg and Wiseman, 1972; Maslow and Mayhew, 1972). This suggests that active micro-filament-dependent motility may be involved, but current uncertainty about the mode of action of these inhibitors leaves open the possibility that they may affect surface properties directly. In one instance, cell-sorting appears to be rather insensitive to inhibition by cytochalasin (Armstrong and Parenti, 1972). Trinkaus (1969) has suggested that motility in cell-sorting may resemble that of the mesenchyme cells of *Arbacia* embryos, in which a cell extrudes a filopodium which adheres to another cell and then contracts, pull-ing the cell body closer to the new contact, rather than the 'amoeboid' loco-motion of cells translocating on a substrate, although these forms of motility may be closely related (Wolpert, 1973). Both are dependent on concomitant adhesion, in a way which has no parallel in the motility of molecules in liquids and so, presumably, cannot be adequately represented in Steinberg's (1970) analogy.

A parallel debate has surrounded the role of adhesion in the translocation of cells on a plane surface. Clearly, any mechanism requires the making and breaking of adhesions to the substrate, but these events could conceivably drive the locomotion so that no other cellular 'motor' is required (Carter, 1967). On a substrate of uniform adhesiveness, the leading edge of a cell is seen as advancing by the same mechanism as the spreading of a liquid drop on a wettable surface, the trailing edge being released by a decrease in its ad-hesion to the substrate, effected by secretion from the cell, and thus dependent on metabolic energy. In a more recent discussion of this mechanism, it has been suggested that the de-adhesion could be caused by hydrolytic enzymes, or by local pH changes (Martz and Steinberg, 1973). In either case, metabolic energy of the cells is transduced via alterations in adhesive properties, and not by active shearing of a system of intracellular filaments. Carter's proposal originated from elegant experiments in which he showed that L cells move up a gradient of adhesiveness between cell and substrate generated by graded deposition of palladium on a surface of cellulose acetate, a phenomenon he termed *haptotaxis*. In a recent extension of this work, Harris (1973b) con-firmed that 3T3 cells 'prefer' a palladium surface to cellulose acetate (*see* Sec-tion 2.5) but reinterpreted the behaviour in terms of active cell locomotion, by which cells are pulled in the direction of the strongest or most long-lasting marginal adhesion. Indeed, there are many descriptive and experimental grounds for the current, rather widely accepted, view that cell locomotion involves a 'motor' which is probably filament-dependent (Porter and Fitzsim-mons, 1973). The question remains as to whether the breakage of trailing adhesions is accompanied by some chemical modification of the contacting surfaces.

Differences in adhesion between cells, and between cells and substrate, have also long been considered important determinants of the social interactions of cells cultured on solid substrates (in particular of contact inhibition of movement) and thus of the morphology of cultures. In terms of the outcome of individual collisions between moving cells, contact inhibition of movement has been defined (Abercrombie and Heaysman, 1954) as the halting, after contact, of continued movement such as would carry one cell over the surface of another. This can be conveniently quantitated in terms of the overlap index, which is the relative frequency with which nuclei overlap. It is now becoming apparent that nuclear overlaps can occur by events other than the failure of contact inhibition of movement. The frequency of these other events may well be determined by differential cell–cell or cell–substrate adhesiveness, but contact inhibition of movement in the strict sense may be more easily understood if the primary effect is indeed inhibition of locomotory activity. Both Weston and Roth (1969) and Martz and Steinberg (1973) have pointed out that if cell–cell adhesions are sufficiently strongly preferred over cell–substrate adhesions, cells will tend to form multilayered clumps, yielding a very high overlap index. The operation of an effect fitting this prediction has been observed (Harris, 1973b), in which cells such as 3T3 and chick heart fibroblasts overlap very extensively if they are grown on substrates to which they adhere weakly. The overlapping appears to result from retraction and clumping of monolayered cell sheets, and not from a failure of contact inhibition of movement. In addition, Harris observed that cell sheets can be caused to clump by jarring the culture so that, in these conditions, the sheet configuration cannot be an 'equilibrium' as proposed by Martz and Steinberg (1973). Overlapping can also occur when a moving cell passes between another and the substrate, usually in a front-side collision (Abercrombie, 1970), and this may be more frequent when cell–substrate adhesion is greater than cell–cell (Weston and Roth, 1969). However, this mechanism has also been proposed for the increased 'underlapping' which occurs when cells grow on a substrate to which they adhere weakly (Abercrombie, 1967), another effect which must be distinguished from real failure of contact inhibition of movement. 'Underlapping' may also contribute to the criss-cross morphology characteristic of many fibroblastic cells which have been transformed by oncogenic viruses, since criss-crossing can occur in cultures of cells that exhibit normal contact inhibition of movement (Bell, 1972a; Guelstein et al., 1973).

What of the role of adhesion in contact inhibition proper? Time-lapse films (Abercrombie, 1970) show that when a moving cell makes contact with another, a contraction of the leading lamella occurs and ruffling activity ceases. This provides evidence that cell–cell adhesion has occurred, because the cell can be rapidly pulled towards the contact by this contraction. Alternatively, retraction of the cell can occur, in which case the appearance of 'retraction fibres' suggests that it may not be adhesion between the membranes which has failed, but the resistance of the lamellar membrane to stretching. It is difficult to believe that adhesion per se stops the forward movement, since the leading lamella is continuously involved in forming new adhesions to the substrate, which are evidently consistent with continued ruffling and locomotion. It also seems unlikely that it is simply the difference in strength between the cell–cell and cell–substrate adhesions which is responsible, since the cell–cell adhesion should be strong enough to support locomotion over the contacted cell if it is strong enough to support the pulling forward of the cell itself by the lamellar contraction.

Martz and Steinberg (1973) have attempted to explain contact inhibition of locomotion in terms of the reluctance of the cell to move onto a less adhesive substrate. This seems better able to describe the behaviour of cells encountering a substrate of low adhesiveness, such as agar. The cells fail to move onto a substrate of this type, and although ruffling activity of the leading lamella continues, no contact contraction is observed (Abercrombie, 1970). A more likely interpretation seems to be that the strength of the cell–cell adhesion is irrelevant, but that it differs qualitatively from cell–substrate adhesion in being accompanied by a signal of some kind (Abercrombie, 1970) which activates the contact contraction and inhibits the spreading component of motility, perhaps by preventing the insertion of new surface in this region (Abercrombie, Heaysman and Pegrum, 1970; Harris, 1973a). Such a mechanism could be related to the observations (Bell, 1972b; Walther, Ohman and Roseman, 1973) that tissue cells settling from suspension onto extended cells do adhere but fail to spread, whereas they would be expected to spread on a non-cellular substrate.

2.4   AVERAGED FORCES VERSUS LOCAL STRUCTURE

The 'fluid' properties of embryonic tissues as seen, for instance, in the spreading of tissue fragments, show that the adhesions in these tissues must be regarded as transient and reversible. This was among the considerations which led Curtis (1960) to examine possible alternative mechanisms of intercellular adhesion distinct from the then-prevalent static mechanisms dependent on cement-like materials.

One of the outstanding achievements of colloid chemistry is the DLVO theory of Derjaguin and Landau (1941) and Verwey and Overbeek (1948) which describes quantitatively how dispersions of lyophobic colloid particles are electrostatically stabilized, i.e. what conditions of ionic environment suppress or permit the adhesion of the particles (see Napper, 1970). The theory postulates the opposition of two forces, a coulombic repulsion between particles which bear like charges, and an attractive force attributed to London dispersion forces. These attractive forces exist between all substances; for example, they are responsible for the liquid state of noble gases at low temperature. They must therefore operate between the surfaces of cells and can be appreciable between macroscopic bodies at distances of several tens of nanometres. In particular, the DLVO theory predicts the existence of two spacings of adhesions, a strong and usually irreversible adhesion with surfaces in molecular contact and a relatively weak, reversible interaction occurring with surfaces separated by a fluid-filled gap of 10–20 nm. This latter arises because the repulsive electrostatic force decreases more rapidly with increasing separation than does the attractive force, yielding a secondary minimum at this separation in the potential energy curve. Its existence is one of the successful predictions of the DLVO theory since it accounts for the weak, readily reversed flocculation of large polystyrene latex beads (Schenkel and Kitchener, 1961) and possibly for the formation of tactoids (Overbeek, 1952).

Curtis (1960) noted that these two spacings appeared to fit rather well the two observed spacings between the twin electron-dense tracks of plasma membranes seen in electron micrographs of cell contacts. In addition to providing a readily reversible mechanism of adhesion in the secondary minimum, and accounting for these spacings in terms of forces which must act between cells, the application of the DLVO theory to cells offered an explanation for the widely reported importance of divalent cations in intercellular adhesion since divalent cations could reduce repulsive electrostatic potential. It also offered a common mechanism for adhesion of cells to each other and to non-cellular substrates, but predicted that differences in adhesiveness between cells would be found to be differences of degree only, and not of specificity. The application of these concepts to cells has recently been continued (Parsegian and Gingell, 1972), using a more sophisticated method of calculating attractive dispersion forces between macroscopic bodies, based on the theory of Lifshitz (1956), and including in the model a layer of carbohydrate 'fuzz' on each side of the the lipid of the plasma membrane. These authors found a secondary minimum at a membrane–membrane spacing of 5–8 nm, the separation between outside surfaces of the 'fuzz', and corresponding to a separation between membrane centres of 13–16 nm. They concluded that an adhesion in the secondary minimum would be maintained in the presence of Brownian movement, but could be broken by contractile forces exerted by the cells themselves, and that variations in the thickness, density or optical properties of the 'fuzz' layer could account for specificity in intercellular adhesion. Protein may contribute to such 'long range' specificity (Nir, Rein and Weiss, 1972), but non-specific mechanisms have also been used to obtain sorting-out of cells within the context of a DLVO model of cell interaction (Good, 1972).

These colloid models of cell adhesion suppose that adhesion is controlled by the bulk properties of cell surfaces, averaged over dimensions considerably greater than the size of individual protein molecules. They therefore contrast sharply with the class of mechanism discussed in Section 2.8, which depend on specific local molecular configurations. The question therefore arises: which mechanism is likely to prove more useful in understanding the relation between cell-surface structure and cell adhesion in animals? The answer is still a matter of subjective attitude rather than evidence, but my own firm preference is for the latter on the following grounds:

1. Ultrastructural studies point to the existence of various specific molecular architectures with qualitatively different properties, associated with at least some kinds of cell adhesion (Section 2.6).
2. The demonstration of a degree of selectivity of both like–like and like–unlike kinds suggests the existence of specific adhesive mechanisms (Section 2.7), as does recent evidence for the involvement of specific carbohydrate residues (Section 2.8).
3. Studies of simpler organisms also seem to point to this answer; for example, discrete contact sites in *Dictyostelium* can be blocked by monovalent fragments of specific antibodies (Beug, Katz and Gerisch, 1973).
4. There seems no good reason why cells should abandon, at the intercellular level, their intracellular means of obtaining genetic control over extended structures through the medium of specific protein configurations.

However, forces involved in colloid stability must apply to cell contacts; the only doubt concerns the importance of their contribution, and it is not easy to see how this can be tested. Weiss (1967) acknowledged the general applicability of the DLVO theory to cells but believed adhesion in the secondary minimum to be too weak to maintain effective adhesion, to achieve which cells must interact more closely. To effect such proximity cells could penetrate the electrostatic repulsive barrier by the extension of processes with low radius of curvature, or by molecular protrusions (Weiss and Harlos, 1972). Perhaps the stability at the secondary minimum is indeed marginal, but sparsely distributed specific bridging macromolecules have evolved to conform with this spacing. The flocculation of colloids by bridging polymers is a well-known phenomenon (Napper, 1970).

2.5   ADHESION TO NON-CELLULAR SUBSTRATES

A problem with the notion that specific molecules are responsible for the adhesion of cells in culture has always been the familiar observation that they adhere to a wide range of non-biological substrates. For example, Parsegian and Gingell (1972) commented that 'for such interactions, evolution did not prepare the cell surface and cell enzymes to bind cells to polystyrene or PTFE'. Attention has been focused on the adhesion of cells to such substrates for a variety of reasons: first, as part of cell culture practice, since many cells will grow only when attached to appropriate solid substrates; secondly, because the cell–substrate interaction seems to offer a useful test system for the investigation of cell adhesion in general; and thirdly, because biomedical technology calls for the selection of surfaces to which cells are either non-adhesive (for blood compatibility) or adhesive (for example, for artificial skin).

Since this chapter deals largely with matters about which there is as yet surprisingly little consensus, it is gratifying to note the emergence of two generalizations about cell–substrate adhesion from widely disparate experimental systems. The first is a correlation between the wettability of surfaces and their adhesiveness for cells, even when adhesiveness is measured in the presence of complex media such as serum or whole blood. Studies on the relation between the critical surface tension (γ_c; a measure of 'wettability') of a variety of surfaces and their resistance towards generation of thrombi when in contact with blood (Baier, 1972) show that surfaces for which γ_c falls between 20 and 30 dyn cm^{-1} exhibit minimal thrombus formation. Baier also noted that the observations of Taylor (1961) on the adhesion of cultured cells show a similar minimum of attachment and spreading between 20 and 30 dyn cm^{-1}. As γ_c increases or decreases from this value, surfaces become increasingly thrombogenic. For example, Baier, Gott and Dutton (1972) found that the blood compatibility of the alloy Stellite 21, which is used in construction of artificial heart valves, depended on a firmly bound adventitious waxy coating derived from a polishing abrasive, which reduced the critical surface tension of the material from about 38 to 22 dyn cm^{-1}. Blood compatibility may be related to the adhesion of platelets to surfaces since the number of platelets adhering to various surfaces at early times after exposure to blood has been correlated with their critical surface tensions (Lyman et al., 1968). Harris (1973b), extending earlier work of Carter (1967), deposited palladium grids on cellulose acetate, polystyrene, glass and wettable (Falconized or sulphonated) polystyrene, and found that cells accumulated preferentially on the metallized areas for the first two substrates, but avoided the metal for the second pair. Carter interpreted these preferences as adhesive, since cells spread less on, and were more easily detached from, the less-preferred substrates and he noted that the preferences are also the order of wettability of the substrates by distilled water. In this work the medium contained serum, as it did in experiments in which rat hepatoma and BHK cells were shown to adhere preferentially to Falconized polystyrene, steel and glass compared with Teflon and polyethylene, further suggesting that biological adhesion may be related to wettability (Grinnell, Milam and Srere, 1972, 1973). It is worth noting that studies of rate of attachment and 'equilibrium configuration', of ease of detachment, and of implantation in the human circulation all point to the same conclusion.

The second generalization is that under most conditions of practical concern, cells adhere not directly to the substrate, but to an adsorbed layer of protein. This is true, for example, of the conditions under which the locomotion and other behaviour of cells is studied in culture. The protein film is adsorbed onto Falconized polystyrene culture dishes from culture media which contain 10 percent of serum and can be seen under the dissecting microscope in solvents of appropriate refractivity (Revel and Wolken, 1973). The film appears as a band 10 nm wide or less after being fixed and embedded for electron microscopy. The protein is eluted from dishes by serum-free medium, but not by normal saline, and may contain a major component detectable by sodium dodecyl sulphate gel electrophoresis. In the biomedical context, optical and contact-angle techniques have been used to show that a germanium surface in contact with blood becomes coated with a film of protein over 12.5 nm thick (possibly fibrinogen) within 60 seconds of exposure to fresh blood (Baier and Dutton, 1969).

I do not wish to suggest that cells cannot adhere to artificial substrates in the absence of an adsorbed protein film. There are now numerous reports that the addition of serum or plasma to media generally decreases the rate of attachment of cells to their culture substrates. The rapid attachment which occurs in the absence of serum differs from attachment in its presence in that it is independent of divalent cations (Takeichi and Okada, 1972; Rabinowitch and DeStefano, 1973) and is less inhibited by low temperature (Nordling, 1967; Rabinowitch and DeStefano, 1973). It seems that cells can themselves supply the material which renders their adhesion dependent on divalent cations since fibroblasts from chick embryonic sclera, plated in serum-free medium, are not detachable by EDTA or trypsin within 60–180 minutes of inoculation, but many cells become so after 24 hours, a change which does not occur in the cold (Takeichi, 1971). When cells are plated in conditioned, serum-free, medium, adhesion is dependent on divalent cations and cells are detachable by EDTA at early times. Similar observations, relating to ease of mechanical detachment, were reported by Daniel (1967).

The divalent cation required has usually been found to be Mg^{2+}, occasionally replaceable by other ions. However, a specific effect of Mn^{2+} at $50 \mu M$ was found in the attachment of ^{51}Cr-labelled sarcoma I cells to serum-coated glass (Rabinowitch and DeStefano, 1973); Mn^{2+} could not be replaced by

Ca^{2+}, Mg^{2+}, Co^{2+}, Ni^{2+} or Zn^{2+} and, indeed, Ca^{2+} competed with Mn^{2+}. It is intriguing that manganese-stimulated adhesion was inhibited by vinblastine, tetracaine and cytochalasin B. The effect of vinblastine recalls the inhibition of the adhesion of 3T3 cells to Falconized plastic by 5×10^{-7} M colchicine (Kolodny, 1972), and that of the cell–cell adhesion of BHK 21 cells by vinblastine, demecolcine and colchicine (Waddell, Robson and Edwards, 1974), although colchicine has been reported to have no effect on the attachment of Ehrlich ascites cells to substrates (Weiss, 1972). The effect of tetracaine could be related to the inhibition of platelet aggregation by local anaesthetics (Mills and Roberts, 1967). Furthermore, manganese reversibly stimulated the formation of cell processes, ruffled membranes and veils and since the inhibitors of adhesion inhibited these activities on both uncoated and coated plastic (Rabinowitch and DeStefano, 1973) the effects could not be interpreted as a consequence of altered adhesion. A similar relationship has been demonstrated between magnesium and spreading of chick embryonic fibroblasts (Takeichi and Okada, 1972).

Assays of this type may depend in part on spreading (Curtis, 1967), but there is no reason, a priori, why assays of intercellular adhesion, for example by aggregation kinetics, should be in any way superior in this respect. This emphasizes again the difficulty in distinguishing 'adhesion' from motility-related cellular activities. There appear to be at least three ways in which such activities of the membrane could determine the number of cells retained in these assays. Spreading may simply enlarge the area of attachment, the extrusion of processes could aid penetration of an electrostatic repulsive barrier (Bangham and Pethica, 1960; Weiss and Harlos, 1972), or the continued operation of the motility-related surface cycle (Abercrombie, Heaysman and Pegrum, 1970; Harris, 1973a) may be required to generate an adhesive surface. Although cells spread on uncoated substrates, this may be a passive effect and only in the presence of adsorbed protein will they exhibit normal behaviour (Witkowski and Brighton, 1972). The idea that active participation of the membrane in one or other of these ways is measured in these assays seems to be contradicted by the observation that the attachment of cells is not inhibited by a variety of metabolic inhibitors (Grinnel, Milam and Srere, 1972).

Consideration of the two generalizations made in this section leads to an enigma, as has been discussed by Baier (1972): since surfaces of both low and high wettability adsorb substantial protein films, why do cells still show preferences for particular underlying substrates? One explanation is that it is possible that proteins are less denatured when adsorbed on surfaces of low wettability. It is not yet established whether any specificity is required in a protein to function in these systems; gelatin can substitute for serum (Rabinowitch and DeStefano, 1973), as can γ-globulin (Takeichi and Okada, 1972). The way should now be open to investigate whether particular structural aspects of adsorbed proteins are required for adhesion of cells.

2.6   THE CELL–CELL ENCOUNTER ASSAYS

The approach referred to in the previous section, that of measuring attachment of cells to non-cellular substrates, has been modified to measure

intercellular adhesions. The surfaces are used to support confluent monolayers of cells, and the rate at which radioactively labelled cells attach to the
monolayers is determined by scintillation counting (Walther, Ohman and
Roseman, 1973). Adhesive specificity occurs in the sense that cells of a given
type adhere faster to monolayers of the same type than to monolayers of
different cells. This is true for the cell pairs mouse kidney with mouse teratoma
and for chick embryonic heart with neural retina, but not for BHK cells with
mouse teratoma. The assay is the most recent, and in some respects the most
promising, of a series of techniques for measuring the rate at which cells in
culture form adhesions, which includes the measurement of the aggregation
kinetics of suspended cells and the collecting-aggregate technique. These experimental methods are derived from the histogenetic aggregation approach
in that all depend on the dispersal of tissues to yield suspensions of viable
single cells which can re-establish adhesions, but they differ in that they
attempt to look at the most immediate outcome of intercellular encounters,
the formation of a stable adhesion in the shortest possible time course. This
is indeed only a limited aspect of the overall process of resynthesis of a tissue,
but by attempting to isolate this aspect these methods aim to reduce the
number of distinct cellular activities involved and so may provide suitable
systems in which to investigate the molecular basis of intercellular adhesion.

Measurement of the short-term kinetics of aggregation of cells in culture
was pioneered by Curtis and Greaves (1965). Curtis (1969) has since emphasized the conditions under which it is possible to determine the actual frequency (collision efficiency) with which cell collisions result in adhesions in
known shear gradients, and thus calculate adhesive energies. The rate of
aggregation of chick embryonic neural retina cells is unaffected by inhibitors
of protein synthesis, metabolic inhibitors or horse serum, but is inhibited by
low temperature, and slightly reduced by the omission of divalent cations
from the medium (Orr and Roseman, 1969). We have found that aggregation
of BHK 21 cells is completely inhibited by low temperature (Edwards and
Campbell, 1971) and is insensitive to inhibitors of protein synthesis unless
the cells are pre-incubated in high concentrations of trypsin (Edwards et al.,
1975). We previously believed aggregation of these cells to be insensitive to
metabolic inhibitors, but now find that cyanide inhibits provided the cells
are pre-incubated in glucose-free medium. Aggregation of BHK cells is also
inhibited by antimitotic alkaloids (Waddell, Robson and Edwards, 1974).
Treatment with neuraminidase promotes the aggregation of BHK cells
(Vicker and Edwards, 1972), an effect which may be electrostatic, although
we have suggested another explanation in terms of sialyl acceptors. Transformation of BHK cells by polyoma virus yields cell lines which consistently
exhibit much less aggregation than the untransformed cells (Edwards, Campbell and Williams, 1971). Aggregation of BHK 21 cells is slightly stimulated
by L-glutamine (Edwards et al., 1975), although in this case glutamine cannot
be replaced by hexosamines, and we suspect it may simply act as an energy
source for the cells. This contrasts with the finding that mouse teratoma cells
(Oppenheimer et al., 1969) and Ehrlich ascites, sarcoma 180 and Taper liver
ascites cells (Oppenheimer, 1973) require L-glutamine for aggregation. In
these instances, glutamine can be replaced by hexosamines, indicating that
synthesis of complex carbohydrate may be required for intercellular adhesion.

The chief disadvantage of the aggregation-kinetic approach is that it cannot readily distinguish the rates of formation of like–like and like–unlike adhesions. To overcome this problem, and others, Roth and Weston introduced the collecting-aggregate technique, in which aggregates or tissue fragments are circulated in suspensions of radioactively labelled cells, and the rate at which cells adhere to the aggregates estimated by autoradiography (Roth and Weston, 1967) or liquid scintillation counting (Roth, McGuire and Roseman, 1971a). This technique involves comparing the rates at which different aggregates collect cells from a given cell suspension. The demonstration that cells of a variety of chick embryonic tissues adhere preferentially to aggregates of like type (Roth, 1968) is the best available evidence for the operation of specific adhesive mechanisms in animal tissues; further results obtained using the technique appear in Section 2.3. The fact that, operationally, adhesion shows specificity leaves open the question as to whether the specificity lies in interactions of surface molecules (stage 2) or depends on some more complex interaction involving the contents of the cells.

Dorsey and Roth (1973) used the collecting-aggregate technique to investigate adhesive properties of normal (3T3) and malignant (SV$_{40}$-transformed 3T3 and 3T12) mouse fibroblasts; aggregates of the two malignant cell types collected either very many cells (SV$_{40}$-transformed 3T3) or very few (3T12), regardless of the identity of the cell suspensions in which they were circulated, whereas 3T3 aggregates collected an appreciable number of SV$_{40}$-transformed 3T3 cells but few 3T12. 3T3 aggregates collected more 3T3 cells if the suspension was prepared from sparse cultures rather than from dense. The authors concluded that malignancy is associated not simply with decreased intercellular adhesiveness, but with decreased intercellular recognition.

Adhesive selectivity occurs between cells of the developing retina and the optic tectum of chick embryos (Barbera, Marchase and Roth, 1973). Cells dissociated from dorsal halves of the neural retina, when suitably pre-incubated, adhere preferentially to ventral halves of the optic tectum, whereas cells from ventral retina adhere preferentially to dorsal tectum, in accordance with the electrophysiological projection. Similar preference is not exhibited by cells from cerebrum, cerebellum, brain stem or optic tectum, but is shared by pigmented retina. The same selectivity is found with tectal halves which have no innervation from neural retina. Whilst the twofold preferences found are, perhaps, rather small it must be remembered that the neural retina contains many cells not destined to interact with the optic tectum. These results may prove to be of great significance if they eventually contribute to an understanding of how specific connections arise between nerve cells.

The relationships between the various types of cell-encounter assays described here are obscure and should be clarified. It is not known, for example, whether the rates of homotypic collection in the collecting-monolayer, collecting-aggregate and aggregation-kinetic methods are determined by the same molecular events, although there is no obvious reason why this should not be so. Neither is it yet clear to what extent these assays depend solely on interactions between stable surface molecules, or simultaneously on motility-related activities of membranes, as discussed in relation to adhesion to surfaces. Indeed, whether they measure adhesive mechanisms which are of significance *in vivo*, and which will prove susceptible to analysis in terms of discrete molecular events, remains to be seen.

2.7   THE RELEVANCE OF JUNCTIONAL ULTRASTRUCTURE

Extensive information is becoming available about the ultrastructure of contacts between animal cells, particularly with the advent of the freeze-cleave technique, and there is a great deal of evidence, much of it admittedly circumstantial, that various specializations of surface membranes detectable by electron microscopy are involved in the adhesion of cells (McNutt and Weinstein, 1972). In many instances, however, the most conspicuous features of junctional specialization are not the intercellular architectures but intracytoplasmic density associated with filaments and their apparent insertion into the surface membrane. Points of strong adhesion between cells which have to sustain tension in tissues may not simply be bonds between plasma membranes, which would presumably too readily stretch or tear, but may be directly linked to filament systems capable of resisting stretch.

A particularly striking example is in the 'intercalated disk' joining mammalian cardiac muscle cells (McNutt and Fawcett, 1969), where the fascia adhaerens component of the junction appears to transmit from cell to cell the tension generated by the myofibrils, and raises the possibility of a molecular relationship between this structure and the Z line. The adhaerens junctions of the epithelial junctional complex appear to be analogous. In other instances it seems possible that the electron microscope is revealing something of the intercellular architecture, for example, the intercellular lamina of the desmosome (Rayns, Simpson and Ledingham, 1969), the 'fibrils' running tangentially in the membrane in freeze-cleave preparations of tight junctions (McNutt and Weinstein, 1972), and the subunit array of gap junctions (McNutt and Weinstein, 1970). In the case of those junctions where the 20-nm intercellular cleft is occluded, it may well be that their physiological functions are not primarily adhesive, but it is difficult to imagine they could serve the other functions proposed, such as tight junctions sealing the intercellular cleft and gap junctions permitting the exchange of small molecules, without incidentally contributing to intercellular bonding. In an investigation of the effects of depletion of divalent cations on the perfused rat heart (Muir, 1967), it was found that the 15-nm appositions, and the bonds between desmosomal and myofibrillar insertion plaques, were disrupted by Ca^{2+} depletion, whereas 'quintuple-layered membrane fusions' (gap junctions) were not disrupted. On prolonged incubation in Ca^{2+}-free medium cells separated completely, but the whole junctional structure remained attached to one cell leaving a hole in the plasma membrane of the adjacent cell. This prior cohesional failure of the non-junctional membrane may be a consequence of its weakness in Ca^{2+}-free medium (McNutt and Weinstein, 1972), since hyperosmotic medium containing 2.5 mM Ca^{2+} appears to separate gap junctions at their central plane (Barr, Dewey and Berger, 1965). Since the gap junction is sufficiently strong to allow its isolation as a fragment containing junctional membrane from two cells, the components of this junction may well prove to be the first molecular species with intercellular adhesive function to be isolated and characterized (Goodenough and Stoeckenius, 1972).

The question arises whether there is any relationship between the findings of the various experimental approaches to cell adhesion outlined in this chapter, and the information available from ultrastructural studies. One view is that the kind of adhesion involved in 'adhesive recognition' precedes the appearance of junctional specializations, and that there may be no molecular relationship between them. For example, 'earlier' adhesions, showing specificity, have been distinguished from 'later' adhesions characterized by desmosomes, tight junctions, etc. (Roth, McGuire and Roseman, 1971b), and a marked lack of specialized junctions has been reported in conventional thin sections made in the early phases of the reaggregation of dissociated cells (Sheffield and Moscona, 1969). It seems possible, however, that molecular species which are components of already identified junctional specializations may play a significant role both *in vivo* and *in vitro* in deciding the outcome of cell encounters. Such species may be present at the free surfaces of moving or dissociated cells, and be able to interact with encountered cells at very early times, even though they are not then ordered sufficiently to be recognizable ultrastructurally. Ordering could then follow as the result of interactions depending on co-operation of junctional components on both cells. This raises the possibility that there may be cell-type specificity vested in components of some of the currently recognized junctional specializations, although it may be entirely a property of contacts such as simple appositions, where electron microscopy presently reveals little.

In connection with his fluid model of embryonic tissues Steinberg (1970) stated, 'Conditions can sometimes arise that "freeze" a pair of apposed tissues and prevent further movement. Most obvious, perhaps, is the development of specialized cell junctions, such as desmosomes, which might act as welds.' However, it is becoming clear that the formation of at least some kinds of specialized junctions is consistent with situations where adhesion is reversible and transient. For example, ultrastructural studies of early contacts between chick heart fibroblasts show that specializations form between overlapping leading lamellae within 20 s of apparent collision (Heaysman and Pegrum, 1973a). After 60 s these junctions are associated with microfilaments inserted in electron-dense plaques underlying the points of closest approach. The association of these specializations with the events of contact inhibition is strengthened by the failure to observe them in contacts between chick heart fibroblasts and mouse sarcoma 180 cells (Heaysman and Pegrum, 1973b). There is other evidence that these effectively transient contacts also involve the formation of gap junctions (Gilula, Reeves and Steinbach, 1972).

2.8   MOLECULAR SPECIES AND MECHANISMS

As long ago as 1946, Weiss proposed that the properties of contacts between cells in animal tissues are controlled by interlocking configurations of surface molecules. For instance one cell could carry a surface component *a* which would bind specifically to a complementary structure *b* located on a second unlike cell. Like cells could be bonded by each possessing both *a* and *b*, raising the problem of how *cis* interaction of the two is to be avoided. Alternatively, like cells could be bonded in a configuration *a–c–a*, where *c* would represent the class of intercellular ligands favoured by Moscona (1968). These ideas were wholly intuitive, and in one sense dangerous since they can be adapted to explain almost any aspect of contact-dependent cell interactions, in the absence of relevant evidence. In seeking examples to illustrate the sort of complementarity he intended, Weiss suggested the interactions of antigen with antibody, hormone with effector cell and enzyme with substrate. Tyler (1946) proposed that the interacting species were antigens and 'auto-antibodies', but Townes and Holtfreter (1955) pointed out that since the 'auto-antibodies' were not elicited by a foreign element, the concept yielded little clarification of the problem. These authors also made passing reference to the linkages responsible for the recombination of dissociated haemoglobins, and it would be as well, in the face of current enthusiasm for the role of carbohydrate, not to ignore the possible involvement of protein–protein interactions such as those responsible for quaternary structure, and for the self-assembly from their monomers of structures such as microtubules. Evidently the postulated molecular species must precede in evolution the machinery of the immune response. Jerne (1971) and Bodmer (1972) have proposed mechanisms by which the immune system may have evolved from molecular species involved in cell recognition. Bodmer (1972) discussed the concept of Jerne (1971) that cell recognition involves a class of molecules already defined immunologically as 'differentiation antigens', and a second hitherto unknown class which act as their 'recognizers', and suggested that large numbers of genes determining these molecular species may be genetically linked to the histocompatibility regions. Goldschneider and Moscona (1972) demonstrated tissue-specific cell-surface antigens in embryonic cells, and consider they may have a role in cell recognition and association. It seems likely that the reason little progress has been made in identifying the molecular species involved has been the lack of suitable assay methods, until the advent of the collection techniques described in Section 2.6.

Roseman and his colleagues (Oppenheimer *et al.*, 1969; Roseman, 1970) have emphasized the possibility that the complex carbohydrates of cell surfaces may be involved in intercellular adhesion. These certainly seem likely candidates, since they are a class of macromolecules which are clearly 'information-rich' as a consequence of the diversity of monomer units and possible linkages between them and yet are not clearly associated with any biological function. Moreover it is already evident that they can be recognized by proteins, the glycosyltransferases responsible for their stepwise synthesis. Roseman (1970) also pointed out that specific interactions are possible between carbohydrates, and instanced the hydrogen-bonded structure of cellulose. Another example is the double-helix formation in sulphonated polysaccharides (Rees, Scott and Williamson, 1969).

Despite its potential importance, the experimental evidence for the involvement of complex carbohydrate in intercellular adhesion in animals is at present slender, and restricted to a very few experiments. In an investigation of the rate of collection of like and unlike (liver) cells by aggregates of chick embryonic neural retina cells, it was found that, at 28 °C, there was a lag of about 25 minutes before control aggregates collected neural retina cells (Roth, McGuire and Roseman, 1971a). However, treatment of aggregates with a crude extract of *Diplococcus pneumoniae*, which contained many glycosidases, increased more than fivefold the number of neural retina cells collected in the same time at the same temperature. The effect was also shown by one β-galactosidase partly purified from the extract, and another purified from *Clostridium perfringens* supernatant. Treatment of neural retina aggregates with these enzymes also resulted in an increase in the (unlike) collection of liver cells, but treatment of liver aggregates had no effect on the collection of either cell type. These effects are large enough to be wholly convincing in spite of the semiquantitative nature of the assay, and the authors concluded that terminal β-galactopyranosyl groups, presumably at the ends of the oligosaccharide chains of glycoproteins, mucins or glycolipids, are involved in the specific adhesion of those cells. On the other hand, the reported inhibitions in this system by low-molecular-weight compounds such a N-acetyglucosamine are much smaller, and must be near the limits of detectability. Evidence was also obtained for the presence, on the outside surface of the plasma membrane of neural retina cells, of galactosyltransferases capable of transferring galactose from UDP–galactose to exogenous and endogenous acceptors (Roth, McGuire and Roseman, 1971b).

These observations formed the basis for the hypothesis (Roseman, 1970) that surface-located glycosyltransferases and their substrates constitute the complementary species in cell adhesion. Among the attractive features of this model are that the *trans*-glycosylation constitutes a novel mechanism of information transfer between cells and that it allows cells to detach by completing the glycosyltransferase reaction, although this important feature seems not to have been demonstrated experimentally. Evidence for the presence of surface glycosyltransferases has since been presented for cells of cultured lines (Roth and White, 1972) and intestinal cells (Weiser, 1973). A UDP–glucose:collagen glucosyltransferase has been found in the membrane of human platelets (Barber and Jamieson, 1971; Bosmann, 1971). A role for this enzyme in the adhesion of platelets to collagen, believed to be the initial event in haemostasis, has been suggested (Jamieson, Urban and Barber, 1971), based on the fact that a number of inhibitors of the transferase, including glycopeptides isolated from collagen, inhibit the adhesion of platelets to collagen. Consistent with this is the demonstration that treatment of collagen with a galactose oxidase inhibits the collagen-dependent aggregation of platelets, an effect reversible by reduction of the oxidized galactose moiety with sodium borohydride (Chesney, Harper and Colman, 1972).

The specificity of the platelet enzyme is for a β-galactosyl residue on ε-hydroxylysine of collagen, so that β-galactosyl residues are involved in both chick embryonic and platelet systems but with the difference that cleavage of the galactosyl residues in neural retina lowers the specificity of adhesion rather than inhibits it. β-Galactosyl residues are also involved in the binding of desialysed glycoproteins to the surface of liver parenchymal cells (Pricer and Ashwell, 1971). In a novel approach to these problems Chipowsky, Lee and Roseman (1973) have shown that SV_{40}-transformed 3T3 cells (but not BHK 21) adhere to Sephadex beads to which galactosyl residues are linked through β-thiogalactosyl linkages, but not to the corresponding β-glucosyl or N-acetyl-β-glucosaminyl derivatives. In addition, cells attached to galactose-substituted beads seem to show an increased propensity to collect further free cells from suspension.

REFERENCES

ABERCROMBIE. M. A. (1964). *Archs Biol.. Paris*, **75**:351.
ABERCROMBIE. M. A. (1967). *Natn. Cancer Inst. Monogr.*, **26**:249.
ABERCROMBIE, M. A. (1970). *Eur. J. Cancer*, **6**:7.
ABERCROMBIE, M. A. and HEAYSMAN, J. E. M. (1954). *Expl Cell Res.*, **6**:293.

ABERCROMBIE, M. A., HEAYSMAN, J. E. M. and PEGRUM, S. M. (1970). *Expl Cell Res.*, **62**:389.
ARMSTRONG, P. B. and NIEDERMAN, R. (1972). *Devl Biol.*, **28**:518.
ARMSTRONG, P. B. and PARENTI, D. (1972). *J. Cell Biol.*, **55**:542.
BAIER, R. E. (1972). *Bull. N.Y. Acad. Med.*, **48**:257.
BAIER, R. E. and DUTTON, R. C. (1969). *J. Biomed. Mater. Res.*, **3**:191.
BAIER, R. E., GOTT, V. L. and DUTTON, R. C. (1972). *J. Biomed. Mater. Res.*, **6**:465.
BANGHAM, A. D. and PETHICA, B. A. (1960). *Proc. R. Soc. Edinb.*, **28**:43.
BARBER, A. J. and JAMIESON, G. A. (1971). *Biochem. biophys. Acta*, **252**:533.
BARBERA, A. J., MARCHASE, R. B. and ROTH, S. A. (1973). *Proc. natn. Acad. Sci. U.S.A.*, **70**:2482.
BARR, L., DEWEY, M. M. and BERGER, W. (1965). *J. gen. Physiol.*, **48**:797.
BELL, P. B. (1972a). *J. Cell Biol.*, **55**:16A.
BELL, P. B. (1972b). *J. Cell Biol.*, **55**:60A.
BEUG, H., KATZ, F. E. and GERISCH, G. (1973). *J. Cell Biol.*, **56**:647.
BODMER, W. F. (1972). *Nature, Lond.*, **237**:139
BOSMANN, H. B. (1971). *Biochem. biophys. Res. Commun.*, **43**:1118.
BURDICK, M. L. and STEINBERG, M. S. (1969). *Proc. natn. Acad. Sci. U.S.A.*, **63**:1169.
CARTER, S. B. (1967). *Nature, Lond.*, **213**:256.
CHESNEY, C. MCL., HARPER, E. and COLMAN, R. W. (1972). *J. clin. Invest.*, **51**:2693.
CHIPOWSKY, S., LEE, Y. C. and ROSEMAN, S. (1973). *Proc. natn. Acad. Sci. U.S.A.*, **70**:2309.
CURTIS, A. S. G. (1960). *Am. Nat.*, **94**:37.
CURTIS, A. S. G. (1961). *Expl Cell Res.*, Suppl. **8**:107.
CURTIS, A. S. G. (1967). *The Cell Surface.* London; Logos Academic.
CURTIS, A. S. G. (1969). *J. Embryol. exp. Morph.*, **22**:305.
CURTIS, A. S. G. and GREAVES, M. F. (1965). *J. Embryol. exp. Morph.*, **13**:309.
DADAY, H. and CREASER, E. H. (1970). *Nature, Lond.*, **226**:970.
DANIEL, M. R. (1967). *Expl Cell Res.*, **46**:191.
DE LONG, G. R. and SIDMAN, R. L. (1970). *Devl Biol.*, **22**:584.
DERJAGUIN, B. V. and LANDAU, L. D. (1941). *Acta phys.-chim. URSS*, **14**:633.
DORSEY, J. K. and ROTH, S. A. (1973). *Devl Biol.*, **33**:249.
DULBECCO, R. and STOKER, M. G. P. (1970). *Proc. natn. Acad. Sci. U.S.A.*, **66**:204.
EDWARDS, J. G. and CAMPBELL, J. A. (1971). *J. Cell Sci.*, **8**:53.
EDWARDS, J. G., CAMPBELL, J. A. and WILLIAMS, J. F. (1971). *Nature, New Biol.*, **231**:147.
EDWARDS, J. G., CAMPBELL, J. A., ROBSON, R. T. and VICKER, M. (1975). *J. Cell Sci.*, **19**:653.
FRYE, L. D. and EDIDIN, M. (1970). *J. Cell Sci.*, **7**:319.
GARBER, B. B. and MOSCONA, A. A. (1972). *Devl Biol.*, **27**:235.
GILULA, N. B., REEVES, O. R. and STEINBACH, A. (1972). *Nature, Lond.*, **235**:262.
GOEL, N., CAMPBELL, R. D., GORDON, R., ROSEN, R., MARTINEZ, H. and YCAS, M. (1970). *J. theor. Biol.*, **28**:423.
GOLDSCHNEIDER, I. and MOSCONA, A. A. (1972). *J. Cell Biol.*, **53**:435.
GOOD, R. J. (1972). *J. theor. Biol.*, **37**:413.
GOODENOUGH, D. A. and STOECKENIUS, W. (1972). *J. Cell Biol.*, **54**:646.
GRINNELL, F., MILAM, M. and SRERE, P. A. (1972). *Archs Biochem. Biophys.*, **153**:193.
GRINNELL, F., MILAM, M. and SRERE, P. A. (1973). *Biochem. Med.*, **7**:87.
GUELSTEIN, V. I., IVANOVA, O. YU., MARGOLIS, L. B., VASILIEV, JU. M. and GELFAND, I. M. (1973). *Proc. natn. Acad. Sci. U.S.A.*, **70**:2011.
GUSTAFSON, T. and WOLPERT, L. (1967). *Biol. Rev.*, **42**:442.
HARRIS, A. K. (1973A). *Ciba Symposium 14*, pp. 3–26. Ed. R. Porter and D. W. FITZSIMONS. Amsterdam; Elsevier.
HARRIS, A. K. (1973a). *Expl Cell Res.*, **77**:285.
HEAYSMAN, J. E. M. and PEGRUM, S. M. (1973a). *Expl Cell Res.*, **78**:71.
HEAYSMAN, J. E. M. and PEGRUM, S. M. (1973b). *Expl Cell Res.*, **78**:479.
JAMIESON, G. A., URBAN, C. L. and BARBER, A. J. (1971). *Nature, New Biol.*, **234**:5.
JERNE, N. K. (1971). *Eur. J. Immun.*, **1**:1.
KOLODNY, G. M. (1972). *Expl Cell Res.*, **70**:196.
LIFSHITZ, E. M. (1956). *J. exp. theor. Phys.*, **2**:73.
LILIEN, J. E. (1968). *Devl Biol.*, **17**:657.
LYMAN, D. J., BRASH, J. L., CHAIKIN, S. W., KLEIN, K. G. and CARINI, M., (1968). *Trans. Am. Soc. artif. internal Organs*, **14**:250.
MARTZ, E. and STEINBERG, M. S. (1973). *J. cell. Physiol.*, **81**:25.
MASLOW, D. E. and MAYHEW, E. (1972). *Science, N.Y.*, **177**:281.
MCNUTT, N. S. and FAWCETT, D. W. (1969). *J. Cell Biol.*, **42**:1.

MCNUTT, N. S. and WEINSTEIN, R. S. (1970). *J. Cell Biol.*, **47**:666.
MCNUTT, N. S. and WEINSTEIN, R. S. (1972). *Prog. Biophys. molec. Biol.*, **26**:47.
MILLS, D. C. B. and ROBERTS, G. C. K. (1967). *Nature, Lond.*, **213**:35.
MOSCONA, A. A. (1957). *Proc. natn. Acad. Sci. U.S.A.*, **43**:184.
MOSCONA, A. A. (1963). *Nature, Lond.*, **199**:379.
MOSCONA, A. A. (1965). *Cells and Tissues in Culture*, pp. 489–531. Ed. E. N. WILLMER. New York; Academic Press.
MOSCONA, A. A. (1968). *Devl Biol.*, **18**:250.
MOSCONA, A. A. (1973). *Cell Biology in Medicine*, pp. 571–591. Ed. E. C. BITTAR. New York; John Wiley.
MOSCONA, M. H. and MOSCONA, A. A. (1963). *Science, N.Y.*, **142**:1070.
MOYER, W. A. and STEINBERG, M. S. (1972). *Am. Zool.*, **12**:335.
MUIR, A. R. (1967). *J. Anat.*, **101**:239.
NAPPER, D. H. (1970). *Ind. Engng Chem. Prod. Res. Dev.*, **9**:467.
NIR, S., REIN, R. and WEISS, L. (1972). *J. theor. Biol.*, **34**:135.
NORDLING, S. (1967). *Acta path. microbiol. scand.*, S192.
OPPENHEIMER, S. B. (1973). *Expl Cell Res.*, **77**:175.
OPPENHEIMER, S. B., EDIDIN, M., ORR, C. W. and ROSEMAN, S. (1969). *Proc. natn. Acad. Sci U.S.A.*, **63**:1395.
ORR, C. W. and ROSEMAN, S. (1969). *J. Membrane Biol.*, **1**:109.
OVERBEEK, J. Th.G. (1952). *Colloid Science*, pp. 302–341. Ed. H. R. KRUYT. Amsterdam; Elsevier.
PARSEGIAN, V. A. and GINGELL, D. (1972). *J. Adhesion*, **4**:283.
PHILLIPS, H. M. and STEINBERG, M. S. (1969). *Proc. natn. Acad. Sci. U.S.A.*, **64**:121.
PORTER, R. and FITZSIMONS, D. W. (Eds) (1973). *Ciba Symposium 14*. Amsterdam; Elsevier.
PRICER, W. E. and ASHWELL, G. (1971). *J. biol. Chem.*, **246**:4825.
RABINOWITCH, M. and DESTEFANO, M. J. (1973). *J. Cell Biol.*, **59**:165.
RAYNS, D. G., SIMPSON, F. O. and LEDINGHAM, J. M. (1969). *J. Cell Biol.*, **42**:322.
REES, D. A., SCOTT, W. E. and WILLIAMSON, F. B. (1969). *Nature, Lond.*, **227**:390.
REVEL, J. P. and WOLKEN, K. (1973). *Expl Cell Res.*, **78**:1.
RICHMOND, J. E., GLAESER, R. M. and TODD, P. (1968). *Expl Cell Res.*, **52**:43.
ROSEMAN, S. (1970). *Chem. Phys. Lipids*, **5**:270.
ROTH, S. A. (1968). *Devl Biol.*, **18**:602.
ROTH, S. A., MCGUIRE, E. J. and ROSEMAN, S. (1971a). *J. Cell Biol.*, **51**:525.
ROTH, S. A., MCGUIRE, E. J. and ROSEMAN, S. (1971b). *J. Cell Biol.*, **51**:536.
ROTH, S. A. and WESTON, J. A. (1967). *Proc. natn. Acad. Sci. U.S.A.*, **58**:974.
ROTH, S. A. and WHITE, D. (1972). *Proc. natn. Acad. Sci. U.S.A.*, **69**:485.
SCHENKEL, J. H. and KITCHENER, J. (1961). *Trans. Faraday Soc.*, **56**:161.
SHEFFIELD, J. B. (1970). *J. Morph.*, **132**:245.
SHEFFIELD, J. B. and MOSCONA, A. A. (1969). *Expl Cell. Res.*, **57**:462.
STEFANELLI, A., ZACCHEI, A. M. and CECCHERINI, U. (1961). *Acta Embryol. Morph. exp.*, **4**:47.
STEINBERG, M. S. (1958). *Am. Nat.*, **92**:65.
STEINBERG, M. S. (1962a). *Proc. natn. Acad. Sci. U.S.A.*, **48**:1769.
STEINBERG, M. S. (1962b). *Proc. natn. Acad. Sci. U.S.A.*, **48**:1577.
STEINBERG, M. S. (1963). *Science, N.Y.*, **141**:401.
STEINBERG, M. S. (1970). *J. exp. Zool.*, **173**:395.
STEINBERG, M. S. and WISEMAN, L. L. (1972). *J. Cell Biol.*, **55**:606.
TAKEICHI, M. (1971). *Expl Cell Res.*, **68**:88.
TAKEICHI, M. and OKADA, T. S. (1972). *Expl Cell Res.*, **74**:51.
TAYLOR, A. C. (1961). *Expl Cell Res.*, Suppl. **8**:154.
TOWNES, P. L. and HOLTFRETER, J. (1955). *J. exp. Zool.*, **128**:53.
TRINKAUS, J. P. (1969). *Cells into Organs*. Englewood Cliffs, N.J; Prentice-Hall.
TYLER, A. (1946). *Growth*, **10**:7.
VERWEY, E. J. W. and OVERBEEK, J. Th. G. (1948). *Theory of the Stability of Lyophobic Colloids*. Amsterdam; Elsevier.
VICKER, M. and EDWARDS, J. G. (1972). *J. Cell Sci.*, **10**:759.
WADDELL, A., ROBSON, R. and EDWARDS, J. G. (1974). *Nature, Lond.*, **248**:239.
WALTHER, B. T., OHMAN, R. and ROSEMAN, S. (1973). *Proc. natn. Acad. Sci. U.S.A.*, **70**:1569.
WEISER, M. M. (1973). *J. biol. Chem.* **248**:2542.
WEISS, L. (1967). *The Cell Periphery, Metastasis and Other Phenomena*. Amsterdam; North-Holland.
WEISS, L. (1972). *Expl Cell Res.*, **74**:21.
WEISS, L. and HARLOS, J. P. (1972). *Prog. Surf. Sci.*, **1**:355.

WEISS. P. (1946). *Yale J. Biol. Med.*, **19**:235.

WESTON, J. A. and ROTH, S. A. (1969. *Cellular Recognition*, pp. 29–37. Ed. R. T. SMITH and R. A. GOOD. New York; Appleton-Century-Crofts.

WISEMAN, L. L.. STEINBERG. M. S. and PHILLIPS, H. M. (1972). *Devl Biol.*, **28**:498.

WITKOWSKI, J. A. and BRIGHTON, W. D. (1972). *Expl Cell Res.*, **70**:41.

WOLPERT, L. (1971). *Curr. Topics dev. Biol.*, **6**:183.

WOLPERT, L. (1973). *Ciba Symposium 14*, p. 345. Ed. R. PORTER and D. W. FITZSIMONS. Amsterdam; Elsevier.

WOLPERT, L. and GINGELL, D. (1969). *Ciba Symposium on Homeostatic Regulators*, pp. 241–259. Ed. G. E. W. WOLSTENHOLME and J. KNIGHT. London; Churchill.

YAFFE. D. and FELDMAN. M. (1965). *Devl Biol.*, **11**:300.

ADDENDUM

Section 2.1   Reports from two laboratories have lent weight to the attempt to analyse formation of intercellular adhesions in terms of a series of temporal stages. Umbreit and Roseman (1975) have shown that during the reaggregation of chick embryonic liver and neural retina cells, the cells first form weak adhesions, in a step insensitive to metabolic inhibition. Subsequent consolidation of these adhesions requires metabolic energy. Jones, Gillett and Partridge (1976) have shown, using electron microscopy, that after initial collision of spontaneously aggregating limpet haemocytes intercellular contact area rapidly increases, apparently through an active response of cellular organelles concerned with motility.

Section 2.2   Hausman and Moscona (1975) have characterized a glycoprotein (mol. wt 50 000) isolated from medium conditioned by cultures of chick embryonic neural retina, which increases the size of 24-hour reaggregates of dispersed primary retinal cells. The authors believe this molecule to be a cell-type-specific cell ligand (Moscona, 1968*), although it has not yet been shown to exhibit cell-type specificity comparable to the crude factor previously reported (Lilien, 1968*).

Section 2.3   Harris (1976) has argued cogently that the transitive property of cells which determines the hierarchy of their sorting behaviour (Steinberg, 1970*), morphogenetic importance notwithstanding, need not be cell adhesion as understood elsewhere in this chapter.

Section 2.7   The isolation of intercellular junctions offers an important possible route to the isolation of molecular species with intercellular adhesive function. Skerrow and Matoltsy (1974) have described the isolation of desmosomes from bovine nasal epithelium. Two of the seven major polypeptides in these preparations are glycosylated, and the authors suggest these may form part of the intercellular lamina of the junctions.

Overton (1975) has reviewed her studies of the formation of intercellular junctions between experimentally juxtaposed chick embryonic cells. In mixtures of 15-day corneal cells with 7-day heart ventricle cells, desmosomes form between like cells, but rarely between unlike. Junctions of the fascia adhaerens type, however, occurred also between unlike cells. Overton

* References marked with an asterisk are to be found in the main reference list.

suggests that as cells differentiate in the developing embryo, those junctions that form early may be less specific than those that form late.

Section 2.8   The hypothesis that cell adhesion involves binding of cell-surface glycosyltransferases to their substrates on other cells (*see* Roth, 1973) seems still to await definitive experimental support.

The inhibition of galactose oxidase of collagen-dependent platelet aggregation (Chesney, Harper and Colman, 1972*) has been reinterpreted by Muggli and Baumgartner (1973), who proposed that for adhesion of platelets, collagen must be in microfibril form, and that the enzyme treatment delays formation of microfibrils.

Glycosyltransferases represent only one possible candidate for a class of proteins able to recognize cell-surface complex carbohydrate. Another intriguing possibility to emerge recently is that tissue cells may possess lectin-like carbohydrate-binding proteins which could fulfil this role. (It must be borne in mind, however, that carbohydrate-binding activity, defined by ability to agglutinate fixed erythrocytes, could be a property of glycosyltransferases.) Carbohydrate-binding proteins, discoidin and pallidin, have been isolated and characterized from two species of cellular slime moulds by Frazier *et al.* (1975). Although these may be involved in intercellular adhesion (Rosen *et al.*, 1974), an unsolved problem is the apparent lack of identity (Huesgen and Gerisch, 1975) between discoidin and its receptr on the one hand, and contact site A, a molecular species implicated in aggregation-phase adhesion of *Dictyostelium* cells by immunological means (Beug, Katz and Gerisch, 1973). Stockert, Morelland Scheinberg (1974) reported that the rabbit hepatic binding protein, implicated in clearance of asialo plasma proteins, has erythrocyte-agglutinating activity, and so is a lectin of mammalian origin. Erythrocyte-agglutinating activity with β-D-galactosyl specificity has since been reported from a variety of animal tissues and cell-culture lines (Teichberg *et al.*, 1975). A high level was present in embryonic muscle tissue, and both Gartner and Podleski (1975) and Nowak, Haywood and Barondes (1976) have suggested the agent could be involved in myoblast fusion, although Den, Malinzak and Rosenberg (1976) failed to obtain inhibition of fusion of chick embryonic myoblasts by disaccharide inhibitors of the agglutination.

Yamada, Yamada and Pastan (1976) have reported that erythrocytes are agglutinated by a cell-surface protein (CSP) isolated from chick-embryo fibroblasts by extraction with urea. CSP may belong to a family of cell-surface proteins of high molecular weight detected by different techniques and in a variety of cells, which includes also the large, external, transformation-sensitive (LETS) protein detected by iodine labelling in permanent lines of mammalian cells (Hynes, 1974) and the surface antigen (SFA) of human fibroblasts. Several workers have suggested that proteins of this family may be involved in intercellular adhesion (Yamada and Weston, 1974; Yamada, Yamada and Pastan, 1975; Rutishauser *et al.*, 1976). Of particular importance is the report that immunoprecipitates of SFA contain a polypeptide of molecular weight 45 000 which co-electrophoreses with fibroblast actin (Wartiovaara *et al.*, 1974), since association with actin could explain the termination of bundles of microfilaments at points of cell–cell and cell–substratum contact in cultured cells (Abercrombie, Heaysman and Pegrum, 1970*; Heaysman and Pegrum, 1973a*). The observation leads to the further speculation that LETS-related proteins could be intercellular components of microfilament-associated junctions of the zonula and fascia adhaerens type in stable tissue. If proteins related to LETS are involved in cell adhesion, transformed cells should be less adhesive than their untransformed counterparts. Most recent studies do not support this (Walther, Ohman and Roseman, 1973*; Dorsey and Roth, 1973*; Cassiman and Bernfield, 1975), although it does hold for the comparison of BHK21 cells and certain polyoma-transformed derivatives (Edwards, Campbell and Williams, 1971*; O'Neill, 1973; Whur *et al.*, 1976).

Two new experimental approaches may shed further light on the molecular basis of cell adhesion: Rutishauser *et al.* (1976) have shown that antibodies directed against specific components of conditioned media inhibit intercellular adhesion of chick embryonic brain and neural retina cells. Pouysségur and Pastan (1976) have isolated mutants of *Balb/c* 3T3 cells defective in adhesiveness to culture substratum.

REFERENCES TO ADDENDUM

BEUG, H., KATZ, F. E. and GERISCH, G. (1973). *J. Cell Biol.*, **56**:647.

CASSIMAN, J. J. and BERNFIELD, M. R. (1975). *Expl Cell Res.*, **91**:31.

DEN, H., MALINZAK, D. A. and ROSENBERG, A. (1976). *Biochem. biophys. Res. Commun.*, **69**:621.

FRAZIER, W. A., ROSEN, S. D., REITHERMAN, R. W. and BARONDES, S. H. (1975). *J. biol. Chem.*, **250**:7714.

GARTNER, T. K. and PODLESKI, T. R. (1975). *Biochem. biophys. Res. Commun.*, **67**:972.

HARRIS, A. K. (1976). *J. theor. Biol.*, in press.

HAUSMAN, R. E. and MOSCONA, A. A. (1975). *Proc. natn. Acad. Sci. U.S.A.*, **72**:916.

HUESGEN, A. and GERISCH, G. (1975). *FEBS Lett.*, **56**:46.

HYNES, R. O. (1974). *Cell*, **1**:147.

JONES, G. E., GILLETT, R. and PARTRIDGE, T. (1976). *J. Cell Sci.*, **22**:21.

MUGGLI, R. and BAUMGARTNER, H. R. (1973). *Int. Soc. Thromb. Haemostas., 4th Congress, Abstracts,* p. 86.

NOWAK, T. P., HAYWOOD, P. L. and BARONDES, S. H. (1976). *Biochem. biophys. Res. Commun.*, **68**:650.

O'NEILL, C. (1973). *Expl Cell Res.*, **81**:31.

OVERTON, J. (1975). *Curr. Topics devl Biol.* **10**:1.

POUYSSÉGUR, J. M. and PASTAN, I. (1976). *Proc. natn. Acad. Sci. U.S.A.*, **73**:544.

ROSEN, S. D., SIMPSON, D. L., ROSE, J. E. and BARONDES, S. H. (1974). *Nature, Lond.*, **252**:128.

ROTH, S. A. (1973). *Q. Rev. Biol.*, **48**:541.

RUTISHAUSER, U., THIERY, J. P., BRACKENBURY, R., SELA, B. A. and EDELMAN, G. M. (1976). *Proc. natn. Acad. Sci. U.S.A.*, **73**:577.

SKERROW, C. J. and MATOLTSY, A. G. (1974). *J. Cell Biol.*, **63**:515.

STOCKERT, R. J., MORELL, A. G. and SCHEINBERG, I. H. (1974). *Science N.Y.*, **186**:365.

TEICHBERG, V. T., SILMAN, I., BEITSCH, D. D. and RESHEFF, G. (1975). *Proc. natn. Acad. Sci. U.S.A.*, **72**:1383.

UMBREIT, J. and ROSEMAN, S. (1975). *J. biol. Chem.*, **250**:9360.

WARTIOVAARA, J., LONDER, E., RUOSLAHTI, E. and VAHERI, A. (1974). *J. exp. Med.*, **140**:1522.

WHUR, P., KOPPEL, H., URQUHART, C. and WILLIAMS, D. C. (1976). *J. Cell. Sci.*, in press.

YAMADA, K. M., YAMADA, S. S. and PASTAN, I. (1975). *Proc. natn. Acad. Sci. U.S.A.*, **72**:3158.

YAMADA, K. M., YAMADA, S. S. and PASTAN, I. (1976). *Proc. natn. Acad. Sci. U.S.A.*, **73**:1217.

3

The role of cell-surface carbohydrates in binding phenomena

Gilbert Ashwell

Laboratory of Biochemistry and Metabolism, National Institute of Arthritis, Metabolism and Digestive Diseases, Bethesda, Maryland

3.1   INTRODUCTION

Despite extensive studies on the nature and role of the various cell-surface constituents, relatively little is known of the mechanism whereby these components effect the specific recognition processes which are characteristic of cellular differentiation and metabolism. However, on a phenomenological level, a considerable body of data has been accumulated although many of the observations recorded are fragmentary and frequently appear to be unrelated. Nevertheless, there has emerged an underlying and constantly recurring theme which points to the predominantly informational role of the glycoprotein constituents of the cell membrane. It is this aspect of membrane physiology which is the focus of the ensuing chapter. Inevitably, the choice of material presented and the emphasis provided represents a subjective selection which defines the interests, prejudices and limitations of the reviewer. For an authoritative and comprehensive treatment of the nature of cell-surface constituents and an examination of the changes occurring during transformation and cell–cell interactions, the reader is referred to reviews by Hughes (1973), Emmelot (1973), Turner and Burger (1973) and Roth (1973).

3.2   BIOLOGICAL EFFECTS OF NEURAMINIDASE ON BINDING

3.2.1   Circulating glycoproteins

During the course of a series of studies on plasma glycoproteins, a new concept was developed for the role of carbohydrates in regulating the serum survival time of these compounds. In essence, the hypothesis was advanced that for many, if not most, of such glycoproteins, sialic acid is essential for their continued viability in the circulation (Ashwell and Morell, 1971, 1974).

These conclusions were prompted by the observation that upon injection into rabbits of a preparation of ceruloplasmin from which the sialic acid had been removed by prior treatment with neuraminidase, the asialoglycoprotein was promptly removed from the circulation within 10 minutes after injection. This is in contradistinction to the behavior of the fully sialylated protein, which exhibited a normal survival time of 56 hours. Evidence in support of the critical role of the thereby exposed galactosyl residues in effecting this rapid clearance was provided by the demonstration that the survival time was markedly increased when the galactose moiety was modified by treatment with galactose oxidase or removal by enzymatic hydrolysis with β-galactosidase (Morell et al., 1968).

The prompt disappearance of desialylated ceruloplasmin from the serum was found to be accompanied by an equally rapid accumulation of this protein within the parenchymal cells of the liver and the principal subcellular site of catabolism was shown to be lysosomal (Gregoriadis et al., 1970). Additional support for the biological significance of these observations was obtained by examination of the behavior of the resialylated protein in vivo. Restoration of the deficient sialic acid residues was accomplished enzymatically by incubation of the partially desialylated ceruloplasmin with a particulate sialyltransferase in the presence of CMP–sialic acid. Injection of this incompletely reconstituted preparation into rabbits led to the recognition of two distinct classes of molecules with markedly different half-lives. Those molecules in which all, or essentially all, of the missing sialic acid had been restored exhibited a normal serum survival time of 56 hours; those in which less than a critical threshold level of sialic acid had been restored disappeared from the circulation within a few minutes after injection (Hickman et al., 1970).

In short, the question of quantitation arose, that is to say, how many galactosyl residues were required to be exposed for the liver to recognize and retain a 'deficient' molecule of ceruloplasmin? By utilizing the estimate of 10 sialic acid residues per mole of ceruloplasmin (Jamieson, 1965), it was possible to calculate the distribution of exposed galactosyl residues per molecule for a known average amount of desialylation and to determine that the exposure of only 2 galactosyl residues was sufficient to mark any given molecule for hepatic destruction (Van Den Hamer et al., 1970).

The generality of the protective property of sialic acid was established by the analogously rapid disappearance from the circulation following injection of asialo derivatives of a number of plasma glycoproteins including orosomucoid, fetuin, haptoglobin, α_2-macroglobulin, human chorionic gonadotropin and follicle-stimulating hormone (Morell et al., 1971).

Since in all cases the major portion of the asialo protein disappearing from the serum was recovered in the liver, the possibility of competitive inhibition was investigated. Thus, upon the injection of tracer amounts of asialoceruloplasmin together with milligram quantities of asialo-orosomucoid, the survival time of the former was dramatically increased whereas injection of orosomucoid itself was without demonstrable effect. Most significantly, this sparing effect was shown to be a property of the carbohydrate chain since the glycopeptides isolated after prolonged digestion with pronase and neuraminidase were similarly effective as competitive inhibitors (Morell et al., 1971).

Subsequent investigation in vitro revealed the primary locus of binding to be localized largely in the plasma membranes of the liver cell. Upon incubation of this material with [^{125}I]asialo-orosomucoid, followed by filtration and counting of the filters, a highly sensitive assay became available to probe the qualitative and quantitative parameters of the binding mechanism (Pricer and Ashwell, 1971; Van Lenten and Ashwell, 1972).

Treatment of the membranes with a small amount of neuraminidase was found to result in the complete loss of their capacity for binding asialoglycoproteins. The destructive action of neuraminidase was shown to be a repairable lesion in that enzymatic replacement of the sialic acid residues restored their binding potency. The latter was accomplished by incubation of the partially desialylated membranes with CMP–sialic acid; the presence of sialyltransferase in the membrane preparation was sufficient to effect incorporation of the essential sialic acid residues. These results confirmed the complex role played by sialic acid in determining the metabolic fate of specific circulating glycoproteins. Effective hepatic uptake requires not only the absence of sialic acid on the glycoprotein to be catabolized but, in addition, the presence of this substituent on the plasma membrane for recognition. In the former case, sialic acid serves to mask the galactosyl residues which are the major determinants of binding and catabolism; in the latter case, the mechanism whereby sialic acid preserves the integrity of the receptor site on the plasma membrane is less clear.

The nature of the receptor for asialoglycoproteins has been considerably clarified by the successful purification of a hepatic protein with retention of the characteristic binding properties associated with the plasma membranes. The isolated material, which is free from lipid, was identified as a sialic-acid-containing glycoprotein in which 10 percent of the dry weight consists of carbohydrate (Hudgin et al., 1974). In aqueous solution, the binding protein exists in a highly aggregated state which was shown to result from the self-associating properties of a single oligomeric protein (Kawasaki and Ashwell, 1976). The tendency towards self-association was promptly and completely reversed by the addition of Triton X-100 which resulted in the appearance of a single component with an estimated molecular weight of 260000. In contrast to the mechanism of binding proposed by Roseman (see Section 3.3), the purified binding protein was devoid of glycosyltransferase activity (Hudgin and Ashwell, 1974).

The presumption that the galactose-specific hepatic receptor participates in the physiological regulation of serum glycoprotein metabolism in mammals has received indirect support by recent studies on avian and reptilian species (Lunney and Ashwell, 1976). In chickens, a high level of circulating asialoglycoproteins has been correlated with a deficiency of the asialoglycoprotein-binding protein. Of greater interest, however, is the identification of an avian hepatic binding activity whereby terminal N-acetylglucosamine residues, rather than galactose residues, serve as the signal for hepatic recognition. It seems reasonable to anticipate that this observation may provide the basis for new insights into the nature and evolutionary origin of hepatic regulatory mechanisms.

3.2.2   Viral attachment to erythrocytes

An early example of such a correlation is provided by the classic investigations of Hirst, Burnet, Gottschalk and others of the Australian school (Burnet, 1951; Gottschalk, 1959). From their studies it became clear that the presence of nonreducing terminal sialic acid residues on the surface of the erythrocyte membrane constituted an absolute requirement for the attachment of influenza virus; prior treatment of the intact cell with neuraminidase completely abolished the binding reaction. The hemagglutination reaction which accompanies successful viral attachment was shown to be inhibited by a variety of naturally occurring sialoglycoproteins. Utilizing this property, Kathan, Winzler and Johnson (1961) succeeded in isolating, in soluble form, a homogeneous protein from human red blood cells which exhibited both M and N blood-group activity as well as a high inhibitory titer for viral hemagglutination. This material, presumably the intact cellular receptor site for myxovirus attachment, was shown to be a glycoprotein with a sialic acid content of 24 percent and a molecular weight of 30000. As would be anticipated, exposure to neuraminidase completely destroyed the inhibitory activity. Somewhat more surprising, however, was the finding that the intact sialyl structure was required for activity. Conversion of the nonreducing 9-carbon sugar acid to the 7- or 8-carbon analogues, by controlled periodate oxidation followed by reduction with borohydride, resulted in an almost complete loss of inhibitory potency (Suttajit and Winzler, 1971). Consequently, it would appear that the polyhydroxy 3-carbon side-chain of sialic acid must be intact in order to bind virus.

In contradistinction to these results, it was found that the analogous conversion of the terminal sialic acid residues of either orosomucoid or ceruloplasmin to the 7-carbon sugar acid had no discernible effect upon the normal survival time in the circulation so long as the modified residue remained glycosidically linked to the underlying galactose moiety (Van Lenten and Ashwell, 1971). The analogy here, of course, is questionable since the properties of the sialic acid on the receptor site are being compared with the properties of the sialic acid on the ligand. The ligand, in the case of viral attachment, is a hemagglutinin and little is known about its chemical structure. It has, however, been isolated from the BEL (AO) strain of influenza virus by disruption with sodium dodecyl sulfate. In the presence of this detergent, the hemagglutinin subunit adsorbed to but did not agglutinate red cells; hemagglutination activity was obtained only after removal of the detergent (Laver and Valentine, 1969).

3.2.3   Circulation of lymphocytes

In a study on the fate of transfused lymphocytes, it was found that the unique pattern of lymphoid circulation was characteristically altered by exposure to neuraminidase prior to transfusion (Woodruff and Gesner, 1969). Upon injection of ^{51}Cr-labeled lymphocytes into allogeneic and syngeneic rats, there was a relatively high concentration of radioactivity in the lymphoid tissue and a low concentration of label in the liver. Within a few hours after transfusion, more than 50 percent of the total radioactivity injected was recovered in the spleen and lymph nodes. This value gradually decreased over a period of 48 hours whereas the liver contained a relatively low and constant percentage of the total radioactivity throughout the entire time course of the experiment.

This pattern of behavior was reversed in the case of lymphocytes which had been previously exposed to neuraminidase. Sacrifice of the recipient animals shortly (1–8 hours) after transfusion revealed a drastic reduction in the amount of radioactivity recovered in the lymphoid tissue together with a striking increase in hepatic radioactivity which amounted, in some cases, to 50–60 percent of the total label injected. However, as the interval between the time of injection and the time of sacrifice increased, the recovery of radioactivity in the lymph nodes increased while the hepatic radioactivity decreased, so that 48 hours after transfusion, the percentage recovery of radioactivity in lymph nodes in recipients of neuraminidase-treated cells was about equal to that obtained in recipients of untreated lymphocytes.

An altogether different circulatory pattern was encountered when the viability of the intact, labeled lymphocytes was destroyed by preincubation for 10 minutes at 45 °C prior to transfusion. When animals injected with these heat-killed cells were sacrificed after 1 hour, virtually no radioactivity accumulated in the lymph nodes and an average of 55 percent of the total injected label was recovered in the liver. Unlike recipients of neuraminidase-treated cells, the concentration of radioactivity in lymph nodes did not rise in those recipients sacrificed at later intervals nor did the hepatic radioactivity decrease nearly so rapidly nor to the same extent.

From these experiments, it would seem reasonable to infer that hepatic tissue is capable of recognizing and binding those lymphocytes in the circulation which are deficient in their normal cell-surface sialic acid residues. Since these cells were shown to retain their viability, and presumably their capacity for regeneration, the restoration of their normal surface structures might, conceivably, provide a signal for their eventual release from the liver. In this case they would be expected, as shown, to migrate normally to the lymphoid tissue. This mechanism could not be operative in the case of the heat-killed but otherwise intact cells. Here, the binding was essentially irreversible and the slow disappearance from the liver, without subsequent migration to the lymph, was altogether consistent with the catabolic turnover of material ingested by the Kupffer cells.

Whether or not this interpretation is valid, the function served by the transitory binding and release of such cells remains obscure. Conceivably, it might reflect nothing more than the nonspecificity of attachment of the desialylated residues to the same hepatic sites involved in the binding and transport of asialoglycoproteins. If this were the case, it would be anticipated that in the presence of added asialoglycoproteins, a marked inhibition of lymphocytic binding to hepatic tissue would be manifested. Alternatively, this phenomenon may represent a biologically significant process utilizing a common mechanism.

3.2.4   Tumor invasiveness

An analogous series of experiments was conducted wherein the fate of circulating neoplastic cells in mice was examined (Gasic and Gasic, 1962). In these studies, 8-week-old mice were inoculated with mouse ascites tumor cells (TA3); after 2 weeks, the animals were killed and the incidence of metastatic lung tumors was determined. On the presumption that the adhesiveness, and hence the virulence, of these cells to the vascular endothelium was influenced by the surface carbohydrates of both surfaces, the effect of neuraminidase was investigated.

Preliminary experiments confirmed the effectiveness *in vitro* of this enzyme in removing sialic acid from the coat of the TA3 cells and the effectiveness *in vivo* of intravenously injected enzyme in removing material staining as carbohydrate from the surface of lung and liver vascular endothelium. The latter effect was almost complete in 3 hours and lasted for 18 hours or more; full recovery of the surface coat of the lung blood vessels required 4 days.

Comparison of the incidence of lung tumors in mice inoculated with control cells versus those in which the cells had been previously exposed to neuraminidase revealed no significant difference. This result was not wholly unexpected since it was known that the uncoated TA3 cells can regenerate their coat *in vitro* within 2 hours and, presumably, comparable recovery *in vivo* could explain the negative results. Quite strikingly different results were obtained when intact TA3 cells were inoculated into mice previously injected with neuraminidase. In this case, there was a highly significant fall in the incidence of metastatic lung tumors; the mean number of tumors decreased by approximately 75 percent. The protective effect of the enzyme was shown to vary according to the time interval between enzyme injection and tumor inoculation (Gasic, Gasic and Stewart, 1968). Protection began as early as 1 hour after treatment, becoming increasingly strong until it reached a maximum at 24 hours and then steadily decreased. The decreasing frequency of metastases with time of exposure to the enzyme, up to 24 hours, corresponded closely to the similar decline in the neuraminidase activity in the blood. The fact that the neuraminidase antimetastatic effect was more intense when there was less circulating enzyme at the time of tumor inoculation was taken as good evidence to favor the hypothesis that neuraminidase reduces the number of metastases by acting mainly on the host.

The totally unexpected finding emerged from this work that neuraminidase *in vivo* produces thrombocytopenia and that a close correspondence exists between the decrease in platelet levels and the decrease in metastases. The time courses of the two phenomena were essentially identical with the maximum thrombocytopenic effect occurring 24 hours after enzyme injection. The infusion of platelets into neuraminidase-treated animals was shown to reverse the antimetastatic effect of the enzyme even under conditions where the augmented platelet count rose to only 20 percent of the normal titer of circulating thrombocytes. When anti-platelet serum, rather than neuraminidase, was injected into mice 24 hours before tumor inoculation, an identical correlation was found between the diminished platelet count and the decreased number of metastatic lesions in the serum-treated as compared with the control animals.

The above findings clearly implicate platelet participation in the formation of metastatic tumors and leave uncertain the involvement of neuraminidase in antimetastases. Conceivably, attachment of the neoplastic cell to the vascular endothelium might reflect a unique aspect of the general role of platelets in adhesive or coagulative processes. However, these observations pose an alternative, if ancillary, question concerning the correlation of neuraminidase activity and the rapid disappearance of platelets from the serum. It might be pertinent to ask whether platelet survival in the circulation is, indeed, a function of their sialic acid content. Secondly, it would be meaningful to know whether the normal pattern of destruction and turnover is merely accelerated by the action of neuraminidase or whether it is altered by hepatic participation, as was discussed earlier in the case of transfused lymphocytes and asialoglycoproteins. The answers to such questions might be anticipated to have immediate relevance for the broader problem of blood coagulation.

In a study on the alteration of tumor histocompatibility induced by neuraminidase, a somewhat different interpretation of the anti tumour effect of the enzyme was suggested (Sandford, 1967). As a consequence of the repeated observation that tumors frequently seem to decrease in specificity during serial transplantation and to develop the ability to grow in foreign strains, the possibility was considered that the diminished specificity might result from a masking of tumor-cell antigens present on the surface of the neoplastic cell. To test the hypothesis that the sialic acid residues might serve this purpose, TA3 ascites carcinoma cells were treated with neuraminidase before intraperitoneal injection into C3H mice. Contrary to the results described above, neuraminidase *in vitro* resulted in a highly significant decrease in the number of lethal tumor 'takes' in allogeneic hosts. An entirely similar protective effect was seen when the recipient animal was given neuraminidase in five intravenous injections at 48-hour intervals beginning 3 hours prior to the inoculation of intact TA3 cells. Thus, in this study, neuraminidase either *in vitro* or *in vivo* was shown to alter the transplantation behavior and it was concluded that the enzymatic removal of sialic acid from the cell surface increased the tumor specificity by exposing histocompatibility antigens previously concealed by the sialomucin.

Additional evidence has been obtained to reinforce the suggestion of an inverse relationship between the presence of cell-surface sialomucins and the absence of antigenic expression (Bagshawe and Currie, 1968). In a study based upon the hypothesis that the sialic acid residues of such privileged cells as trophoblasts and tumor cells act as a barrier to prevent their intimate contact with lymphocytes, Bagshawe and Currie examined the immunogenicity of L 1210 murine leukemia cells which had been exposed to neuraminidase prior to injection into mice.

Following the inoculation of 40 000 L1210 cells intraperitoneally into C57Bl/DBA2 mice, 100 percent of the animals (9 out of 9) were dead within 20 days. When the experiment was repeated with cells which had been treated briefly with *Vibrio cholerae* neuraminidase, 100 percent of the animals were still alive after 100 days. The viability of the treated cells was demonstrated *in vivo* by their growth in irradiated recipients. Upon injection of either intact or enzyme-treated cells into mice 2 hours after exposure to 600 rad of whole body irradiation, 100 percent mortality resulted within 20 days and the ascitic fluid, recovered before death, was shown to contain large numbers of viable and virulent leukemia cells.

The growth of the neuraminidase-treated L1210 cells in irradiated recipients but not in intact recipients was interpreted as suggesting that an intact immune system was necessary to prevent their growth. This presumption was tested by challenging the surviving recipients of the enzyme-treated cells. Thirty days after receiving 40000 neuraminidase-treated cells, the apparently healthy animals were reinoculated with an equal number of intact leukemia cells; all of the animals were alive when the experiment was terminated 60 days later. Direct confirmation of an increased immune titer was obtained by serum agglutination tests which revealed that sera from two immune mice agglutinated L 1210 cells at a dilution of 1 : 1024 whereas the sera from tumor-bearing animals were inactive at dilutions greater than 1 : 8. It was concluded that the development of immunity after exposure to neuraminidase-treated tumor cells was consistent with the concept that sialic acid groups in the cell periphery can mask cell-surface antigens from detection by the immune system.

3.2.5   Trophoblast implantation

Continuing their interesting studies in this general area, Gasic and Gasic (1970) turned their attention to the adhesive and invasive phenomenon of ovum implantation attendant upon successful mating. At varying time intervals for several days *post coitum*, mice were given 50 units of *Vibrio cholerae* neuraminidase by tail vein and pregnancy was determined visually after 14 days and confirmed at delivery. In the control group, 78 percent (25 out of 32) of the mice became pregnant, whereas in the experimental group receiving neuraminidase on day 4 and day 6 *post coitum*, there was complete suppression of pregnancy (0 out of 34). Here, too, the time of treatment was critical with maximum effectiveness obtained when the enzyme was injected between 4 and 7 days after the time of implantation of the ovum at the onset of pregnancy.

To evaluate the contribution of neuraminidase-mediated thrombocytopenia on pregnancy, mice were injected with antiplatelet serum either 4 or 6 days after mating. Although this treatment did reduce the number of successful pregnancies to 44 percent (20 out of 45), it was markedly less effective than neuraminidase.

When autopsies were performed shortly after the onset of pregnancy, a high frequency of free blastocysts and implant resorptions was observed in the neuraminidase-treated mice. From this it would appear that the enzyme acts by interfering with both the implantation and the development of the already implanted ovum. Whether this disruption in the orderly sequence of events results from partial stripping of the endometrial surface is not clear, but it is significant that luteal function is highly dependent on sialic-acid-rich gonadotropins which are produced by the trophoblast in the earliest stages of implantation. Cleavage of the sialic acid residues of these hormones has been shown to result in their prompt removal from the circulation and subsequent destruction in the liver (Morell *et al.*, 1971).

An alternative, or possibly supplementary, mechanism whereby sialic acid may participate in the sequence of events culminating in pregnancy relates to the role of this compound on the zona pellucida of rabbit ovum (Soupart and Clewe, 1965). Following brief incubation of freshly isolated ova with either intact or heat-inactivated neuraminidase, the ova were replaced in the oviducts of recipient rabbits which were then mated with fertile males. After 22 hours, the recipient animals were killed, the ovaries and oviducts were

removed, and the recovered ova were examined for evidence of sperm penetration. A dose–response curve clearly indicated a relationship between
neuraminidase treatment of the zona pellucida and the inhibition of sperm
penetration. In all cases, penetration was significantly reduced, and in 25
percent of the experiments, the inhibition was 100 percent. Corresponding
deformational changes were noted in the shape of the enzyme-treated ova
and it was suggested that the specific enzymatic removal of sialic acid residues
from the zona pellucida and the consequent alteration in mechanical and
elastic properties might explain the failure of sperm penetration. In light of
the present discussion, the possibility should also be considered that the initial
attachment of the sperm to the zona pellucida may be dependent upon the
presence of surface sialic acid residues and that the binding reaction is abolished when they are removed.

3.2.6   Transport

Prompted by earlier observations that tumor strains were characterized by
increased levels of surface sialic acid and that, in general, such cells seemed
to concentrate K^+ better than normal cells, Glick and Githens (1965) investigated regulation of this ion by sialic acid. Employing L 1210 mouse leukemia
cells which had been maintained in the ascitic form, they monitored the
changes in both K^+ and Na^+ in normal and neuraminidase-treated cells.
Different incubation media were utilized in order to regulate the direction
of flow. In K^+-free medium, K^+ release from the cells was accompanied by
uptake of Na^+; with K^+-low medium, K^+ uptake accompanied release of
Na^+. Under these conditions, K^+ transport, regardless of the direction of
flow, was significantly inhibited in the sialic-acid-deficient cells. In contrast
to the relatively sharp response of K^+ to the removal of sialic acid, Na^+ transport was not significantly affected. The relative specificity of this phenomenon for K^+ was shown by the fact that sialic acid was found to be without
effect on either glucose or lysine uptake in the presence or absence of K^+,
thereby indicating that sialic acid does not influence transport of positively
charged substances in general. These results indicated that sialic acid
apparently mediates both the inward and outward diffusion of K^+ in a leukemia cell.

Based on earlier reports from several laboratories that ascites tumor cells
release large quantities of enzymes into the surrounding medium *in vitro*,
Glick, Goldberg and Pardee (1966) extended these observations on control
and neuraminidase-treated L 1210 leukemia cells. Following a 90-minute incubation at 37 °C, the cells were recovered by centrifugation, monitored for
the extent of cell lysis, and the supernatant fluid examined for sialic acid content by means of the colorimetric resorcinol procedure of Svennerholm. An
aberrant ratio in the absorption spectrum suggested that the neuraminidase-
treated cells, in addition to sialic acid, released an unidentified substance or
substances which reacted in the resorcinol assay. This material was identified
as protein by chromatographic and spectrophotometric procedures. Quantitative studies, employing disk gel electrophoresis, revealed the presence of
nine major protein bands in the supernatant fluid of both control and experimental cells with no discernible differences in the number and location of

the bands. However, there was a marked diminution in the amount of protein released in three of the bands, no significant change in five others and an increase in one. The total amount of protein released by the neuraminidase-stripped cells was decreased by 40 percent. When this figure was corrected for protein resulting from lysis during the incubation, it was shown that net protein release was inhibited 66 percent in the sialic-acid-deficient cells. On the basis of these results, it was speculated that the enzymatic removal of membrane-bound sialic acid might cause the release of a 'transport protein' which, in turn, would block the secretion of intracellular proteins. Such a mechanism might be advantageous for a tumor cell when characteristically releasing large quantities of intracellular enzymes since the production and leakage of these enzymes might be related to the invasiveness of the tumor cell. Thus, in this manner it was suggested that alterations in the surface properties of tumor cells might contribute to the unchecked growth of tumors.

3.2.7   Phagocytosis

Recent studies have revealed a new and unsuspected role for the participation of cell-surface sialic acid residues in the phagocytic response of polymorphonuclear leukocytes (Constantopoulos and Najjar, 1973). In the system employed, the isolated leukocytes were incubated with *Staphylococcus aureus* at 37 °C for 30 minutes. At the end of that time, stained smears of the incubated cells were prepared and counted; the phagocytic index was defined as the percentage of cells counted (400) which contained ingested bacteria. The inclusion in the incubation mixture of a particular cytophilic γ-globulin, termed *leukokinin*, had previously been shown to bind specifically to the leukocytes and to stimulate their phagocytic activity several-fold. It was later found that the whole molecule was not necessary for this biological effect and that all of the stimulatory activity resided in a small peptide cleaved from the parent molecule by a membrane enzyme, leukokininase. The resulting, highly basic tetrapeptide, named *tuftsin*, was shown to consist of L-threonyl-L-lysyl-L-prolyl-L-arginine.

Prior incubation of the polymorphonuclear cells with small amounts of neuraminidase preparations from either *Vibrio cholerae* or *Clostridium perfringens* completely abolished the stimulatory effect of added tuftsin. Viral neuraminidase was somewhat less effective in that the reduction of stimulation was approximately 50 percent. When mucin, a substrate for the bacterial neuraminidase, was present during incubation of cells with the enzyme, the cells responded to tuftsin with an increase in the phagocytic index equal to that observed with control cells. The addition of tuftsin in high concentrations also protected the cells. Thus, cells incubated with 1 milliunit of bacterial neuraminidase and 4.2 μmol of tuftsin, after washing, responded to 1 nmol of tuftsin to essentially the same degree as did cells never exposed to the enzyme. It was postulated that the protective effect of tuftsin may result from its ability to bind to membrane sialic acid residues and thereby make them inaccessible to the enzyme. The protective action of tuftsin does not, of course, identify the sialic acid groups as the actual receptor of the tetrapeptide; they may simply bind the activator in order to provide a high local concentration in the vicinity of the ultimate protein receptor which then mobilizes the cell membrane for more effective phagocytosis, pinocytosis and cell mobility (Nishioka *et al.*, 1972).

3.3   ROLE OF GLYCOSYLTRANSFERASES IN BINDING

3.3.1   **Intercellular adhesion**

In a stimulating attempt to rationalize the constantly increasing evidence for cell-surface carbohydrate participation in recognition processes, Roseman (1970) proposed the hypothesis that intercellular recognition and communication arise from the interaction between cell-surface glycosyltransferases and their appropriate cell-surface glycosyl acceptors. This concept has provided a marked stimulus for further experimentation designed to test its validity. As a consequence, meaningful new information has become available which bears directly on the central problem of the role of complex carbohydrates in intercellular recognition and binding phenomena.

At the outset, it was realized that the accumulated documentation of complex carbohydrates as constituents of a broad spectrum of cell surfaces related, in all probability, to their diverse roles in a variety of biological functions. Consequently, attention was directed toward a single major function of cell surfaces, intercellular adhesion, in an attempt to determine whether or not complex carbohydrates were involved in this process. With the development of a quantitative assay to measure the adhesive process (Orr and Roseman, 1969), an examination of the metabolic requirements for cellular aggregation of a mouse teratoma tumor was undertaken (Oppenheimer *et al.*, 1969). In this system cells of the ascites-grown form of mouse teratoma adhered to each other when grown in a complex tissue culture medium; omission of a single constituent, L-glutamine, prevented agglutination. Only two compounds were found to replace L-glutamine, namely the hexosamines D-glucosamine and D-mannosamine. The response to L-glutamine and the two amino sugars was completely inhibited by a variety of metabolic inhibitors including sodium cyanide, sodium azide, dinitrophenol and sodium fluoride. That the inhibitory effect was not the result of cell death was shown, in the case of sodium fluoride, by the prompt restoration of agglutinating ability upon removal of the inhibitor from the incubation mixture. Furthermore, two metabolic antagonists of L-glutamine (6-diazo-5-oxonorleucine and *O*-diazoacetylserine) blocked the action of the amino acid but did not inhibit the effect of D-glucosamine or D-mannosamine.

Since L-glutamine functions as a nitrogen donor in the conversion of fructose-6-phosphate to glucosamine-6-phosphate, a block in the synthetase reaction would prevent the formation of all amino sugars. Furthermore, the fact that this block was overcome uniquely by the addition of glucosamine and mannosamine (but not galactosamine) is in complete accord with the known metabolic pathways whereby these amino sugars are incorporated into the oligosaccharide chains of complex carbohydrates and provides strong inferential support for the thesis that at least this cell type requires such complex carbohydrates for adhesiveness.

Buttressed by the above results, Roseman and his colleagues set about to determine whether or not evidence could be obtained to indicate that plasma membranes did, in fact, contain glycosyltransferases and whether they might play a role in intercellular adhesion. The methodology adopted involved the adhesion of radioactively labeled cells to unlabeled aggregates of the same (specific) or different (nonspecific) types of tissue as determined by radioactive monitoring of the number of cells so collected (Roth, McGuire and Roseman, 1971a). Exposure of embryonic chicken neural retina cells to β-galactosidase (but not to trypsin or collagenase) resulted in a significantly diminished specificity of attachment and it was concluded that terminal β-galactopyranosyl groups were involved in specific intercellular adhesion of these cells. The intact cells, in the presence of labeled UDP–galactose, catalyzed the transfer of galactose to endogenous acceptors of high molecular weight as well as to exogenous acceptors. A variety of controls established that the cells remained viable during the course of the reaction, that the sugar nucleotide did not enter the cells and that, after centrifugation, the supernatant solution was essentially devoid of transferase activity. Of critical significance was the finding that the incorporation of galactose by whole cells comprised 75–90 percent of the transferase activity of cell homogenates, and that the addition of appropriate galactosyl acceptors interfered with the adhesive specificity of the retina cells under conditions where nonacceptors were without effect. These results were interpreted as indicating that the cell-surface galactosyltransferase did play a role in cellular recognition (Roth, McGuire and Roseman, 1971b). In this context it is perhaps meaningful to recall that sialyltransferase has been identified on liver plasma membranes and that, in the presence of added CMP–sialic acid, it is capable of restoring binding activity to the neuraminidase-inactivated membranes (Pricer and Ashwell, 1971). Whether or not this glycosyltransferase is identifiable with the binding protein, or actually participates in the membrane binding of desialylated glycoproteins, has not been established.

Additional supportive evidence has been obtained to indicate the presence of glycosyltransferases on normal (Roth and White, 1972) as well as virus-transformed fibroblasts (Bosmann, 1972a, b). It has been a consistent finding that normal cells in confluent cultures have less endogenous transferase activity per cell than when they are present in sparse cultures whereas transformed cells do not show this concentration-dependent activity. These results are compatible with the idea that glycosyltransferases on normal cells are capable of glycosylating acceptors on adjacent cells and that, as cell contact increases, the number of transfer reactions increases with a consequent reduction in the number of available acceptor sites; the latter effect explains the diminished endogenous transferase activity (Roth, 1973). In the case of the virally transformed fibroblasts at confluency, the levels of several glycosyltransferases were shown to be significantly elevated compared with normal controls whereas, in sparse cultures, the transferase activity of both normal and transformed cells was essentially equivalent (Bosmann, 1972a).

Complementary studies on the glycolipid composition of normal hamster cells indicated a more complex structure at confluency than was seen during the early growth phase. Similarly, upon transformation by polyoma virus, the glycolipids of the transformants were shown to be less complex than those of normal cells. However, regardless of the degree of confluence of the culture, the transformed cells contained mostly simpler glycolipids such as ceramide lactose and ceramide glucose (Hakamori, 1970). The generality of these findings has been confirmed for glycolipids (Cumar *et al.*, 1970; Robbins and Macpherson, 1971) as well as for glycoproteins (Meezan *et al.*, 1969).

It is clear from the above that the application of the enzyme–substrate analogy to rationalize the behavior of cells in growth and differentiation has potentially far-reaching consequences in terms of providing a unifying hypothesis for the explanation of such diverse phenomena as regulatory control mechanisms, intercellular recognition, cell-surface modification and contact inhibition. Whether, or to what extent, subsequent investigation will affirm or negate this concept is almost irrelevant; its heuristic value has been already established.

3.3.2   Platelet–collagen adhesion

The conceptual basis of intercellular adhesion, as outlined in the preceding section, was quickly recognized by Jamieson and his colleagues as being potentially applicable to the fundamental problem of hemostasis, that of platelet–collagen adhesion (Jamieson, Barber and Urban, 1971; Jamieson, Urban and Barber, 1971). The platelet plasma membranes having previously been isolated and characterized, a means had become available to distinguish between the membrane-associated phenomena of collagen adhesion and the intracellular events associated with platelet aggregation (Barber and Jamieson, 1970).

Upon incubation of the isolated platelet plasma membranes with labeled UDP–glucose and a collagen receptor, radioactive glucose was transferred to an acid-insoluble fraction wherein the locus of incorporation was demonstrated to be on the glucosylgalactosylhydroxylysine moiety of collagen (Barber and Jamieson, 1971a). The activity of the membrane-bound collagen:glucosyltransferase represented a twentyfold purification and a recovery of 55 percent of the initial enzymatic activity of the intact platelets. Examination of the specificity of the reaction revealed that none of 12 potential glycoprotein acceptors tested exhibited significant activity; in the absence of any added receptor, the endogenous activity was negligible (Barber and Jamieson, 1971a).

Most significant, however, was the correlation observed between inhibitors of the collagen:glucosyltransferase reaction and the adhesion of platelets to collagen. Thus, glucosamine, aspirin, chlorpromazine and sulfhydryl antagonists, all inhibitors of the glucosyltransferase, were similarly effective in blocking aggregation; unsubstituted ε-amino groups on the hydroxylysine residues were essential for the activity of both systems.

Based upon the identification of a specific collagen:glucosyltransferase on the outer surface of the platelet membrane, the incorporation of glucose from UDP–glucose exclusively into the heterosaccharide units of collagen, and the identity of inhibitory effects shown by a wide range of substances in each reaction, the conclusion was drawn that the carbohydrate portion of collagen was clearly implicated in platelet adhesion and the suggestion was made that this reaction might represent a special case of the general mechanism for intercellular adhesion (Barber and Jamieson, 1970).

Collagen:galactosyltransferase, the second enzyme implicated in the biosynthesis of the disaccharide units of collagen, was similarly identified as a constituent of the platelet plasma membranes (Barber and Jamieson, 1971b). The activity of this enzyme, as contrasted to that of the glucosyltransferase, was markedly stimulated by the inclusion of Triton X-100 in the incubation mixture. Since it is believed that the detergent stimulates enzymes located on the inner surface of the membrane vesicle, as opposed to those on the exterior, this behavior was considered as suggesting that the galactosyltransferase might be buried in the interior of the membrane or on the inner surface. A second major difference between the two enzymes emerged from a consideration of their behavior toward the inhibitors glucosamine, aspirin and sulfhydryl blocking agents, none of which exerts a significant effect upon the reactivity of the galactosyltransferase. In view of these differences, it was considered as unlikely that this enzyme was involved in the phenomenon of platelet:collagen adhesion.

In an independent and concurrent study, Bosmann (1971) identified and characterized four glycoprotein:glycosyltransferases, including glucosyl- and galactosyltransferase, from platelet membranes. Subsequent studies by the same author (Bosmann, 1972c) resulted in the identification of a glycoprotein:sialyltransferase together with a membrane-associated neuraminidase. The sequential activity of these two enzymes, acting in concert upon the surface sialic acid residues, was suggested as a general model for cell–cell or cell–macromolecule adhesiveness.

3.4   CONCLUSION

As was stated at the outset, the purpose of this brief and eclectic review has been to attempt to correlate certain biologically significant aspects of cell and membrane physiology with the current rapidly developing appreciation of the role of cell-surface carbohydrates in recognition phenomena. No attempt has been made to cover the problems associated with changes in cell-surface topography during normal and neoplastic growth, the mechanism of contact inhibition, the metabolic *sequelae* of lectin induction or the ubiquitous role of adenyl cyclase, since these topics are treated in detail elsewhere in this series. It is hoped, however, that the subject matter actually covered, although far from comprehensive and all too frequently speculative in content, may provide at least the outlines of a conceptual basis for subsequent and systematic investigations encompassing all of the above areas.

REFERENCES

ASHWELL, G. and MORREL, A. G. (1971). *Glycoproteins of Blood Cells and Plasma*, p. 173. Ed. G. A. JAMIESON and T. J. GREENWALT. Philadelphia; J. B. Lippincott.
ASHWELL, G. and MORELL, A. G. (1974). *Adv. Enzymol.*, **41**:99.
BAGSHAWE. K. D. and CURRIE. G. A. (1968). *Nature. Lond.*, **218**:1254.
BARBER. A. J. and JAMIESON. G. A. (1970). *J. biol. Chem.*, **245**:6357.
BARBER, A. J. and JAMIESON. G. A. (1971a). *Biochim. biophys. Acta*, **252**:533.
BARBER, A. J. and JAMIESON, G. A. (1971b). *Biochim. biophys. Acta*, **252**:546.
BOSMANN. H. B. (1971). *Biochem. biophys. Res. Commun.*, **43**:1118.
BOSMANN. H. B. (1972a). *Biochem. biophys. Res. Commun.*, **48**:523.
BOSMANN. H. B. (1972b). *Biochem. biophys. Res. Commun.*, **49**:1256.
BOSMANN, H. B. (1972c). *Biochim. biophys. Acta*, **279**:456.

BURNET. F. M. (1951). *Physiol. Rev.*, **31** : 131.

CONSTANTOPOULOS. A. and NAJJAR. V. A. (1973). *J. biol. Chem.*, **248** : 3819.

CUMAR, F. A., BRADY, R. O., KOLODNY, E. H., MCFARLAND, V. W. and MORA, P. T. (1970). *Proc. natn. Acad. Sci. U.S.A.*, **67** : 757.

EMMELOT, P. (1973). *Eur. J. Cancer*, **9** : 319.

GASIC. G. J. and GASIC. T. B. (1962). *Proc. natn. Acad. Sci. U.S.A.*, **48** : 1172.

GASIC. G. J. and GASIC. T. B. (1970). *Proc. natn. Acad. Sci. U.S.A.*, **67** : 793.

GASIC. G. J., GASIC. T. B. and STEWART. C. C. (1968). *Proc. natn. Acad. Sci. U.S.A.*, **61** : 46.

GLICK, J. L. and GITHENS. S. (1965). *Nature, Lond.*, **208** : 88.

GLICK. J. L., GOLDBERG. A. R. and PARDEE. A. B. (1966). *Cancer Res.*, **26** : 1774.

GOTTSCHALK. A. (1959). *The Viruses*, p. 51. Ed. F. M. BURNET and W. M. STANLEY. New York ; Academic Press.

GREGORIADIS, G.. MORELL. A. G.. STERNLIEB. I. and SCHEINBERG. I. H. (1970) *J. biol. Chem.*, **245** : 5833.

HAKAMORI. S. (1970). *Proc. natn. Acad. Sci. U.S.A.*, **67** : 1741.

HICKMAN, J.. ASHWELL. G., MORELL. A. G.. VAN DEN HAMER. C. J. A. and SCHEINBERG, I. H. (1970). *J. biol. Chem.*, **245** : 759.

HUDGIN, R. L. and ASHWELL, G. (1974). *J. biol. Chem.*, **249** : 7369.

HUDGIN, R. L., PRICER, W. E., JR., ASHWELL, G., STOCKERT, R. J. and MORELL, A. G. (1974). *J. biol. Chem.*, **249** : 5536.

HUGHES. R. C. (1973). *Prog. Biophys. molec. Biol.*, **26** : 189.

JAMIESON. G. A. (1965). *J. biol. Chem.*, **240** : 2019.

JAMIESON. G. A.. BARBER. A. J. and URBAN. C. (1971). *Glycoproteins of Blood Cells and Plasma*, pp. 219–232. Ed. G. A. JAMIESON and T. J. GREENWALT. Philadelphia ; J. B. Lippincott.

JAMIESON. G. A.. URBAN. C. L. and BARBER. A. J. (1971). *Nature. New Biol.*, **234** : 5.

KATHAN. R. H.. WINZLER. R. J. and JOHNSON. C. A. (1961). *J. exp. Med.*, **113** : 37.

KAWASAKI. T. and ASHWELL. G. (1976). *J. biol. Chem.*, **251** : 1296.

LAVER. W. G. and VALENTINE. R. C. (1969). *Virology*, **38** : 105.

LUNNEY, J. and ASHWELL. G. (1976). *Proc. natn. Acad. Sci. U.S.A.*, **73** : 341.

MEEZAN. E.. WU. H. C.. BLACK. P. H. and ROBBINS. P. W. (1969). *Biochemistry*, **8** : 2518.

MORELL. A. G.. VAN DEN HAMER. C. J. A.. SCHEINBERG. I. H. and ASHWELL. G. (1968). *J. biol. Chem.*, **241** : 3745.

MORELL. A. G.. GREGORIADIS. G.. SCHEINBERG. I. H.. HICKMAN, J. and ASHWELL. G. (1971). *J. biol. Chem.*, **246** : 1461.

NISHIOKA. K.. CONSTANTOPOULOS. A.. SATOH. P. S. and NAJJAR. J. A. (1972). *Biochem. biophys. Res. Commun.*, **47** : 172.

OPPENHEIMER. S. B.. EDIDIN. M.. ORR. C. W. and ROSEMAN. S. (1969). *Proc. natn. Acad. Sci. U.S.A.*, **63** : 1395.

ORR. C. W. and ROSEMAN. S. (1969). *J. Membrane Biol.*, **1** : 109.

PRICER. W. E. and ASHWELL. G. (1971). *J. biol. Chem.*, **246** : 4285.

ROBBINS, P. and MACPHERSON. T. (1971). *Nature. Lond.*, **229** : 569.

ROSEMAN. S. (1970). *Chem. Phys. Lipids*, **5** : 270.

ROTH. S. (1973). *Q. Rev. Biol.*, **48** : 541.

ROTH. S.. MCGUIRE. E. J. and ROSEMAN. S. (1971a). *J. Cell Biol.*, **51** : 525.

ROTH. S. MCGUIRE. E. J. and ROSEMAN. S. (1971b). *J. Cell Biol.*, **51** : 536.

ROTH. S. and WHITE. D. (1972). *Proc. natn. Acad. Sci. U.S.A.*, **69** : 485.

SANDFORD, B. H. (1967). *Transplantation*, **5** : 1273.

SOUPART. P. and CLEWE. T. H. (1965). *Fert. Steril.*, **16** : 677.

SUTTAJIT. M. and WINZLER. R. J. (1971). *J. biol. Chem.*, **246** : 3398.

TURNER, R. S. and BURGER, M. M. (1973). *Ergebn. Physiol.*, **68** : 121.

VAN DEN HAMER. C. J. A.. MORELL. A. G.. SCHEINBERG. I. H.. HICKMAN. J. and ASHWELL. G. (1970). *J. biol. Chem.*, **245** : 4397.

VAN LENTEN. L. and ASHWELL. G. (1971). *J. biol. Chem.*, **246** : 1889.

VAN LENTEN. L. and ASHWELL. G. (1972). *J. biol. Chem.*, **247** : 4633.

WOODRUFF, J. J. and GESNER. B. M. (1969). *J. exp. Med.*, **129** : 551.

4

The antigenic nature of mammalian cell surfaces

Michael J. Weiss and J. V. Klavins

Long Island Jewish–Hillside Medical Center, Queens Hospital Center Affiliation, Jamaica, New York

4.1 INTRODUCTION

Until recently, the plasma membrane was viewed as a passive barrier between the cell interior and its external environment. Early models of the plasma membrane (Danielli and Davson, 1956) envisaged a lipid bilayer sandwiched between two layers of protein. This view was later modified to include pores in an attempt to attain a better agreement with observed function.

More recently a different view of the plasma membrane has been proposed (Singer, 1971) and is more consistent with the observed data. The suggestion is that the membrane consists of a lipid bilayer with proteins interdispersed throughout the lipid. The membrane proteins can then be classified in two broad categories: those which are easily extracted from the membranes and those which cannot be extracted without the complete disruption of the plasma membrane. It must be borne in mind that membrane proteins consist of a hydrophobic portion embedded in the lipid layer, and a hydrophilic portion. There is evidence (*see* Bretscher, 1973) that in human red blood cells certain membrane proteins extend through the lipid bilayer and have a hydrophilic moiety extending into the external environment and another hydrophilic moiety extending into the cytoplasm. The major glycoprotein of the red cell membrane has been isolated from human red blood cells utilizing mild conditions (Winzler, 1969). This glycoprotein has a large hydrophilic moiety containing all of the carbohydrate residues and a small hydrophobic portion which can be shown by the use of proteolytic enzymes to be embedded in the lipid portion of the membrane.

Another very important feature of Singer's model is the hypothesis that the proteins are as 'icebergs in a sea of lipid' and are free to migrate. This is supported by experiments (Frye and Edidin, 1970) involving mouse–human cell fusion; immediately afterwards, species-specific antigens were located in discrete areas of the surface of the new hybrid cell but eventually mouse-specific and human-specific antigens became intermixed and randomly distributed. This was true even under the influence of a variety of inhibitors of protein synthesis, used so as to rule out synthesis of antigens *de novo*.

4.2   BLOOD-GROUP SUBSTANCES

The blood-group substances are the most extensively studied alloantigens and our discussion of them as a model system will be brief and limited to the ABO (H) MN systems.

Since the classic ABO system was first described (Landsteiner, 1900), the number of known human blood-group substances has increased to over 60. Blood-group substances have been demonstrated in the tissue and body fluids in a large number of individuals, as well as on the erythrocyte membrane. The term 'secretor' has been used to describe those individuals who secrete the blood-group substances in their saliva. It has been shown (Morgan and Watkins, 1948) that secretors of the O type possess an antigen designated the 'H' blood group.

The existence of two blood-group antigens which are termed the 'M' and 'N' antigens was demonstrated in 1927 by Landsteiner and Levine. These antigens were independent of the ABO system. In addition, rabbit antiserum to rhesus-monkey blood was found to be immunologically active against an antigen (now known as the RH antigen) in 85 percent of the humans tested (Landsteiner and Wiener, 1940).

It was clearly shown (Poulik and Bron, 1969) that the A, B, M and N antigens can be extracted from the red blood cell ghosts in the aqueous phase, using the butanol extraction procedure of Maddy as modified by Rega *et al.* (1967). The fact that these materials remain in the aqueous phase, even when subjected to repeated butanol extractions, provides substantial evidence of their glycoprotein nature.

A glycoprotein has been isolated from erythrocyte ghosts using 50% aqueous phenol (Winzler, 1969), and found to consist of 87 amino acid residues and 100 carbohydrate residues. All of the carbohydrate residues are located on one portion of the molecule in 10 side-chains while the remainder of the molecule, the hydrophobic portion, is buried in the lipid bilayer. This glycoprotein contains the MN antigenic site and the influenza virus receptor site, as well as other antigenic determinants. Approximately 90 percent of the total sialic acid content of the red blood cell plasma membrane is accounted for by this molecule, as is approximately 80 percent of the membrane carbohydrate.

Over the past two decades, the immunogenic determinants of the ABO (H) system have been determined principally by the work of Morgan and Watkins in England and Kabat and his group in the United States. The materials used for these studies were the soluble glycoproteins from ovarian cysts. The sequence of the carbohydrate core common to all of the blood-group substances is shown in *Figure 4.1*. The structures of the immunogenic determinants of the blood-group substances are given in *Table 4.1*.

It has been proposed that the *A*, *B* and *H* genes control the final conversion

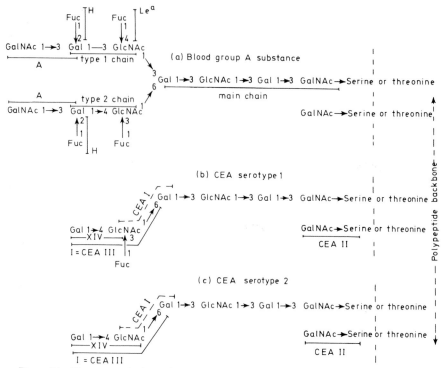

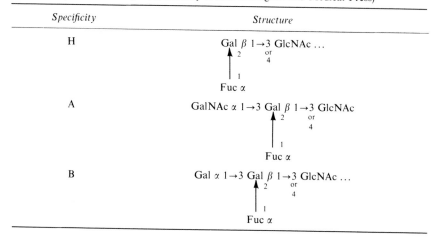

Figure 4.1 The structural relationship of blood-group A substance and carcinoembryonic antigens (CEAs) of serotypes 1 and 2. The structure of the blood-group A glycoprotein has been adapted from the studies of Feizi et al. (1900), but the 'optional' secondary side-chains that are present on some primary chains have been omitted in the interests of clarity. (- - - -) Blood-group and CEA determinants that have been adequately characterized; (- - - -), incompletely characterized CEA I. All the fucosyl and N-acetylgalactosaminyl residues are believed to be of the α-anomeric form and all the galactosyl and N-acetylglucosaminyl residues were thought to be of the β form. Fuc, L-fucose; Gal, D-galactose; A, H and Le^a, determinants A, H and Le^a; XIV, factor XIV; I, antigenic complex I (= CEA III). From Simmons and Perlmann, 1973, courtesy of Cancer Research Inc.)

Table 4.1 STRUCTURES RESPONSIBLE FOR ABH IMMUNE RESPONSE
(From Watkins, 1966, courtesy of ASP Biological and Medical Press)

Specificity	Structure
H	Gal β 1→3 GlcNAc ... ↑2 or 4 │1 Fuc α
A	GalNAc α 1→3 Gal β 1→3 GlcNAc ↑2 or 4 │1 Fuc α
B	Gal α 1→3 Gal β 1→3 GlcNAc ... ↑2 or 4 │1 Fuc α

of an existing precursor substance into final blood-group substances (Watkins, 1966). It is now well established that the various genes code for specific carbohydrate transferases. The H gene codes for the fucosyltransferase with the following specificity:

H enzyme

$$\text{GDP–Fuc} + \text{GalNAc}\cdots\text{R} \rightleftharpoons \text{Fuc } \alpha\ 1\rightarrow2\ \text{GalNAc}\cdots\text{R} + \text{GDP}$$

(for this and subsequent reactions R = the GlcNAc $1\rightarrow6$ Gal branch of the structure shown in *Figure 4.1*).

The product of gene A is an N-acetylgalactosaminyltransferase with the specificity:

A enzyme

$$\text{UDP–GalNAc} + \text{GalNAc}\cdots\text{R} \rightleftharpoons \text{GalNAc } 1\rightarrow3\ \text{GalNAc}\cdots\text{R} + \text{UDP}$$

$$\begin{array}{ccc} \uparrow^2 & & \uparrow^2 \\ \mid_1 & & \mid_1 \\ \text{Fuc } \alpha & & \text{Fuc } \alpha \end{array}$$

Gene B codes for galactosyltransferase with the specificity:

B enzyme

$$\text{UDP–Gal} + \text{Gal } \beta\ 1\rightarrow3\ \text{GalNAc}\cdots\text{R} \rightleftharpoons \text{Gal } \alpha\ 1\rightarrow3\ \text{Gal } \beta\ 1\rightarrow3\ \text{GalNAc}\cdots\text{R}$$
$$+ \text{UDP}$$

$$\begin{array}{ccc} \uparrow^2 & & \uparrow^2 \\ \mid_1 & & \mid_1 \\ \text{Fuc } \alpha & & \text{Fuc } \alpha \end{array}$$

It is obvious from the preceding discussion that the precursor of the ABO (H) system having the structure shown in *Figure 4.1* is the substrate for the transferases specified by the corresponding genes. The H-, A- and B-specific transferases use only the Gal 1–4 branch of the basic precursor. The H-specific structure then acts as a substrate for A and B transferases. This has been confirmed (Tuppy and Schenkel-Brenner, 1969) using the 'A' enzyme described by Hearn, Smith and Watkins (1968). These workers converted the O-type erythrocytes to A-type erythrocytes. The O-type erythrocytes have also been converted to B-type erythrocytes under the influence of the B enzyme.

4.3   HISTOCOMPATIBILITY ANTIGENS

Histocompatibility antigens are plasma-membrane components, expressed on the surface of the cells. They have been studied in man and in animals, particularly chick, guinea-pig, mouse, rat, dog, rhesus monkey and chimpanzee. In each species there is a major histocompatibility system which can be identified by allograft rejection or serologically demonstrable antigens.

The most extensive studies of histocompatibility antigens have been carried out in mouse and man. In the mouse there are at least 30 genetic loci which determine the antigens detectable by graft rejection. Between the two histocompatibility antigen loci (H-2D and H-2K) is a locus which determines a variation in the serum protein. These two distinct types of antigen can be detected serologically. A gene or gene complex determining the intensity of the immune response to antigen, *Ir-1*, is located between the serum protein locus and *H-2K*. In the mouse there is a high degree of genetic polymorphism.

A high degree of genetic polymorphism appears also in man. At least five thousand phenotypes have been identified in the HL-A system, in which there are two complexes of antigens, the 'LA series' and the '4 series' which are comparable to H-2D and H-2K. These antigens correspond to separate loci consisting of gene complexes. The two loci recombine with an estimated frequency of between 0.5 and 1 percent, a recombination frequency similar to that observed between the mouse *H-2D* and *H-2K* regions. The LA and '4' series determinants appear as multiple alleles at a single genetic locus, each consisting of approximately 20 specificities. The two loci are closely linked on the same chromosome (Ceppellini *et al.*, 1967). Although HL-A antigens are not expressed on the surface of unfertilized or fertilized eggs, they are present on spermatozoa (Fellous and Dausset, 1970). While they are present in most tissues of early embryos, the HL-A antigens appear not to be present in the syncytiotrophoblast of the placenta (Van Werf, 1971), so preventing placental reactivity with maternal HL-A antibodies and avoiding embryo rejection.

It has not, as yet, been definitely determined whether the HL-A antigens are distributed uniformly over the cell surface or whether they are concentrated in patches. The antigens are proteins with molecular weights between 30000 and 50000.

The HL-A antigens of man cross-react with the H-2 system of mouse. Probably a wide range of cross-reactivity exists among phylogenetically distant species. For example, there is evidence of shared determinants between mammalian histocompatibility antigens and such bacterial antigens as *Streptococcus* (Rapaport *et al.*, 1968).

The nature of no individual histocompatibility antigen has yet been clearly defined. Groups of antigens can be isolated by pressure homogenization, sonication, salt extraction (3 M KCl) and solubilization with papain and detergents. It is not known how these procedures modify the structure of the antigens. To test biological activity and specificity, such procedures as mixed leukocyte reaction, complement fixation, microcytotoxicity assays using ^{51}Cr labeling, delayed-type hypersensitivity assays and other serological methods have been used. However, these methods do not allow the extraction and isolation of specific gene products. With the chemical characterization of these antigens there will eventually emerge a better understanding of their biological significance.

The histocompatibility antigens are important not only in transplantation, but may be significant in immune surveillance, protecting the host from malignant mutations (Burnet, 1970) or other undesirable external or internal effects. They have provided antigenic polymorphism throughout evolution although the significance of this phenomenon is not, as yet, understood.

4.4   ANTIGENICALLY ALTERED CELL SURFACES IN
MALIGNANT NEOPLASMS

4.4.1   Introduction

Alterations in the antigenic composition of plasma membranes—both qualitative and quantitative—are known to accompany malignant transformation, and the immunological responses elicited by cells are altered accordingly.
There are now specific substances, to some degree characterized, which
can be classified in the following two groups: qualitatively altered antigens
('neoantigens') and quantitatively altered antigens, including a large sub-
group, the so-called 'oncofetal antigens'.
The neoantigens, found on malignant cells but not on normal adult cells
in comparable tissues, also include groups of substances which, strictly speak-
ing, are not components of the plasma membrane. These are various cell products which may reside within the cell and reach the outside either by reverse
pinocytosis or by passing through the membrane itself. Such substances have
been identified by immunological methods as antigens but their precise
chemical characteristics are not known. Some of the antigens consist of peptide and polysaccharide components. To this group belong tumor-specific
transplantation antigens, which are new surface components produced following changes in the cell genome. Such changes may be induced by viruses
or may result from a resistable susceptibility for certain specific mutations
or by a breakdown in suppression mechanisms. The mode by which virus-
induced tumor-specific transplantation antigens are expressed will be discussed more specifically in Vol. 5 of the present series.
The second group of antigens, the quantitatively altered type, are normal
cell components expressed in abnormal amounts in malignant cells. They may
have enzymatic as well as antigenic activity. Some of these components have
been completely deleted from the plasma membrane of malignant cells.
To this group belong the oncofetal antigens, which are expressed in malignant neoplasms and normal embryonic cells, whereas in adult normal tissues
they are present in small or undetectable amounts. This group of antigens
has been suggested to represent retrogenetic expression. Many of these components represent fetal isoenzymes which disappear during differentiation.
These substances are evidently not components of the plasma membrane, but
they pass through it at some time to become detectable on the cell surface
by methods specific for the determination of antigenic components.
In animal neoplasia there is ample evidence for the presence of tumor-
specific transplantation antigens specified by particular types of virus; this has
practical implications in diagnosis, prevention and treatment of tumors. In
human neoplasia no such antigens are known, although there are some indications that portions of viral genome might be associated with some malignant human tumors. Consequently, such neoantigens are currently of limited
practical significance in detection, therapy or prevention of human neoplasia.
In fact, it is not clear whether at the present time any specific antigen in
humans has been demonstrated which is not a normal component expressed
in altered quantity. The authors' own preliminary work, using an aqueous
extract of the whole human fetus, indicates that specific antibodies can be
produced against such fetal extracts. The precise localization and the

constitution of these antigens in our experiments is not known nor do we know whether with more sensitive methods we would be able to demonstrate such antigens in non-neoplastic, normal or otherwise pathologically altered tissues.

4.4.2 Qualitatively altered membrane antigens (neoantigens)

The neoantigens include substances which are not detectable, using present methods, in normal or non-neoplastic tissues.

4.4.2.1 TUMOR-SPECIFIC TRANSPLANTATION ANTIGENS

The tumor-specific transplantation antigens (TSTAs), also known as 'tumor rejection antigens', 'tumor-specific surface antigens' and 'tumor-distinctive histocompatibility antigens', as already mentioned are neoantigens produced by the neoplastic cell.

TSTAs are present on the external surface of plasma membranes, and may be removed by trypsin treatment (Dickinson, Caspary and Field, 1972), but are not present in intracellular fractions (Baldwin and Moore, 1969). These antigens can be induced by viruses and chemical carcinogens.

(a)   *Virus-induced TSTAs.* Many RNA as well as DNA viruses induce TSTAs, but the chemical nature of these virus-specific transplantation antigens is not known. It is also not known whether they are induced by viral genetic or host genetic determinants.

Several virus-specific surface antigens in leukemic cells of mice have been demonstrated by cytotoxicity methods and immunofluorescence microscopy. According to these demonstrations *in vitro*, the following antigen groups are found: the E antigens in C57B1 leukemia; the FMR antigens in neoplasms induced by Friend, Moloney and Rauscher viruses; and the G antigen groups induced by Gross virus in many spontaneous leukemias and in apparently normal tissues of those strains which have high susceptibility to leukemia.

Cell-surface antigens which are virus-specific have been demonstrated by transplantation rejection procedures. It is not known how the antigens demonstrated *in vitro* are related to TSTAs *in vivo*.

There is a difference in antigenic response to mature DNA virus and to the TSTAs. Immunity to neutralizing virus is distinct from immunity to the TSTAs. In some RNA tumor viruses the situation is similar, as in rodent tumors induced by Graffi leukemia virus and by Rous sarcoma virus.

In all virus-induced tumors in animals a common TSTA appears (Deichman, 1969), whose identity remains unaltered in different species if induced by the same virus. For example, cross-reaction between chick embryo cells and mouse cell lines is evident after transformation (Gelderbom and Bauer, 1973). Similarly, a common antigen was found on the surfaces of several strains of cell lines transformed by mouse sarcoma virus (Aoki, Stephenson and Aaronson, 1973). The antigens cross-reacted within cell lines from rats as well as mice.

TSTAs can be induced *in vitro* by infecting BALB/3T3 cells with SV_{40} virus (Tevethia and McMillan, 1973).

(b)   *Chemically induced TSTAs.* Chemically induced carcinomas as well as naturally occurring tumors, such as Ehrlich ascites tumors, have demonstrable plasma membrane antigens (Fitch, 1962). Cell-surface components which are immunogenic can be isolated from aminoazo-dye-induced rat hepatomas and these are unique to this particular tumor (Baldwin and Glaves, 1972). Such preparations include a tumor-specific antigen with a molecular weight of approximately 55000 as well as other antigens of higher molecular weight (Baldwin, Harris and Price, 1973). These were solubilized from tumor cell surfaces after digestion with papain.

Chemical carcinogenesis induces unique TSTAs in the resulting malignant cells which vary from one case to another, both in the amount and the species of antigen observed. The same carcinogen, inducing tumors in more than one representative of the same strain of animal, may not produce identical TSTAs (Sjogren, 1965).

A review of 3-methylcholanthrene-induced murine sarcomas revealed only rare cross-reactions among TSTAs (Basombrio, 1970). Weak cross-reactions were observed (Hellström, Hellström and Sjogren, 1970) when cell-mediated immunity was studied using an assay of inhibition of colony formation. There were fewer cross-reacting antigens when mice were immunized with multiple tumors (Reiner and Southam, 1969). Cross-reactivity has also been demonstrated among methylcholanthrene-induced sarcomas and tumors induced by benzpyrene (Koldovsky, 1961).

Distinct individual antigens were found among rat hepatomas induced by aminoazo dyes and observed in syngeneic recipients (Baldwin and Barker, 1967a, b). Lymphocytes from a recipient have been shown to inhibit colony formation by hepatoma cells (Baldwin and Embleton, 1971). Using immunofluorescence microscopy with immune serum raised against the tumors, these antigens were demonstrated to have individuality (Baldwin *et al.*, 1970), and were clearly hepatoma plasma membrane components. The antibodies reacting with these antigens were absorbed with hepatoma cell plasma membrane fractions, using immunofluorescence tests (Baldwin and Moore, 1969) as a further demonstration of their surface-associated nature.

Cross-reactivity has been observed between the X antigens in leukemias induced by 3-dimethylbenzanthracene (Gorner, Tuffrey and Batchelor, 1962). However, the question as to whether or not antigens are unusual expressions of normal transplantation antigens has not been clearly answered.

Ultraviolet radiation induces tumors having a TSTA which is not cross-reacting (Graffi, Pasternak and Horn, 1964). Radiation-induced tumors have not been studied to the same extent as virally or chemically induced neoplasms.

TSTAs can be induced *in vitro*, for example by 3-methylcholanthrene in mouse prostate cells (Mondal *et al.*, 1970).

The localization of TSTA determinants is not known. The thymus leukemia (TL) antigen (Old, Boyse and Stockert, 1963) is genetically determined by a locus close to *H-2*. It is carried by certain (TL$^+$) inbred mice but not by other strains (TL$^-$). When some TL$^-$ strains develop radiation-induced leukemias, they have TL antigen. In the sera of TL$^-$ mice, immunized with allogeneic TL$^+$ leukemia cells, cytolytic antibodies for syngeneic TL$^+$ tumor cells are found *in vitro*, but *in vivo* these mice have no resistance. In such immune animals, the transplanted cells lose the TL antigen and regain it only when transplanted to nonimmunized TL⁻ animals. This loss of TL antigens is accompanied by an increase in the level of H-2 antigens; it represents a kind of 'antigenic modulation' (Old *et al.*, 1968). This has been shown *in vitro* by incubating the TL⁺ cells with TL antibody. The process is blocked by actinomycin D and occurs at 37 °C but not 40 °C. Thus it would appear that the antibody on the cell surface inhibits the synthesis of TL antigen and stimulates the synthesis of H-2. This is an example of an immunological reaction participating in cell regulatory processes. It has been observed that the induction of tumor-associated antigen on the cell surface is associated with concomitant loss of normal liver-specific antigens (Baldwin, Glaves and Pimm, 1971). In the context of the present discussion a genetic mutation may not be necessary for the expression of one antigen and the regression of the other (Old *et al.*, 1968). Further, this mode of action is supported by observations that in aminoazo-dye-induced hepatomas, induction of tumor-associated antigen in the plasma membrane is associated with concomitant loss of normal liver-specific antigens. Phenotypic expression of the Gross leukemia virus can likewise be lost and recovered, and the same type of modulation phenomenon can be observed in other systems.

It is possible that chemical carcinogens attack the sites in the genome which govern the normal transplantation antigens, possibly by a chemical change in DNA or a repression–derepression phenomenon. In the first case there would be an appearance of true neoantigens such as TSTA and in the second case, as a result of repression–derepression, normal antigens would appear or disappear or be expressed in varying amounts. These would then belong to the second group of antigens discussed in this review. The persistence of these antigens in 3-methylcholanthrene-induced papillomas of mice is indicative of genetic change. The carcinogen may directly affect the operating segment of the genome and, by induction of random events, produce changes specific for that particular tumor in a particular individual. A close association of polycyclic hydrocarbons with cellular macromolecules has been repeatedly observed. On the other hand, there is evidence that the capacity of chemical carcinogens to react with DNA *in vitro* does not correlate with the carcinogenic potential of the compound *in vivo* (Raymond and Dipple, 1973).

(c) *TSTAs and related antigens in human malignant neoplasms.* In human malignant neoplasms, unlike animal tumors, there is no well-established agent—chemical or biological—that is known to induce malignant transformation. There is evidence of tumor cell surface changes which induce an altered cell-mediated immune reaction. In many cases lymphocytes from cancer patients are cytotoxic to cultured autochthonous tumor cells. Cross-reaction has been shown within certain histologically definable groups of tissues but not between different groups. Such groups with inherent cross-reactivity are melanomas, colon carcinomas, breast carcinomas, bladder carcinomas, endometrial carcinomas, ovarian carcinomas, seminomas and sarcomas (Hellström *et al.*, 1971).

No antigen common to all human neoplasms has been found (Korosteleva, 1957), although there are some human malignant neoplasms which have tumor-specific components.

Tumors of ectodermal origin show some evidence of common specific antigenic components. For example common antigenic material has been observed in human melanomas (Carrel and Theilkaes, 1973). It was common not only to melanomas, but also to two neuroblastomas and one ganglioneuroma. All these tumors are suggested as being derived from neural tube. The antigenic material has a molecular weight of between 42000 and 60000. On isoelectric focusing the fractions with activity were discovered between pH 4.5 and 5.5. All materials, even with different isoelectric points, contained the same antigenic determinants. This material was also excreted in urine. Evidence of human melanoma-specific antigen has also been derived from studies on the induction of specific antibodies in monkeys (Metzger *et al.*, 1973).

Meningiomas and glioblastomas producing an isoantibody have common antigens. Furthermore, serum from a patient with osteogenic sarcoma was found to react with three meningioma cell cultures and sera from meningioma patients with cells from osteogenic sarcoma. Also, normal sera in 30 percent of cases reacted with meningioma cells.

Cell-bound and humoral immunity has been demonstrated against neuroblastoma cells (Hellström *et al.*, 1968) and specific antigens in human breast cancer (Edynak *et al.*, 1971). Soluble membrane antigens from carcinomas of the lip and cervix have been reported to react with antiserum to herpes nonvirion antigens (Hollinshead and Tarro, 1973). These soluble antigens may be markers of the presence of a viral genome in tumor cells.

Among the endodermal derivatives, tumor-specific antigens have not been reported frequently. A tumor-associated antigen with beta mobility (greater than carcinoembryonic antigen) was found in gastric cancer secretions in humans (Dentsch *et al.*, 1973). It has an immunologically reactive carbohydrate moiety, and is different from blood-group substances. This antigen is present in gastric secretions in 91 percent of patients with stomach carcinoma. A sulfomucoprotein present in gastric juice (Häkkinen, 1966) of carcinoma patients is probably not a plasma-membrane component.

Among mesodermal derivatives, a common antigen has been reported in eight osteogenic sarcomas, two rhabdomyosarcomas, one chondrosarcoma and two giant-cell tumors (Priori, Wilbur and Dmochowski, 1971). Sera from 50 percent of patients and 30 percent of their relatives contain antibodies to sarcoma cells. In tissue cultures the antigen can be lost or gained during subculture. In early passages of two osteosarcomas the antigen was undetectable but appeared in the fifth and sixth passages and remained detectable for three more passages before disappearing. In two other sarcomas the antigen was present in early passages, but was not detected subsequently.

There is evidence (Cohen, Ketcham and Morton, 1973) that in human sarcomas the host develops a kind of transplantation resistance to a common sarcoma antigen. Peripheral lymphocytes from sarcoma patients (14 out of 18 cases) were cytotoxic to an allogeneic osteosarcoma cell line but not to fibroblasts from the same patient.

In human nephroblastomas (Wilms' tumors) a common antigen (W antigen) has been discovered (Burtin and Gendron, 1973) which was absent from fetal organs. In addition a similar antigen was present in one renal cell carcinoma, a pool of mammary carcinomas and two gastric carcinomas. This antigen was soluble in phytic acid, destroyed by perchloric acid, and altered by periodic acid. It is a glycoprotein or a polysaccharide. It was not detectable in all nephroblastomas, probably owing to lack of sensitivity in the methods used.

A specific tumor-associated antigen has been isolated in a human serous cystoadenocarcinoma of the ovary (Bhattacharya and Barlow, 1973). Specific anti-tumor antibodies were present. The antigen was a mucoprotein of high molecular weight with beta electrophoretic mobility which gave a single peak on Sephadex G-200. It was not related to carcinoembryonic antigen or histo-compatibility antigens and appeared to be a common antigen for such ovarian tumors.

Antigens on human leukemia cells can induce specific antibodies in monkeys, and these can be used to distinguish specific types of leukemia cells (Metzgar, Mohanakumar and Miller, 1972). A common antigen for acute and chronic lymphocytic leukemia cells, capable of being destroyed by neuramini-dase or trypsin, is suggested. Antigens associated with acute leukemia have been demonstrated (Bentwich et al., 1972) in a purified plasma-membrane component from Burkitt's lymphoma cells in tissue culture. An antibody which combines with Burkitt's lymphoma cells is present in patients as well as in healthy persons. The antigen is thought to be present specifically on the tumor cell surface only and is a transplantation antigen.

Unique components consisting of lipid complexes have been isolated from human tumors: a phospholipid malignolipin (Kosaki et al., 1958) and lipid haptens (Rapport, Graf and Alonzo, 1958a) including cytolipin H (Rapport et al., 1958b). Cytolipin H is antigenic in test animals but not in the host.

HL-A antigens are extractable in soluble form with salt (Reisfeld, Pellegrino and Kahan, 1971), and this method has been applied to the extraction of soluble tumor-associated antigens, for example in chemically induced hepatoma in guinea-pigs (Meltzer et al., 1971). A similar method has been used for the extraction of tumor-associated antigens from a human adeno-carcinoma of the colon. Such antigens inhibit migration of patients' leukocytes, and produce delayed hypersensitization in skin tests. The lymphocytes of patients suffering from the carcinoma gave a blastogenic response, which was not shown by normal lymphocytes, when stimulated with low doses of tumor antigens. However, whole tumor cells stimulate blastogenesis of allo-geneic normal lymphocytes (Anderson, McBride and Hersch, 1972).

4.4.2.2   NATURE AND FUNCTION OF NEOANTIGENS

Qualitative differences in plasma-membrane antigens may reside in the carbohydrate portion rather than in the peptide sequences. For example chick muscle and embryonic cells transformed by Bratislava 77 and Fuji-nami viruses differ from nontransformed cells in the neutral sugar and sialic acid contents of their plasma membrane (Perdue, Warner and Miller, 1973).

The absence of appropriate transferases can lead to the incomplete synthesis of glycoproteins and glycolipids (Hakomori and Murakami, 1968), which may be responsible for the observed decrease in sialic acid content of the plasma membranes of some transformed cells.

In the fluid-mosaic membrane model the carbohydrate moieties are oriented toward the outside of the plasma membrane (Hirano et al., 1972). These cell-surface moieties are obviously major determinants in immunologi-

cal reactions and play a physiological and pathological role in immune reactions.

In the pathogenesis of malignant neoplasms, the plasma membranes constitute an important element. They may act primarily as important mediators
of tumor immunology or in many other phenomena such as lack of density-
dependent inhibition of growth. Ultimately, the plasma membrane may be
responsible for the 'success' or 'failure' of a tumor by evading or by succumb-
ing to the immune surveillance of the host.

All of these examples might indicate that, if there is no expression of common endogenous antigen, the neoantigen is derived either from an altered
cellular genome or by externally induced new information, from a virus for
example. In many cases cross-reactivity exists within a group of specific histo-
logic types.

The question of what the TSTAs represent in virus-induced tumors is un-
resolved. When two different gs-negative cell lines derived from tumors induced by virus were incubated with ferritin-labeled anti-avian sarcoma virus
antiserum, the plasma membranes were labeled uniformly. These common
antigens did not represent any known structural viral antigens.

It is not known what specific transplantation antigens induced by visna
and progressive pneumonia viruses represent (Law and Takemoto, 1973).
These are strong membrane-bound tumor antigens, having no cross-reactivity; they are resistant to X-irradiation, and are effective immunogens even
toward irreversibly damaged tumor cells. After repeated culture *in vivo* or
in vitro in the presence of virus-neutralizing antibody, polyoma tumor
sublines which do not produce infective virus can be established. No sublines
lacking TSTA have been produced (Sjogren, 1964a, b). In DNA viruses the
immunity against the viral coat antigens is distinct from the TSTAs; the virus-
neutralizing immunity does not induce resistance against tumor grafts. However, graft immunity can be induced by tumor cell lines which do not release
demonstrable virus.

It is entirely possible that what have been considered as neoantigens
(TSTAs) may turn out to be embryonal antigens which result from derepression to such an extent that they are detected in tumors. However, studies to
confirm this assumption have yet to be carried out..

(a)   *Virus-specific antigenic changes.* Virus-induced modifications of membranes and incorporation of virus-specific proteins will be discussed in more
detail by H.-D. Klenk in Vol. 5, Chapter 9 of this series.

Both the RNA and DNA virion antigens can be found on cell surfaces.
In some tumors induced by DNA viruses virus-specific cell-surface antigens
have been detected by transplantation rejection techniques, but these antigens
are not found as frequently as with tumors induced by RNA viruses. It is
not known whether all the antigens induced by viruses are altered proteins
or whether alterations occur only in carbohydrate structures, as in bacteria
infected by phage.

In association with RNA viruses, besides virus-specific transplantation
antigen, a viral envelope antigen may be found on the cell surface if the cell
is producing virus. On the other hand, there is no information that the viral
coat antigens would be present in the membranes of the transformed cell
except during the passage of the virus. Type-specific (Ve) viral envelope

antigens have been reported in chick embryo cells that have been transformed or infected by avian RNA tumor viruses (Kurth and Bauer, 1972).

There is some evidence (Defendi, 1967) that transplantation antigen is synthesized in SV_{40}-induced tumors during primary infection and that this synthesis does not require cellular or viral DNA replication. The situation is similar for T antigens, which are cellular antigens that do not require DNA replication itself, but probably require some of the enzymes involved in DNA replication. Furthermore, it has been suggested that these surface antigens represent the expression of the viral cistron or cistrons. Also, a strain of herpes virus induces altered antigenic reactivity of infected cell surfaces (Roizman and Roane, 1963). It has also been suggested that the new immunologic specificity on the plasma membrane is conferred by the virus, that is, it is genetically determined.

Rat leukemia virus gs-1 antigen specific for hamster C-type virus present in transformed hamster cells has been reported (Freeman et al., 1971). In addition, the induced complement-fixation antigen or T antigen, residing in the nucleus, may be determined by a viral gene, since, for example, polyoma viral DNA has been demonstrated in transformed cells (Benjamin, 1966). It would be of great interest to determine whether or not the products coded by virus-specific DNA sequences represent TSTAs or other kinds of complexes.

The surface antigens may play an important role in the immunological surveillance of malignant neoplasms. Depending on the strength of these antigens stimulating the immunological reaction, as well as the condition of the hosts, the ultimate outcome of infection with an oncogenic virus may be either a malignant neoplasm or elimination of the thus-transformed cell. It has also been postulated that some extracellular regulatory process and cell division may be dependent on surface alterations.

4.4.3   Quantitatively altered plasma-membrane antigens

4.4.3.1   CARCINOEMBRYONIC ANTIGEN IN HUMAN MALIGNANT NEOPLASMS

The oncofetal antigens have evoked a great deal of interest and controversy. Carcinoembryonic antigen (CEA), a member of this group (Gold and Freedman, 1965), has been used in the diagnosis and prognosis of a wide variety of human malignant neoplasms, but particularly those of the gastrointestinal system. Laurence and Neville (1972) have reviewed the topic of CEA in great depth. CEA is found in carcinomas of the gastrointestinal tract, pancreas, bronchus, neuroblastoma, testis, breast, urinary bladder and less commonly in other neoplasms.

CEA can be demonstrated to be present in human mucin-producing colonic tumors when such tumors are grown in hamster cheek pouches (Goldenberg and Hansen, 1972).

Collatz, von Kleist and Burtin (1971) found no circulating autoantibodies against CEA. Elevated values of CEA are found in 30 percent of diseases associated with regeneration or inflammation (Laurence et al., 1972). Though not useful as a screening test, this is valuable in following up patients after therapy. A good prognosis is associated with a fall in concentration of CEA, while elevated values indicate metastases.

CEA is present in meconium and the antigen content of amniotic fluid is proportional to the antigen content of meconium. Thus CEA levels could be used as a diagnostic test for meconium and consequently of fetal well-being. CEA is also present in the colonic mucosa of children up to 7 years old. It is found in normal adult colon and plasma but in a lower concentration than in carcinoma patients. Recently an antigen has been identified in normal human tissues, a glycoprotein, which cross-reacts with CEA (von Kleist, Chavanel and Burtin, 1972).

CEA is associated with plasma membranes. It was shown electron-microscopically (Gold, Krupey and Ansari, 1970) that cell surfaces, especially the glycocalyx, contain CEA. This localization was still found after culturing tumor cells *in vitro* for 7 years (Burtin *et al.*, 1970). On the other hand, immunofluorescence microscopic studies showed preeminent localization of CEA in the glycocalyx and dependence on a certain degree of tumor differentiation (more fluorescence in better-differentiated tumors), suggesting (Denk *et al.*, 1972) that CEA is not a constituent of plasma membranes but is a secretory product.

CEA is a glycoprotein which is soluble in 50%-saturated ammonium sulfate solution and in perchloric acid at concentrations up to 1.0 M, has a sedimentation coefficient of 7–8 S (Krupey, Gold and Freeman, 1968), and differs from blood-group substances by the absence of N-acetylgalactosamine and the presence of mannose.

An ion-sensitive antigenic site, found in perchloric acid extracts from normal lung and colon, is shared with CEA, which is tumor-specific. This site is named the *tumor-associated antigenic site* (TAA) to differentiate it from CEA (Lo Gerfo and Herter, 1972). Also, a glycoprotein with a sedimentation constant of between 3 and 4 S is similar to CEA (von Kleist, Chavanel and Burtin, 1972), indicating that CEA-like substances are present in various tissues (Lo Gerfo, Krupey and Hansen, 1971). False positive results may be found in situations where CEA is present in low concentration. The glycoprotein is present in normal tissues such as colon, spleen and lung and is similar to CEA except for its sedimentation constant.

It is possible that α- and β-TAA may represent another glycoprotein reported from different laboratories (von Kleist, Chavanel and Burtin, 1972). The two glycoproteins, α-TAA and β-TAA in colonic adenocarcinoma, have the ion-sensitive site and differ in carbohydrate quantity. An antiserum reacting with these antigens allowed the detection of antigens in patients' sera from a wide variety of neoplasms, pulmonary emphysema and alcoholic cirrhosis; the antiserum also reacted with CEA. Additionally, α-TAA has been isolated from meconium.

An antigen with β-mobility from colonic tumors was found to be identical to CEA. In addition, this antigen shares sites with another β-antigen which is present in nontumorous tissues (Orjasaeter, Fredriksen and Liavag, 1972).

In human breast carcinomas there is a glycoprotein which shares antigenic determinants with CEA. Thus it can be explained that CEA levels as measured by radioimmunoassay techniques are raised in 50 percent of breast cancer patients. This glycoprotein can be isolated from human breast cancer cell surface preparations using lithium diiodosalicylate, followed by phenol partition (Kuo, Rosai and Tillack, 1973).

Several CEA preparations have identical amino acid sequences (Terry *et al.*, 1972), indicating a common polypeptide component. On the other hand, ultracentrifugation of $[^{125}I]$CEA in CsCl (Turner *et al.*, 1972) gave three distinct fractions. One had blood-group H determinant, but not CEA. Another fraction had CEA and possibly A determinant. The third appeared to be a low-molecular-weight component with no antigenic activity. Several antibodies have been observed after immunization of rabbits with CEA. The heterogeneous nature of CEA may explain variability among different laboratories in respect to immunological reactions and quantitative determinations. Extensive studies of the carbohydrate composition of various carcinoma fractions and comparisons with blood-group substances revealed that the qualitative composition was identical (Simmons and Perlmann, 1973). The variation between tumor-extract glycoproteins and blood-group substances was a quantitative one, involving sugar levels. There was less polysaccharide per molecule in tumor glycoproteins (40–50 percent) than in blood-group substances (70–80 percent). There was a similar pattern in the behavior of *N*-acetylgalactosamine in periodate oxidation among these glycoproteins (50 percent was protected, the rest rapidly destroyed), suggesting two differently linked residues. These similarities and quantitative differences suggest that tumour glycoproteins could represent precursors of blood-group substances.

Although the CEA fraction with specificity I (reacting intensely and specifically with CEA antiserum) has not been investigated as to the molecular sequence of major determinants, the other two specificities of CEA (II and III) show striking similarities with the central sequences of blood-group substances which lack the peripheral sequences with determinants of A, B, H and Le^a specificities (*Figure 4.1*) (Simmons and Perlmann, 1973).

The suggestion that there are missing transferases for completing the synthesis of blood-group substance is very attractive. An accumulation of unincorporated type I sidechains with Le^a-specific pentasaccharide ceramide (Hakomori and Jeanloz, 1969) and type 2 chain represented by a lacto-*N*-fucopentosyl-(III)-ceramide (Yang and Hakomori, 1971) has been shown in some adenocarcinomas.

In this context it is possible that in studies by Lo Gerfo, Krupey and Hansen (1971) and others on the presence of CEA in non-neoplastic tissues, the ion-sensitive determinant may not be the CEA I component (Simmons and Perlmann, 1973) but CEA II, a suggested blood-group precursor.

Supportive evidence for the precursor nature of certain tumor antigens is the absence of blood-group substances in malignant tumors. On the other hand, direct measurements of transferases for A and B isoantigens do not correlate with determinations of these isoantigens in colon carcinomas. Soluble fractions of colonic cancer and fetal intestinal cell membranes produce delayed skin-hypersensitivity reactions in intestinal cancer patients (Hollinshead *et al.*, 1972). These antigens are different from CEA, having a lower molecular weight. CEA does not elicit a skin reaction.

There is some evidence of an immunological relationship between CEA and fetal sulfoglycoprotein antigen (FSA) of gastric juice. The latter increases with increasing age in both males and females. It is suggested that FSA might be an early indicator of alimentary canal cancers (Terry *et al.*, 1972). FSA occurs in the gastric juice of 96 percent of gastric cancer patients (Häkkinen and Viikari, 1969). Also, 14 percent of cases with peptic ulcers have been found to have this antigen. Its presence in plasma membranes has not been clearly demonstrated.

4.4.3.2   OTHER ONCOFETAL ANTIGENS IN HUMAN MALIGNANT NEOPLASMS

There has been less interest in other oncofetal antigens in human malignant neoplasms. A tumor-associated antigen in Hodgkin's disease has been reported (Order, Porter and Hellman, 1971). This antigenic component is not specific for Hodgkin's disease, but is increased in quantity by comparison with normal tissues or in such situations as lymphosarcoma, sarcoidosis and infectious mononucleosis (Order and Hellman, 1973). The antigenic complex consists of slow (S) and fast (F) migrating antigens. It is not known whether these substances are autoantigenic. The location of this antigen is not known, and the possibility exists that it is not a plasma-membrane antigen. It has been shown (Chism, Order and Hellman, 1973) that these antigens are present in embryonic hematopoietic tissues (liver, spleen and thymus) and in some non-neoplastic adult tissues. Hodgkin's tumor-associated antigens give an elution pattern on Sephadex G-200 chromatography similar to that of normal fetal liver, the F and S antigens being in the peak I fraction. It was suggested that the S may be a de-differentiation antigen expressed in immature lymphocytes (Katz et al., 1973).

In patients with various types of leukemia, a serum component is found in one-third of cases, which is also present in embryonic tissues and in embryonic serum (Harris, Harrell and Anderson, 1971). This leukemia-associated antigen (LAA) has been demonstrated to be a tumor cell surface component and is also present in the sera of some patients with Hodgkin's disease. It is not known whether an antigen with γ-globulin electrophoretic mobility, which has been found in fetal sera and sera of tumor and leukemia patients (Edynak et al., 1972), is a membrane component. This antigen is also present in benign human neoplasms (Edynak et al., 1972) and in fetuses but not in tumors of other species. Using the immunoprecipitation method of Ouchterlony, it has been demonstrated also in 11 percent of sera of patients with malignant neoplasms and only in 2 out of 101 patients with non-neoplastic diseases. It has been suggested (Alexander, 1972) that it may be related to LAA.

A globulin (T globulin) is present in sera of patients with various malignant neoplasms and in some sera from pregnant women. It is specific to ceramide-lactoside receptors on plasma membranes. Ceramide-lactoside is cytolipin H, a cell-surface factor which probably becomes expressed in neoplastic cell transformation. It is present in trophoblastic tissues (Tal, 1965). The hapten ceramide-lactoside is not soluble in an aqueous medium.

There is a large macromolecular complex in fetal tissues in bronchial, gastric, pancreatic, kidney and liver carcinomas (Yachi et al., 1968). This complex is not present in fluids, and is not soluble in perchloric acid or alcohol. It can be stained with periodic acid–Schiff's reagent and is heat-labile. As determined by gel filtration, its molecular weight is larger than that of IgG. It

In brain, it was found that glioblastomas and astrocytomas contain traces of brain-specific fetal antigen (Delpech et al., 1972).

Originally, it was suggested that sarcoma antigens might be related to viruses. Later studies (Mukherji and Hirshaut, 1973) indicated that such antigens are present in embryonic tissues. There are antibodies against such antigens in patients with a wide variety of neoplasms, in their relatives and in some (10 percent) normal individuals. The antigen was also present in adult skin.

In the future, it can be expected that many more oncofetal plasma-membrane antigens will be detected. Some such antigens may be present, but not accessible to immune sera. For example, trypsinization increases the number of combining sites for anti-tumor antibody on embryonic cells. The reactivity of anti-HeLa immunoglobulin G is abolished by absorption with untrypsinized cells.

4.4.3.3   UNIVERSAL ONCOFETAL ANTIGENS IN HUMAN MALIGNANT NEOPLASMS

In our laboratories we demonstrated the presence of embryonic antigens in human carcinomas (Klavins, Mesa-Tejada and Weiss, 1971). We obtained a 6–7-week-old intact human fetus from an ectopic pregnancy. Saline extracts of the whole fetus were used to immunize a rabbit. Using the Ouchterlony gel diffusion technique, we demonstrated that antiserum raised to this fetal tissue extract, after absorption with a mixture of human adult tissues, reacted by producing precipitin bands with saline extracts of a wide variety of malignant neoplasms representing all three germinal layers. With indirect immunofluorescence microscopy it was seen that epidermis of normal adult skin also reacted with the antiserum. We have not yet tested whether absorption of immune serum with human skin extracts would remove the reactivity to carcinoma antigen as it did in experiments on mice (Stonehill and Bendich, 1970).

To date we have examined 31 carcinomas representing each of the three germinal layers: 6 from ectoderm, 4 (kidneys) from mesoderm and 21 from endoderm. Using the Ouchterlony technique, 26 of these tumors reacted with rabbit antisera against the saline extract of human embryos. In the subsequent studies we used an aqueous extract of a mixture of embryonic tissues, obtained by suction biopsy, to immunize the rabbits. Those malignant neoplasms which did not react were the following: 2 colon carcinomas, 2 lung carcinomas and 1 kidney carcinoma. Of these cases, 1 adenocarcinoma of colon and 1 bronchogenic carcinoma were examined using immunofluorescence microscopy. The γ-globulins of the immune rabbit serum were bound to the tumor cells. In addition, 4 other malignant neoplasms were examined using this technique only. They were: 1 adenocarcinoma of breast, 1 undifferentiated large-cell bronchogenic carcinoma and 2 adenocarcinomas of colon. Again, the anti-embryonic antibodies were localized on the tumor cells, but not on the normal tissues or the epithelial cells of a fibroadenoma of the breast, adenomatous polyp of the colon, cervical epithelium of the uterus, follicular epithelium of the ovary, epithelium of the oviduct and follicular adenoma of the thyroid. The precise localization of the antibodies on the cells could not be ascertained, although the fluorescence appeared to be restricted to the cell periphery. This finding indicates a possible association of the embryonic antigens with the plasma membrane. In the sections of fetal lung and colonic cancer, the immunofluorescence was present most strikingly in the glycocalyx in a pattern identical to that described for CEA (Gold, Krupey and Ansari, 1970). A less potent antibody complex has been raised by a placenta from a fetus of estimated 8 or 9 weeks' gestation. This antiserum reacted with fewer malignant neoplasms than the one raised by a mixture of fetal tissues (Mesa-Tejada *et al.*, 1972a, b).

We have now isolated a relatively homogeneous component which does not cross-react with CEA or α-fetal protein. It is a glycoprotein with a molecular weight of approximately 60000 and an isoelectric point at pH 4.80. The distribution of this antigen among various human malignant neoplasms and other tissues is not yet known.

We do not know whether, in the mixture of oncofetal antigens in human fetuses, there is a single universal component, present in malignant neoplasms. Neither do we know whether such a component, if present, would be present in small amounts in other than fetal or malignant neoplastic tissues.

4.4.3.4   DELETION OF NORMAL ANTIGENS IN HUMAN MALIGNANT NEOPLASMS

Loss of normal plasma-membrane antigens has been observed on several occasions in human malignant neoplasia, for example, in stomach (Nairn *et al.*, 1962) and in colon (Burtin, von Kleist and Sabine, 1971). A and B blood-group antigens were reported to be lost in human bladder carcinomas (Kay, 1957). Such losses in a wide variety of tumors (Davidson, 1972) can be of significant diagnostic importance, particularly in distinguishing some malignant entities from unique ones (Davidson *et al.*, 1973). This loss has not been related to changes in transferase activity (Kiang, Edstrom and Kennedy, 1972). It has also been suggested that the loss of surface glycoprotein may be related to the metastasizing capacity of tumors (Kim and Carruthers, 1972).

The loss of blood-group substances in the presence of CEA may indicate the precursor nature of this component. Such correlative studies would be of great interest. On the other hand, loss of some terminal carbohydrate residues might result in appearance of CEA due to conditions other than malignant neoplasia, such as by autolysis or other chemical degradations during isolation.

Of interest is the finding that the skin from some patients with advanced cancer, when transplanted to normal recipients, will not be rejected as quickly as normal skin and can remain up to 60 days while normal skin is rejected within 11 days (Amos, Hattler and Shingleton, 1965). The question arises, is there some loss of antigens, or is there some effect on the host immune system? Is there a transmissible virus? Perhaps blocking antibodies in skin present in a cancer patient may protect the graft from rapid rejection.

In human malignant neoplasia, there is a decrease in the reactivity of HL-A antigens. In a patient with a malignant lymphoma there was an initial loss of HL-A3, 7 and 8 reactivity. Subsequently the reactivity for antisera to HL-A1 was lost, along with morphological and biological malignant transformation of lymphocytic cells. This occurred during a period of exacerbation of the disease. There was no concurrent loss of reactivity from fibroblasts of the same patient (Seigler et al., 1971). Another instance has been reported (Tsuji et al., 1972) where a patient with ovarian adenocarcinoma had lost all reactivities for 15 antisera to HL-A antigens.

When cell lines from human tumors were examined (Kersey et al., 1973), there was absence of HL-A reactivity in four of eight such lines. No loss of reactivity was observed when human fibroblasts were transformed by the small-plaque mutant of SV_{40} or by the Kirsten strain of murine sarcoma virus.

4.4.3.5   ONCOFETAL ANTIGENS IN ANIMAL MALIGNANT NEOPLASMS

There are numerous oncofetal antigens in animals. Some in this group, although not specifically those of embryonic tissues, have been associated with growing and rejuvenating tissues (Darcy, 1957). One such is an α-migrating mucoprotein in the serum of rats, always found in higher concentrations in animals with tumors than in normal ones. Sera from pregnant and young (1-week-old) rats also have high values.

The embryonic (oncofetal) antigens in animals can be induced by both viral and chemical carcinogenesis. It has been demonstrated repeatedly that embryonic antigens appear in association with SV_{40}-induced hamster tumor cells. Immunization with embryonic tissues (Coggin, Ambrose and Anderson, 1970) produced significant tumor transplantation immunity against SV_{40}-transformed hamster cells.

The polyoma-virus-induced transplantation antigen in transformed hamster cells is similar to antigen on embryonic hamster cells, but not to that on adult cells (Pearson and Freeman, 1968), and in cells infected with avian leukosis virus an antigen appears to be a component of embryonic chick cells (Daugherty and DiStefano, 1966).

Hepatomas induced by aminoazo dyes and sarcomas induced by 3-methylcholanthrene develop cell-surface changes which are reflected in the expression of at least two classes of neoantigens (Baldwin, Glaves and Vose, 1972; Baldwin et al., 1972). Tumor-specific antigens are detectable by tumor rejection reactions against transplanted tumor cells in syngeneic hosts and by analysis in vitro of the cell-mediated and humoral immune responses brought about by this rejection. These antigens are diverse and are different for each tumor.

Another group of antigens have been detected on the cell surface of rat hepatomas and sarcomas and are also embryonic components. They react with the serum and lymphocytes from multiparous rats. These antigens are common to both tumor types. Both antigenic components—the embryonic antigens and those of the tumor cell—are associated with cell surfaces. The fetal antigens, however, are soluble and they may be secretory products.

In other experiments (Le Mevel and Wells, 1973), mice were immunized with irradiated embryonic cells and so protected from the challenge with methylcholanthrene-induced sarcoma cells. On the other hand, lymphocytes sensitized against embryonic cells were cytotoxic for mouse fibrosarcomas induced by 7,12-dimethylbenz[a]anthracene (Ménard, Colnaghi and Della Porta, 1973). In experimentally induced hepatomas in rat and also in human liver tumors, embryonic antigens have been detected (Uriel, 1969).

Immunofluorescence microscopy (Baldwin, Glaves and Vose, 1972; Baldwin *et al.*, 1972) showed that such antigens appeared on the surface of carcinogen-induced rat hepatoma and sarcoma cells. In these studies it appeared that there was a common embryonic antigen, since mouse sera reacted with different tumors. Similarly, in lymphocyte cytotoxicity studies, a single preparation of lymph nodes reacted with more than one tumor type. These antigens may also be intracellular, although expressed at the cell surface. A fetal-type antigen on rat hepatoma cells (Woo and Cater, 1972) appeared to contain a sialic acid and to be sited on the cell surface.

An antigen complex with universal reactivity was reported (Stonehill and Bendich, 1970) in murine embryos, cross-reacting with all malignant mouse neoplasms tested. There were three fractions of antigens after separation on DEAE-Sephadex. Some relationship with α-fetal protein was noted (Stonehill *et al.*, 1972). It is not known whether these antigens are membrane components. They are soluble in water but not in trichloroacetic acid. They disappear during incubation in saline, but reappear after incubation in 25% mouse serum. Similar findings were observed using immune mouse sera raised against embryonic tissues (Ting *et al.*, 1972). Using the capacity of transplanted fetal liver cells to form spleen colonies in lethally irradiated recipients, it was found that there was suppression of fetal colonies in spleen, when animals were immunized with a wide variety of malignant neoplasms (Salinas and Hanna, 1973). This suggests immunological cross-reactivity of fetal antigens on tumor cell surfaces.

4.4.3.6   DELETION OF H-2 ANTIGENS IN TRANSFORMED CELLS

As with human malignant neoplasms, various antigens expressed on normal animal cells are lost when these undergo malignant transformation. For example, there is evidence of decreased H-2 antigen reactivity following malignant transformation of mouse cells (Haywood and McKhann, 1971).

Loss of specific transplantation antigens is difficult to detect, if only a few among many antigens are lost, but there is evidence that such a loss occurs (Hellström and Möller, 1965). Deletion of sets of H-2 antigens in hybrid mice has been demonstrated with concomitant increase of the other set. No variants lacking all H-2 antigens have been found. A loss of expression of one *H-2* allele in a tumor originating in F_1 hybrids was reported (Klein and Hellström, 1960).

In the mouse there is a loss of specificities at the K end of the *H-2* locus in cell variant populations. Specificities 12, controlled at the D end of the *H-2* locus, and the specificities in the translocation were not lost simultaneously in tumors of the mouse.

Organ-specific antigens may not necessarily be plasma-membrane antigens, but they can also be lost or reduced. Deletions of organ-specific antigens have been reported in malignant neoplasms; for example, there was deletion of surface antigens in 4-dimethylaminoazobenzene-induced rat hepatoma cells.

Whether the deletions are accompanied by the appearance of embryonal antigens (precursors of the deleted antigens) is not known; however, there is some indication that this can occur. Cells in culture derived from murine lymphomas induced by Moloney leukemia virus (MLV) progressively lose H-2 antigens and gain MLV-determined cell-surface antigens (Cikes, Friberg and Klein, 1973). This is compatible with the modulation theory. Finally, it may appear that there is no true disappearance of an antigen, but only blockage of exposure. For example, it was shown that in murine tumor cell lines TA3/Ha considerable amounts of H-2^a were synthesized, but were not exposed on the surface of the cell (Friberg and Lilliehook, 1973).

4.4.3.7   NATURE OF ONCOFETAL ANTIGENS

The cross-reactivity of embryonic tumor antigens can occur even between different species. There is a common embryonic antigen among SV$_{40}$-transformed hamster cells and human embryonic cells (Ambrose, Anderson and Coggin, 1971). The sera from Burkitt lymphoma and infectious mononucleosis patients cross-react with hamster fetal cells of 10 days' gestation, but not with cells of 14 days' gestation or adult cells (Harris, Harrell and Anderson, 1971).

Using immune mouse sera raised by embryonic tissues (Ting et al., 1972), it was observed that the fetal antigens were different from tissue-specific cell-surface antigens of the papova-virus-induced tumors. This would suggest that such antigens (TSTAs) are coded for by the viral genome and incorporated into the transformed cells. Against this concept are the findings that immunization with embryonic tissues (Coggin, Ambrose and Anderson, 1970) produces tumor transplantation immunity against SV$_{40}$-transformed hamster cells. It would appear that tumor-specific antigens may be related to embryonic antigens. Similarly, mice immunized with irradiated embryonic cells are protected from challenge with methylcholanthrene-induced sarcoma cells (Le Mevel and Wells, 1973).

The cell-surface antigens can be modified by different factors. For example, a female hormone complex can shorten the persistence of fetal surface antigens required for the development of cell-mediated immunity (Ballard and Tomkins, 1969). It has been shown that an antigen in cells infected with avian leukosis virus is a component of embryonic chick cells (Daugherty and DiStefano, 1966). The COFAL antigen disappears from chick embryonic cells in tissue culture, but reappears again on infection with avian leukosis sarcoma virus. This might indicate derepression of the cells by the virus.

The question can be raised whether the oncofetal antigens could be related to viruses, particularly some of the antigens which are present in a wide variety of tissues and species. For example, the gs-antigen of mouse leukemia virus is present in certain strains of mouse embryo (Huebner, Kellogg and Sarma, 1970), and these gs-antigens are not expressed in adult mice. In leukemia of such strains these antigens again appear.

We feel that there is a very wide variability in a single tumor specimen in its quantitative expression of different components, therefore it may not be possible to differentiate two similar tumors on the basis of their quantitative composition. Those tumor cells which are in a more highly differentiated state can produce more of a substance which is similar to the normal adult pattern, whereas cells which are less differentiated may produce substances of embryonic nature.

4.4.3.8   COMMENTS ON ANTIGENICALLY ALTERED CELL SURFACES

The nature of the tumor antigens is not clear. One concept is that the cells already have different antigens normally, but these become manifested only when the antigenic pattern of the cell line from which the tumor arises produces sufficient concentration by monoclonal proliferation of the tumor. This hypothesis is not supported by experiments where monoclonal cell lines produce distinct rejection antigens after treatment *in vitro* with chemical carcinogens.

Another hypothesis is that the chemical or other carcinogen may induce mutation-like heritable alterations which then express new antigens on the cell surface. Some tumors, however, have no significant change in antigenicity by comparison with the tissue of origin.

The production of new cell-surface antigens by carcinogens does not necessarily coincide with malignant transformation.

One of the factors in the autonomy of neoplasia may be a lack of certain normal regulatory components. For example, in rat mammary carcinoma (R3230AC), which does not regress after removal of the ovaries, there is lack of a specific cytoplasmic β-estradiol-binding protein which normally facilitates binding of estradiol to cell chromatin (McGuire *et al.*, 1972).

It is also possible that viruses, for example, may induce the synthesis of normally repressed proteins (the embryonic proteins) in the target cells which then may alter the plasma membrane and induce altered cellular behavior related to neoplasia.

Transformed cells are agglutinated by some plant lectins, such as wheat-germ agglutinin or concanavalin A. Masking the agglutinin-binding sites of the transformed cells causes reversion to a normal pattern of growth (Burger and Noonan, 1970); the contact inhibition of cell division is restored. This subject has been discussed in more detail by D. F. Smith and E. F. Walborg in Vol. 3, Chapter 5 of this series. It is of interest that concanavalin A can bind CEA (Chu, 1973). It has been suggested (Pitot, 1969) that stability of mRNA may be an important factor in carcinogenesis. The term *membron* has been introduced to represent membrane-associated mRNA template. mRNA expression can be regulated by the cellular environment. Deviations in genetic expressions of malignant cells have been related to alterations in the stability of mRNA. Heritable changes in mRNA may be induced by alteration in membrane mosaic structures. It has been suggested (Coggin, 1973) that the soluble antigens of the tumor may in fact represent a mechanism designed to evade the host's immune response. It has been shown that the soluble antigens are released from the tumor cell surface in a quantity large enough to overwhelm the humoral arm of this immune response. In this condition there is an antigen excess which now acts to paralyze the cellular arm of the host's immune response. This is accomplished by coating the killing lymphocytes with excess antigens, reducing their cytotoxicity. Leukophoresis and subsequent washing of the white cells removes the antigens and renders the leukocytes cytotoxic. A similar study (Kim, 1973) involving animal tumors has shown a basic difference between metastasizing and nonmetastasizing tumor cells. Nonmetastasizing tumor cells have a thick glycoprotein 'coating' around the plasma membrane which does not dissociate from it, whereas metastasizing tumor cells do not have this bound surface antigen, but instead 'shed' large quantities of soluble glycoprotein antigens. Bearing these two independent observations in mind, it may well be that the so-called oncofetal antigens represent, aside from a general derepression, the tumor's adaptive response to overcome the host's immune system.

REFERENCES

ALEXANDER. P. (1972). *Nature, Lond.*, **235**:137.

AMBROSE, K. R., ANDERSON, N. G. and COGGIN, J. H. (1971). *Nature, Lond.*, **233**:194.

AMOS, D. B., HATTLER, B. G. and SHINGLETON, W. W. (1965). *Lancet*, i:414.

ANDERSON, R. J., MCBRIDE, C. M. and HERSH, E. M. (1972). *Cancer Res.*, **32**:988.

AOKI, T., STEPHENSON, J. R. and AARONSON, S. A. (1973). *Proc. natn. Acad. Sci. U.S.A.*, **70**:742.

BALDWIN, R. W. and BARKER, C. R. (1967a). *Int. J. Cancer*, **2**:355.

BALDWIN, R. W. and BARKER, C. R. (1967b). *Br. J. Cancer*, **21**:793.

BALDWIN, R. W. and EMBLETON, M. J. (1971). *Int. J. Cancer*, **7**:17.

BALDWIN, R. W. and GLAVES. D. (1972). *Clin. exp. Immun.*, **11**:51.

BALDWIN, R. W., GLAVES, D. and PIMM, M. V. (1971). *Progress in Immunology*, p. 919. Ed. B. AMOS. New York; Academic Press.

BALDWIN. R. W.. GLAVES, D. and VOSE, B. M. (1972). *Int. J. Cancer*, **10**:233.

BALDWIN, R. W., HARRIS, J. R. and PRICE. M. R. (1973). *Int. J. Cancer*, **11**:385.

BALDWIN. R. W. and MOORE, M. (1969). *Int. J. Cancer*, **4**:753.

BALDWIN, R. W., BARKER, C. R., EMBLETON, M. J. and MOORE, M. (1970). *Immunity and Tolerance in Oncogenesis*, pp. 11–24. Ed. L. SEVERI. Perugia, Italy; University of Perugia.

BALDWIN, R. W., GLAVES, D., PIMM, M. V. and VOSE, B. M. (1972). *Annls Inst. Pasteur, Paris*, **122**:715.

BALLARD, P. L. and TOMKINS, G. M. (1969). *Nature, Lond.*, **222**:344.

BASOMBRIO. M. A. (1970). *Cancer Res.*, **30**:2458.

BENJAMIN, T. L. (1966). *J. molec. Biol.*, **16**:359.

BENTWICH, Z., WEISS, D. W., SUBITZEAN, D., KEDAR, E., IZAK, G., COHEN, I. and EYAL. O. (1972). *Cancer Res.*, **32**:1375.

BHATTACHARYA, M. and BARLOW, J. J. (1973). *Cancer, N. Y.*, **31**:588.

BRAWN, R. J. (1970). *Int. J. Cancer*, **6**:245.

BRETSCHER, M. S. (1973), *Science, N. Y.*, **181**:622.

BURGER, M. M. and NOONAN, K. D. (1970). *Nature. Lond.*, **228**:512.

BURNET. F. M. (1970). *Nature, Lond.*, **226**:123.

BURTIN, B., VON KLEIST, S. and SABINE, M. C. (1971). *Cancer Res.*, **31**:1038.

BURTIN, P. and GENDRON, M. C. (1973). *Proc. natn. Acad. Sci. U.S.A.*, **70**:2051.

BURTIN, P., SABINE. M. C. and CHAVANEL, G. (1972). *Int. J. Cancer*, **10**:72.

BURTIN, P., BUFFE, D., VON KLEIST, S., WOLFF, E. and WOLFF, E. (1970), *Int. J. Cancer*, **5**:88.

CARREL, S. and THEILKAES, L. (1973). *Nature, Lond.*, **242**:609.

CEPPELLINI, R.. CURTONI, E. S., MATTIUZ, P. L., MIGGIANO, V., SCUDELLER, G. and SERRA, A. (1967). *Histocompatibility Testing, 1967*, pp. 149–185. Ed. E. S. CARTONI, P. L. MATTIUZ and R. M. TOSI. Copenhagen; Munksgaard.

CHISM, S. E., ORDER, S. E. and HELLMAN, S. (1973). *Am. J. Roentg*, **117**:5.

CHU. T. M. (1973). *Fedn Proc. Fedn Am. Socs exp. Biol.*, **32**:859.

CIKES. M.. FRIBERG, S., JR. and KLEIN, G. (1973). *J. natn. Cancer Inst.*, **50**:347.

COGGIN, J. A. (1973). Mechanisms employed by tumors to escape immunologic rejection by their hosts. *3rd International Conference on Embryonic and Fetal Antigens In Cancer, Knoxville, Tennessee*. Ed. N. G. ANDERSON, J. H. COGGIN, JR. and E. B. COLE. Springfield, Virginia; National Technical Information Service, U.S. Department of Commerce. Conf. 731141.

COGGIN, J. A., AMBROSE, K. R. and ANDERSON, N. G. (1970). *J. Immun.*, **105**:525.

COHEN, A. M., KETCHAM, A. S. and MORTON, D. L. (1973). *J. natn. Cancer Inst.*, **50**:585.

COLLATZ, E.. VON KLEIST, S. and BURTIN, P. (1971). *Int. J. Cancer*, **8**:298.

DANIELLI, J. F. and DAVSON, H. (1956). *J. cell. Physiol.*, **5**:495.

DARCY. D. A. (1957). *Br. J. Cancer*, **11**:137.

DAUGHERTY, R. and DISTEFANO. H. S. (1966). *Virology*, **29**:586.

DAVIDSON, I. (1972). *Am. J. clin. Path.*, **57**:715.

DAVIDSON, I., NORRIS, H. T., STEJSKAL, R. and LILL, P. (1973). *Archs Path.*, **95**:132.

DEFENDI, V. (1967). *Cross-Reacting Antigens and Neoantigens*, pp. 81–82. Ed. J. J. TRENTIN. Baltimore; Williams & Wilkins.

DEICHMAN, G. I. (1969). *Adv. Cancer Res.*, **12**:101.
DELPECH, B., DELPECH, A., CLEMENT, J. and LANMONIER, R. (1972). *Int. J. Cancer*, **9**:374.
DENK, H., TAPPERNER, G., ECKERSTORFER, R. and HOLFNER, J. H. (1972). *Int. J. Cancer*, **10**:262.
DENTSCH, E., APFFEL, C. A., MORI, H. and WALKER, J. E. (1973) *Cancer Res.*, **33**:112.
DICKINSON, J. P., CASPARY, E. A. and FIELD, E. J. (1972). *Nature. New Biol.*, **239**:181.
EDYNAK, E. M., HIRSHAUT, Y., OLD. L. J. and TREMPE. G. L. (1971). *Proc. Am. Ass. Cancer Res.*, **12**:75.
EDYNAK, E. M., OLD, L. J., VRANA, M. and LARDIS, M. (1972). *N. Engl. J. Med.*, **286**:1178.
FEIZI *et al.* (1900).
FELLOUS, M. and DAUSSET, J. (1970). *Nature, Lond.*, **225**:191.
FITCH, F. W. (1962). *Archs Path.*, **73**:144.
FREEMAN, A. E., KELLOFF, G. J., GILDEN, R. V., LANE, W. T., SWAIN, A. P. and HUEBNER, R. J. (1971). *Proc. natn. Acad. Sci. U.S.A.* **68**:2386.
FRIBERG, S., JR. and LILLIEHOOK, B. (1973). *Nature, Lond.*, **241**:112.
FRYE, C. D. and EDIDIN, M. (1970). *J. Cell Sci.*, **7**:319.
GELDERBOM, H. and BAUER, H. (1973). *Int. J. Cancer*, **11**:466.
GOLD. P. and FREEDMAN, S. O. (1965). *J. exp. Med.*, **122**:467.
GOLD, P., KRUPEY, J. and ANSARI. H. (1970) *J. natn. Cancer Inst.*, **45**:219.
GOLDENBERG, D. M. and HANSEN. H. J. (1972). *Science, N.Y.*, **175**:1117.
GOLDENBERG, D. M., NUBAR, G. O. T., HANSEN, H. J. and VANDEVOORDE, J. P. (1972). *Am. J. Obstet. Gynec.*, **113**:66.
GORNER, P., TUFFREY, M. A. and BATCHELOR. J. R. (1962). *Ann. N.Y. Acad. Sci.*, **101**:5.
GRAFFI, A., PASTERNAK, G. and HORN, K. H. (1964). *Acta biol. med. germ.*, **12**:726.
HÄKKINEN. I. P. T. (1966). *Scand. J. Gastroenterol.*, **1**:28.
HÄKKINEN, I. P. T. and VIIKARI, S. (1969). *Ann. Surg.*, **169**:277.
HAKOMORI, S. and JEANLOZ, R. W. (1969). *J. biol. Chem.*, **239**:3606.
HAKOMORI, S. and MURAKAMI, W. T. (1968). *Proc. natn. Acad. Sci. U.S.A.*, **59**:254.
HARRIS, W., HARRELL, B. W. and ANDERSON, N. G. (1971). *Embryonic and Fetal Antigens in Cancer*, pp. 129–137. Ed. N. G. ANDERSON and J. H. COGGIN, JR. U.S. Department of Commerce, Conf. No. 71-0527.
HAYWOOD, G. R. and MCKHANN, C. F. (1971). *J. exp. Med.*, **133**:1171.
HEARN, V. M., SMITH. Z. G. and WATKINS, W. M. (1968). *Biochem. J.*, **109**:315.
HELLSTRÖM. I., HELLSTRÖM. K. E. and SJOGREN, H. O. (1970). *Cell. Immun.* **1**:18.
HELLSTRÖM, I., HELLSTRÖM, K. E., PIERCE, G. E. and BILL, A. H. (1968). *Proc. natn. Acad. Sci. U.S.A.*, **60**:1231.
HELLSTRÖM, K. E. and MÖLLER, G. (1965) *Prog. Allergy*, **9**:158.
HELLSTRÖM, K. E., HELLSTRÖM, I., SJOGREN, H. O. and WARNER. G. A. (1971). *Progress in Immunology*. pp. 939–949. Ed. B. AMOS. New York; Academic Press.
HIRANO, H., PARKHOUSE, B., NICOLSON, G. L., LENNOX. E. S. and SINGER, S. J. (1972). *Proc. natn. Acad. Sci. U.S.A.* **69**:2945.
HOLLINSHEAD. A. C. and TARRO, G. (1973). *Science, N.Y.*, **179**:698.
HOLLINSHEAD, A. C., MCWRIGHT, C. G., ALFORD, T. C., GLEW, D. H., GOLD, P. and HERBERMAN, R. B. (1972). *Science, N.Y.*, **177**:887.
HUEBNER. R. J., KELLOGG, G. F. and SARMA, P. S. (1970). *Proc. natn. Acad. Sci. U.S.A.*, **67**:366.
KATZ, D. H., ORDER, S. E., GRAVES, M. and BENECERAF, B. (1973). *Proc. natn. Acad. Sci. U.S.A.*, **70**:396.
KAY, H. E. M. (1975). *Br. J. Cancer*, **11**:409.
KERSEY, J. H., YUNIS, E. J., TODARO, G. J. and AARONSON, S. A. (1973). *Proc. Soc. exp. Biol Med.*, **143**:453.
KIANG, D. T., EDSTROM, R. D. and KENNEDY, B. J. (1972). *Proc. Am. Ass. Cancer Res.*, **13**:93.
KIM, U. (1973). Mechanisms employed by tumors to escape immunologic rejection by their hosts. *3rd International Conference on Embryonic and Fetal Antigens in Cancer, Knoxville, Tennessee.* Ed. N. G. ANDERSON, J. R. COGGIN, JR. and E. B. COLE. Springfield, Virginia; National Technical Information Service, Conf. 731141.
KIM, U. and CARRUTHERS, C. (1972). *Proc. Am. Ass. Cancer Res.*, **13**:69.
KLAVINS, J. V., MESA-TEJADA, R. and WEISS, M. (1971). *Nature, New Biol.*, **234**:153.
KLEIN. E. G. and HELLSTRÖM, K. E. (1960). *J. natn. Cancer Inst.*, **25**:271.
KOLDOVSKY, B. (1961). *Folia biol., Praha*, **7**:162.
KOROSTELEVA, V. S. (1957). *Byull. eksp. Biol. Med.*, **44**:987.
KOSAKI. T., IKODA. T., KOTANTI, Y., NAKAGWA, A. and SAKA. T. (1958). *Science, N.Y.*, **127**:1176.
KRUPEY, J., GOLD, P. and FREEDMAN, S. O. (1968). *J. exp. Med.*, **128**:387.
KUO. T., ROSAI, J. and TILLACK. T. W. (1973). *Fedn Proc. Fedn Am. Socs exp. Biol.*, **32**:859.
KURTH. R. and BAUER, H. (1972). *Virology*, **47**:426.

LANDSTEINER, K. (1900). *Zentbl. Bakt. ParasitKde*, **27**:357.
LANDSTEINER, K. and LEVINE, P. (1927). *Proc. Soc. exp. Biol. Med.*, **24**:94.
LANDSTEINER, K. and LEVINE, P. (1928). *Proc. Soc. exp. Biol. Med.*, **24**:600.
LANDSTEINER, K. and WIENER, A. S. (1940). *Proc. Soc. exp. Biol. Med.*, **43**:233.
LANGVAD, E. (1968). *Int. J. Cancer*, **3**:17.
LAPPE, M. A. (1969). *Nature, Lond.*, **233**:82.
LAURENCE. D. J. R. and NEVILLE, A. M. (1972). *Br. J. Cancer*, **26**:335.
LAURENCE. D. J. R.. STEVENS, U., BETTELHEIM. R., DARAY, D., LEESE, C., TURBERVILLE. C., ALEXANDER, P., JOHNS. E. W. and NEVILLE. A. M. (1972). *Br. med. J.*, **3**:605.
LAW. L. W. and TAKEMOTO. K. K. (1973). *J. natn. Cancer Inst.*, **50**:1075.
LEHRS, H. (1930). *Z. ImmunForsch.*, **66**:175.
LE MEVEL, B. P. and WELLS, S. A., JR. (1973). *Nature. New Biol.*, **244**:183.
LO GERFO, P. and HERTER. F. P. (1972). *Oncology*, **4**:1.
LO GERFO, P., KRUPEY, J. and HANSEN, H. J. (1971). *New Engl. J. Med.*, **285**:138.
MCGUIRE, W. L., HUFF, K., JENNINGS, A. and CHAMNESS, G. C. (1972). *Science, N Y.*, **175**:335.
MELTZER, M. S., LEONARD, E. J.. RAPP. H. J. and BORSOS. T. (1971). *J. natn. Cancer Inst.*, **47**:703.
MÉNARD, S., COLNAGHI, M. I. and DELLA PORTA, G. (1973). *Cancer Res.*, **33**:478.
MESA-TEJADA, R., FIERER, J. A., KLAVINS, J. V., WEISS, M. and BERKMAN, J. I. (1972a). *Embryonic and Fetal Antigens in Cancer*, pp. 177–180. Ed.   N. G. ANDERSON. J. H. COOGIN JR.. E. COLE and J. W. HOLLIMAN. Springfield, Virginia: National Technical Information Service, U.S. Department of Commerce. Conf. No. 720208.
MESA-TEJADA, R., WEISS, M., KLAVINS, J. V. and BERKMAN, J. I. (1972b). *Fedn Proc Fedn Am. Socs exp. Biol.*, **31**:639.
METZGAR, R. S., MOHANAKUMAR, T. and MILLER, D. S. (1972). *Science, N. Y.*, **178**:986.
METZGAR, R. S., BERGOC, P. M., MORENO, M. A. and SIEGLER, H. F. (1973). *J. natn. Cancer Inst.*, **50**:1065.
MONDAL. S.. IYPE, P. T.. GRIESBACH. L. M. and HEIDELBERGER. C. (1970). *Cancer Res.*, **30**:1593.
MORGAN, W. T. J. and WATKINS, W. M. (1948). *Brit. J. exp. Path.*, **29**:157.
MUKHERJI. B. and HIRSHAUT, Y. (1973). *Science, N.Y.*, **181**:440.
NAIRN, R. C., FOTHERGILL, J. E., MCENTEGARD, M. G. and RICHMOND, H. G. (1962). *Br. Med. J.*, **1**:1791.
OLD. L. J. and BOYSE. E. A. (1964). *A. Rev. Med.*, **15**:167.
OLD. L. J., BOYSE, E. A. and STOCKERT E. (1963). *J. natn. Cancer Inst.*, **31**:977.
OLD. L. J., BOYSE, E. A.. CLARKE. D. A. and CARSWELL, E. A. (1962). *Ann. N.Y. Acad. Sci.*, **101**:80.
OLD. L. J., BOYSE, E. A. and STOCKERT, E. (1963). *J. natn. Cancer Inst.*, **31**:977.
ORDER. S. E. and HELLMAN S. (1973). *J. Am. med. Ass.*, **233**:174.
ORDER. S. E. PORTER M. and HELLMAN, S. (1971). *N. Engl. J. Med.*, **285**:471.
ORJASAETER H.. FREDRIKSEN, G. and LIAVAG, I. (1972). *Acta path. microbiol. scand.*, **80B**:599.
PEARSON. G. and FREEMAN. G. (1968). *Cancer Res.*, **28**:1665.
PERDUE. J. F., WARNER, D. and MILLER, K. (1973). *Biochim. biophys. Acta*, **298**:817.
PITOT, H. C. (1969). *Archs Path.*, **87**:212.
POULIK, M. D. and BRON, C. (1969). *Red Cell Membrane Structure and Function*, pp. 131–154. Ed. G. A. JAMIESON and T. J. GREENWALT. Philadelphia; J. B. Lippincott.
PRIORI, E. S., WILBUR, J. R. and DMOCHOWSKI, L. (1971). *J. natn. Cancer Inst.*, **46**:1299.
RAPAPORT, F. T., DAUSSET. J., LEGRAND, L., BARGE, A.. LAWRENCE, H. S. and CONVERSE. J. M. (1968). *J. clin. Invest.*, **47**:2206.
RAPPORT, M. M., GRAF L. and ALONZO, V. P. (1958a). *Cancer, N.Y.*, **11**:1136.
RAPPORT, M. M.. GRAF L. SKIPSKI, V. P. and ALONZO, V. P. (1958a). *Nature, Lond.*, **181**:1803.
RAYMOND M. P. and DIPPLE. A. (1973). *Biochemistry*, **12**:1538.
REGA, A. F., WEED. R. I.. REED, C. F. BERG G. G. and ROTHSTEIN A. (1967). *Biochim. biophys. Acta* **147**:297.
REINER, R. and SOUTHAM, C. M. (1969). *Cancer Res.*, **29**:1814.
REISFELD R. A., PELLEGRINO, M. A. and KAHAN B. DG. (1971). *Science, N.Y.*, **172**:1134.
ROIZMAN B. and ROANE P. R. JR. (1963). *Virology*, **19**:198.
SALINAS. F. A. and HANNA. M. G.. JR. (1973). *Proc. Am. Ass. Cancer Res.*, **14**:116.
SCHENKEL-BRUNNER, H. and TUPPY, H. (1969). *Nature, Lond.*, **223**:1272.
SEIGLER, H. F., KREMER, W. B., METZGAR, R. S., WARD, F. E., HAUNG, A. T. and AMOST, C. B. (1971). *J. natn. Cancer Inst.*, **46**:577.
SIMMONS. D. A. and PERLMANN. P. (1973). *Cancer Res.*, **33**:313.
SINGER, S. J. (1971). *The Molecular Organization of Biological Membranes. Structure and Function of Biological Membranes*, Ed. L. I. ROTHFIELD. New York; Academic Press.
SINGER. S. J. and NICOLSON G. L. (1972). *Science, N. Y.*, **175**:720.
SJOGREN. H. O. (1964a). *J. natn. Cancer Inst.*, **32**:645.

SJOGREN. H. O. (1964b). *J. natn. Cancer Inst.*, **33**:661.

SJOGREN, H. O. (1965). *Prog. exp. Tumor Res.*, **6**:280.

SJOGREN. H. O. and JONNSON. N. (1966). *J. exp. Med.*, **123**:487.

STONEHILL, E. H. and BENDICH, A. (1970). *Nature, Lond.*, **228**:370.

STONEHILL. E. H., BENDICH. A., HIGGINS. P. J. and BORENFREUND, E. (1972). *Proc. Am. Ass. Cancer Res.*, **13**:43.

TAL, C. (1965). *Proc. natn. Acad. Sci. U.S.A.*, **54**:1318.

TARRO, G. (1973). *Proc. natn. Acad. Sci. U.S.A.*, **70**:325.

TERRY. W. D.. HENKART. P. A.. COLIGAN, J. E. and TODD. C. W. (1972). *J. exp. Med.*, **136**:200.

TEVETHIA, S. S. and MCMILLAN. V. L. (1973). *Proc. Am. Ass. Cancer Res.*, **14**:57.

TING. R. C.. LAVRIN. D. H.. SHIU. G. and HERBERMAN. R. B. (1972). *Proc. natn. Acad. Sci. U.S.A.*, **69**:1664.

TSUJI. K.. ITO, M., MIYAMOTO. H. and YAMASHITA. H. (1972). *Gann*, **63**:495.

TUPPY. H. and SCHENKEL-BRUNNER, H. (1969). *Eur. J. Biochem.*, **10**:152.

TURNER. M. D.. OLIVARES, T. A.. HASWELL, L. and KLEINMAN. M. S. (1972). *J. Immun.*, **108**:1328.

URIEL. J. (1969). *Sem. Hôp. Paris (Path. Biol.)*, **17**:877.

VAN WERF. A. J. M. (1971). *Lancet*, **i**:595.

VON KLEIST. S.. CHAVANEL, G. and BURTON, P. (1972). *Proc. natn. Acad. Sci. U.S.A.*, **69**:2492.

WATKINS. W. M. (1966). *Glycoproteins: Their Composition Structure and Function*, pp. 462–515. Ed. A. GOTTSCHALK. Amsterdam; Elsevier.

WINZLER, R. J. (1969). *Red Cell Membrane Structure and Function*, pp. 157–171. Ed. G. A. JAMIESON and T. J. GREENWALT. Philadelphia; J. B. Lippincott.

WOO J. and CATER. D. B. (1972). *Biochem. J.*, **128**:1273.

YACHI. A.. MATSUMURA. Y.. CARPENTER C. M. and HYDE. L. (1968). *J. natn. Cancer Inst.*, **40**:663.

YANG. H. and HAKOMORI. S. (1971). *J. biol. Chem.*, **246**:1192.

5

Surface immunoglobulins: characteristics, mobility and role in immune phenomena

Milton Kern
National Institute of Arthritis, Metabolism and Digestive Diseases, Bethesda, Maryland

To account for the substantial production of serum antibody by animals immunized with an antigen, Burnet (1959) proposed the clonal selection theory. This theory predicts that certain lymphoid cells, each capable of producing antibody of a single specificity, are specifically and selectively stimulated by the appropriate antigen to proliferate and produce antibody, and that the antibody produced by such cells is specific for the stimulating antigen. In this view the cells are genetically committed with regard to their ability to produce antibody with specificity for a single antigenic determinant and the antigen represents the stimulating or triggering agent. While Burnet originally proposed that the immunological recognition occurs through antibody-like sites on the cell surface, there is now abundant evidence for the presence of antibody, or, in more general terms, immunoglobulins, on the surface of lymphoid cells.

5.1   EVIDENCE FOR CELL-SURFACE IMMUNOGLOBULINS

The presence of immunoglobulins associated with the surface plasma membrane of lymphoid cells has been demonstrated by a variety of techniques. In most of the studies, advantage has been taken of the fact that such immunoglobulins can combine either with antigen or with anti-immunoglobulin antisera which have specificity for a variety of structural sites on the relevant immunoglobulin molecule itself.

Reiss, Mertens and Ehrich (1950) showed that lymph node cells, derived from rabbits immunized with bacteria, were capable of binding bacterial cells to their surface and that the greatest number of plasma cells binding bacteria appears 5–7 days after immunization. Further studies (Mäkelä and Nossal, 1961) showed that some of the lymphoid cells that did not secrete antibody were capable of binding bacteria on their surface.

The binding of red blood cells to lymphocytes can also be used to demonstrate immunoglobulin determinants on the cell surface. Rosette formation, in which red blood cells cluster around individual lymphocytes, has been observed, for example, with spleen cells from mice immunized with red blood cells (McConnell *et al.*, 1969). However, observations of this type in which bacteria or red blood cells bind to lymphoid cells, are subject to the reservation that soluble, secreted immunoglobulins could have been fortuitously bound to the cell surface.

The first evidence that the immunoglobulin which is present on the surface of lymphoid cells does not represent some artifactual process was provided by the study of Sell and Gell (1965), who showed that rabbit peripheral-blood lymphocytes, in the presence of anti-rabbit immunoglobin antisera, exhibited a marked stimulation of [³H] thymidine incorporation into DNA and that as many as 49 percent of the cells were induced to differentiate to blast cells. They concluded that the small lymphocytes were the most likely source of the blast cells. Moreover, the mitogenic as well as blastogenic activity of the anti-immunoglobulin antisera indicated that the structure of the cell-surface immunoglobulin was similar to that of the serum immunoglobulin from which the anti-immunoglobulin antisera were prepared, and also that the presence of immunoglobulin in the surface membrane was related somehow to biological function.

The clonal selection theory predicts that a relatively small proportion of cells from nonimmunized animals should bind antigen. This hypothesis was tested directly (Naor and Sulitzneau, 1967) by studying the binding of ^{125}I-labeled bovine serum albumin to spleen cells from nonimmunized mice. The distribution of grain counts upon examination by autoradiography revealed that a very small proportion of the cells showed uptake in excess of the background level observed for most of the cells.

^{125}I-labeled flagellin and hemocyanin have also been shown to react, *in vitro*, with about 1 in 5000 of lymphocyte-like cells from mouse spleen, thoracic duct, lymph node, bone marrow and peritoneal exudate (Byrt and Ada, 1969). On the other hand, few if any reactive cells were observed in thymus suspensions. Confidence that the antigen uptake reflected surface binding was afforded by the findings that the extent of antigen uptake was not increased by purposely damaging the cells and that the reaction was not inhibited by azide, which does inhibit phagocytosis by macrophages.

Some of the cells which react with antigen have been shown to play a role in the immune response (Ada and Byrt, 1969). Spleen cells were briefly incubated *in vitro* with ^{125}I-labeled flagellin, washed and then stored at 4 °C for 20 hours to permit radiation damage to those cells which had been coated with the labeled antigen. The cells were then injected into syngeneic mice that had previously been subjected to X-ray treatment sufficient to prevent any immune response to injected antigen. Under these circumstances the X-irradiated mice functioned as a form of tissue culture. After a suitable delay, the mice which had received the cells were injected with antigen and the titer of serum antibody measured. Cells treated with unlabeled antigen, or antigen labeled with carrier iodine rather than ^{125}I, or cells incubated in the absence of antigen, all yielded antibody specific for the serological type of flagellin used. However, the cells previously incubated with ^{125}I-flagellin either did not produce serum antibodies or produced antibodies at a very low rate. The fact that the ^{125}I-labeled flagellin of a given serological type did not inhibit serum antibody production by cells in the same cell suspension that were capable of secreting antibody to a serologically unrelated flagellin, indicates that the radiosensitivity was restricted to those cells bearing the appropriate antibody on their surface. The immunoglobulin nature of the antigen receptor site was suggested by the observation that prior treatment of cells with anti-immunoglobulin antisera inhibited the subsequent binding of antigen (Byrt and Ada, 1969). Furthermore, mouse spleen cells treated in this manner and then injected into irradiated mice showed suppressed antibody production (Warner, Byrt and Ada, 1970).

Antigen-coated glass beads have been used to deplete mouse spleen cells of populations that show a high rate of antibody secretion as well as those cells that exhibit immunological memory (Wigzell and Andersson, 1969). Memory cells are defined as those which upon injection into irradiated animals show no secretory response unless the animals are injected with the appropriate antigen. It should be noted that both memory cells and the cells that exhibited a high rate of antibody secretion readily passed through antigen-coated columns if soluble antigen was also present. In an extension of this study, Walters and Wigzell (1970) found that the selective retention of cells on such columns was blocked if the cells were first incubated with anti-immunoglobulin antisera. The possibility that this effect could be caused by any antibody capable of combining with any part of the cell surface was excluded because prior incubation of cells with antiserum directed against lymphocytes rather than against immunoglobulins did not alter the selective retention of the cells on antigen-coated columns.

It is known that antibody molecules can be covalently linked to a variety of compounds without substantially inactivating the antibody site that combines with antigen. Anti-mouse immunoglobulin antisera which had been labeled with ^{125}I or fluorescein were used to assess the percentage of immunoglobulin-bearing cells in various lymphoid tissue sources of mouse; autoradiography and fluorescence analysis (Raff, Sternberg and Taylor, 1970) gave results that were essentially the same. The amounts of lymphocytes bearing immunoglobulin on their surface for spleen, lymph node, thoracic duct and bone marrow cells were 40, 20, 15 and 15 percent, respectively. About 0.2 percent of the thymus cells bore immunoglobulins on their surface that were detected by autoradiography but not by the fluorescence technique, presumably because of the difference in sensitivity of the two techniques. The possibility that surface immunoglobulins arise as a consequence of being absorbed onto cells was excluded because thymus cells, washed after being incubated with mouse serum, still failed to stain for surface immunoglobulins.

The presence of surface immunoglobulins as detected by immunofluorescent staining has also been reported for rabbit (Pernis, Forni and Amante, 1970). It was observed that about 50 percent of lymphocytes derived from spleen and peripheral blood reacted with anti-immunoglobulin antisera which had been covalently linked to fluorescent dyes, and that thymus cells all gave negative results by the immunofluorescence assay.

Ultrastructural studies have also been used to demonstrate cell-surface immunoglobulins. Electron microscopy provides the resolution required to distinguish whether the test indicator reagent is in or on the pertinent cell. Jones, Marcuson and Roitt (1970) showed by autoradiography that anti-immunoglobulins labeled with ^{125}I were associated with the periphery of the cell. Similar results were achieved (Mandel, Byrt and Ada, 1969) with [^{125}I]-hemocyanin; that is, using an antigen as indicator. The presence of surface immunoglobulins on human lymphoblasts from Burkitt-lymphoma-derived cell cultures and leukemic blood (Hammond, 1970) and the distribution of immunoglobulin on the surface of mouse lymphoid cells (de Petris and Raff, 1972) were shown by improving the visualization of immunoglobulin by use of ferritin conjugated to anti-immunoglobulin antisera.

The iodination of immunoglobulins and other serum proteins, as catalyzed by lactoperoxidase (Marchalonis, 1969), has been used effectively in studying the surface of red blood cells (Phillips and Morrison, 1970) and lymphocytes (Marchalonis, Cone and Santer, 1971). Autoradiographic studies of lymphocytes showed that ^{125}I became bound to surface proteins of living cells from murine spleen and tumors. Labeled immunoglobulins could be isolated following lactoperoxidase iodination of living murine spleen cells (Baur *et al.*, 1971). Evidence that these were not intracellular immunoglobulins that became labeled by the lactoperoxidase technique was based on the findings that the cell sap from such cells contained less than 3 percent of the total radioactivity and that lactic and isocitric acid dehydrogenases, two intracellular enzymes, were not labeled. The isolation of surface immunoglobulins from murine lymphocytes and human neonatal thymus has been reported employing essentially the same techniques (Marchalonis, Atwell and Cone, 1972a, b).

It is exceedingly difficult to deny the various lines of evidence detailed here indicating the presence of immunoglobulins associated with the cell surface.

5.2   THE DIFFERENT KINDS OF SURFACE IMMUNOGLOBULINS

All classes of immunoglobulins possess light and heavy polypeptide chains and have a minimum of 2 heavy and 2 light chains or multiples of this basic structure. Immunological distinctions with regard to the classes of immunoglobulins (IgG, IgA, IgM, IgD and IgE) are based on the antigenic determinants of their constituent heavy chains. For example, rabbit IgG injected into goats results in the production of antibody molecules, some of which react specifically with the heavy-chain moieties of IgG but none of which reacts with the heavy chain of any of the other classes of immunoglobulin. Since the mixture of antibodies produced in response to IgG immunization contains some which are specific for determinants on the heavy chain and some specific for determinants on the light chain, techniques for the selective removal of anti-light-chain antibodies from such antisera are required to prepare class-specific anti-immunoglobulins. Similarly, using appropriate immunization or selective absorption techniques or both, antibodies specific for the light chains of immunoglobulins can also be prepared.

Since anti-immunoglobulin antisera usually contain antibodies specific for both heavy and light polypeptide chains, the detection of surface immunoglobulins with such a reagent may reflect either light- or heavy-chain determinants or both. The presence of light-chain determinants on the surface of rabbit lymphoid cells has been shown by using anti-light-chain antisera to induce blast cell formation *in vitro* (Sell and Gell, 1965). At low concentrations anti-light-chain antisera are not themselves mitogenic for normal lymphoid cells but can inhibit the mitogenicity induced by tuberculin preparations (Greaves, Torrigiani and Roitt, 1969). Furthermore, anti-light-chain antisera bind to cells as detected by high-resolution autoradiography (Jones, Marcuson and Wright, 1970) and the binding of antigen by mouse splenic cells is inhibited in the presence of anti-light-chain antisera (Warner, Byrt and Ada, 1970).

By studying either the binding of class-specific anti-heavy-chain antisera, or by using such antibody preparations to inhibit the binding of an antigen, several kinds of heavy-chain determinants have been detected on the surface of lymphocytes. Thus the presence of IgM heavy-chain determinants on mouse splenic cells was demonstrated by blocking antigen receptor sites (Warner, Byrt and Ada, 1970) and both IgM and IgG heavy-chain determinants were shown to be present on such cells by inhibition of rosette formation (McConnell *et al.*, 1969). Similarly, immunofluorescent techniques have been used to establish the presence of IgM and IgG heavy-chain determinants on rabbit spleen and peripheral blood lymphocytes (Pernis, Forni and Amante, 1970), IgA heavy-chain determinants on mouse spleen cells (Rabellino *et al.*, 1971), IgD heavy-chain determinants on human peripheral blood lymphocytes (Van Boxel *et al.*, 1972) and IgE on human basophil cells (Ishizaka, Tomioka and Ishizaka, 1970).

That an immunoglobulin-positive lymphocyte carries a single class of immunoglobulin on its surface is suggested by the finding that the sum of the total number of cells that react with different class-specific reagents is approximately equal to that found with an antiserum containing antibodies for all known classes of heavy chain (Rabellino *et al.*, 1971). If a reasonable proportion of the cells possessed more than one class of immunoglobulin on the surface, then the sum of the number of cells reacting with sera having individual specificities would be greater than the sum observed by using a polyvalent antiserum. Fröland and Natvig (1972) employed fluorescein-labeled anti-IgG and rhodamine-labeled anti-IgM to establish that no doubly labeled cells could be detected. Their experiments also established that membrane-associated immunoglobulins of human peripheral blood lymphocytes are restricted not only to a single class but to a single subclass.

Contrary findings suggest that some guinea-pig lymph node cells (Biozzi *et al.*, 1969) and some mouse spleen cells (Lee, Paraskeras and Israels, 1971) bear more than one class or subclass of immunoglobulin simultaneously. Unfortunately there is no simple explanation for this discrepancy, which may be related to the substantial differences in methodology used by the different laboratories or, alternatively, to differences in the stages of differentiation of the cells studied. Regardless of how this issue is resolved, it does not alter the fact that lymphocytes which bear IgG, IgM, IgA, IgD and IgE have been observed.

It must be emphasized that the combining site of an antibody molecule has a relatively small volume (Kabat, 1956; Farah, Kern and Eisen, 1960) and, therefore, reacts specifically with only a small part of the surface of the appropriate antigen. In this sense the reactivity of cell-surface immunoglobulins with anti-immunoglobulin antisera reflects only the reactivity of antigenic determinants and does not provide direct information concerning overall structure.

Several studies have provided evidence that the immunoglobulins derived from cell surfaces which have been labeled with ^{125}I are composed of polypeptide chains that have an electrophoretic mobility like that of authentic heavy and light chains (Baur et al., 1971; Marchalonis, Atwell and Cone, 1972a, b). Since heavy chains of IgM and IgG can be distinguished by acrylamide-gel electrophoresis in sodium dodecyl sulfate. Baur et al. (1971) were able to show that some mouse spleen cells bear IgM heavy chains and are devoid of IgG heavy chains. Furthermore, only IgG heavy chains are found on the surface of a line of murine myeloma cells known to secrete solely IgG, and only IgM on a line of human Burkitt lymphoma cells that synthesize, but do not secrete, IgM.

In an extension of this study (Vitetta, Baur and Uhr, 1971), the surface immunoglobulins of mouse spleen lymphocytes were shown to comprise principally the monomeric or subunit form of IgM, rather than the pentameric structure characteristic of serum IgM. Several observations indicated that the monomeric IgM was not an artifact of the isolation and characterization procedures. Monomers of IgM are linked exclusively by disulfide bonds to form the pentamer whereas their constituent light and heavy chains are linked both by disulfide bonds and noncovalent forces. It was therefore necessary to prove that the techniques used did not induce either the dissociation of the pentamer, by reduction of disulfide bonds, or the association of heavy and light chains to form the monomer, by oxidation of sulfhydryl bonds. Dissociation of the pentamer was ruled out because authentic IgM carried through the same procedure did not degrade to form monomer while the assembly of heavy and light chains was excluded because the IgM monomer was obtained whether or not the cells were lysed in the presence or absence of agents that alkylate sulfhydryl groups, which effectively precludes the formation of disulfide linkages. Finally, the identification of the IgM monomer by its mobility on sodium dodecyl sulfate polyacrylamide gels eliminated the possibility that the monomer was linked solely by noncovalent forces.

Murine thymus cells also bear IgM heavy and light chains (Marchalonis, Atwell and Cone, 1972a, b) as judged by these procedures. The data for human thymus were not so conclusive because the ^{125}I-labeled heavy chains migrated on sodium dodecyl sulfate polyacrylamide-gel electrophoresis at a rate between that of authentic heavy chains of IgM, on the one hand, and the heavy chains of IgG on the other. However, for both human and murine thymus, analysis of the cell-surface product by gel filtration on Sepharose 6B revealed that it migrated more slowly than pentameric IgM and more rapidly than IgG. Its elution position was consistent with the elution position expected for the monomeric form of IgM.

It is important to note that the monomer occurs only in trace amounts in normal mammalian sera. In this context the surface IgM is an incomplete immunoglobulin even though the available evidence, based on acrylamide-gel electrophoresis, suggests that the constituent heavy and light chains of immunoglobulins are linked together by disulfide bonds. The observation that IgM occurs as monomers suggests that other surface immunoglobulins detected by anti-immunoglobulin antisera are in the form of associated heavy and light chains.

5.3   IMMUNOGLOBULINS ON T AND B LYMPHOCYTES

A variety of lines of evidence has led to a division of lymphocytes into two broad groups termed 'T' (thymus-derived cells) and 'B' (bursa- or bone-marrow-derived cells) according to their site of differentiation. In mice, T cells have been characterized in part by the presence of an antigenic marker, θ, on their surface, which is not observed on B cells (Raff, 1970). As judged by immunofluorescent techniques, θ-bearing cells contain little or no surface immunoglobulin whereas many non-θ-bearing cells have an abundance of such immunoglobulins. In addition, some cells have been observed that have neither θ nor immunoglobulins on their surface, but it is not known whether these are T cells, B cells or a third type (Stobo, Rosenthal and Paul, 1973). As might be expected from the foregoing, the distribution of lymphocytes which bear surface immunoglobulins in different lymphoid tissues varies markedly, as assessed by immunofluorescence (Raff, Sternberg and Taylor, 1970). When calculated on the basis of the total number of lymphocytes present, the relative proportions of immunoglobulin-bearing cells in spleen, bone marrow, peripheral blood, lymph node, peritoneal exudate and thymus are 49, 47, 14, 7, 10 and 0.1 percent, respectively (Rabellino et al., 1971). The very low percentage of immunoglobulin-bearing cells detected in thymus is consistent with the fact that 95–100 percent of thymus cells bear θ surface determinant (Raff, 1970).

The report that T cells have as much immunoglobulin on their surface as B cells (Marchalonis, Atwell and Cone, 1972a, b) has been the subject of considerable dispute. Marchalonis and co-workers suggest that the immunoglobulin on the surface of T cells may not be so accessible as that on B cells. In this sense, the detection of surface immunoglobulins using anti-immunoglobulin techniques such as immunofluorescence is subject to restrictions imposed by the size and shape of the indicator molecules while steric problems are brought to a near minimum in the surface iodination technique. Nevertheless, using the same surface iodination technique, Vitetta et al. (1972) were unable to detect surface immunoglobulins on T cells from a variety of sources, although they claimed to be able to detect as little as one B cell per 400 thymocytes. Furthermore, they believed that they could detect as few as 250 molecules of immunoglobulin per cell. Also Grey, Kubo and Cerottini (1972) were unable to demonstrate hidden immunoglobulin determinants on T cells insofar as variations in the method of cell disruption did not show any disproportionate increase in T cell immunoglobulin compared with that observed when B cell populations were similarly treated. Furthermore, reduced and alkylated immunoglobulin:anti-immunoglobulin precipitates, and controls consisting of unrelated immune precipitates which had been formed in the presence of lysates of surface-iodinated thymus cells both yielded essentially the same pattern on sodium dodecyl sulfate polyacrylamide-gel electrophoresis. Spleen cells processed in this manner showed readily discernible heavy- and light-chain peaks with anti-immunoglobulin antisera but not with control antisera. Vitetta *et al.* (1972), who could not detect immunoglobulins on T cells, explained the contradictory results of Marchalonis, Atwell and Cone (1972a, b) by possible contamination of the thymus cell preparations with B cells and by the lack of suitable controls for non-specific immunoprecipitation in the studies of the latter workers.

A new dimension was added to this controversy by the findings that rat T cells, stimulated with tuberculin to undergo blast cell transformation, retain the antigenic markers for T cells on their surfaces and, in addition, that 92 percent of such blast cells then possess readily detectable surface immuno-globulin (Goldschneider and Cogen, 1973). Blast cell formation, of itself, is not responsible for the appearance of surface immunoglobulins because blast cells induced by concanavalin A have the T cell antigenic markers on their surface although they are essentially free of surface immunoglobulins as judged by the immunofluorescence technique. From Goldschneider and Cogen's study it becomes clear that some 10–20 percent of the cells of suitably induced populations are T cells which have immunoglobulins on their surface in sufficient abundance as to be easily detectable by the immunofluorescence techniques used for B cells. Since blast cell transformation from small lymphocytes involves a considerable synthesis of cellular materials it is not known whether the surface receptors observed result from unmasking of cryptic sites in the plasma membrane or by synthesis *de novo*. It will be of considerable interest to see if this phenomenon is observed in other species as well as in rats and, in particular, whether surface immunoglobulins become more dense on T cells that are subjected to antigenic stimulation.

The presence or absence of immunoglobulins on T cells is an issue which is central to a more complete understanding of the processes that lead to the production of antibodies. Unlike B cells, T cells do not actively secrete any of the known classes of immunoglobulins (Miller and Mitchell, 1965), but the ultimate production of antibody by B cells is very much dependent on collaboration with T cells. If thymus cells or bone marrow cells alone are injected into X-irradiated animals, small amounts of antibody are produced in response to the subsequent injection of antigen, whereas the injection of both cell types results in substantial antibody production on challenge with antigen (Clamen, Caperson and Triplett, 1969). The specificity of this so-called 'helper' function of T cells has been shown by the finding that both T and B cells are inactivated by ^{125}I-labeled antigen but that T cells can be protected by anti-light-chain antiserum, suggesting that the receptor is an immunoglobulin (Basten *et al.*, 1972).

The absence, or marginal detectability, of immunoglobulins on freshly isolated T cells in many investigations has led to the suggestion that the antigen receptor sites which have been demonstrated on T cells are either a novel immunoglobulin molecule or even a molecule unrelated to immunoglobulins but having the capability of recognizing antigens (Greaves, 1970; Vitetta *et al.*, 1972; Nossal *et al.*, 1972). The possibility that the antigen recognition site on the T cell is not an immunoglobulin carries with it the substantial problem that an entirely new set of molecules having the same specificity as their counterparts on B cells would be required. It is interesting to note in this regard that an antigen-specific factor released by T cells can be used to replace the helper function of T cells (Munro *et al.*, 1974). Although the factor specifically binds antigen, it is not an antibody *per se* because it does not react with antisera against immunoglobulins, whereas it does react with antisera against H2 determinants, that is, the major histocompatibility complex of the mouse. Unfortunately, the available data do not permit a conclusion as to whether or not the solubilized factor and the T cell receptor site are one and the same.

The studies involving a search for immunoglobulins on T cells can be considered to suffer a general defect in that they offer a rather restricted view. For example, the usual emphasis on the relatively small number of immunoglobulin molecules on T cells would be unwarranted if the proportion of T cells which are reactive in immune phenomena is quite small in comparison to the number of T cells that have not yet differentiated to, or achieved, this state. If one assumes that such fully reactive T cells are very low in number, then the characteristics of the general T cell population with regard to surface immunoglobulins or any other parameter could well be irrelevant. In this context, it should be noted that subpopulations of T cells which differ both in functional (Mosier and Cantor, 1971) and physical (Konda, Nakao and Smith, 1972) properties are known.

It has been calculated that there are some 10^5 immunoglobulin molecules per B cell (Rabellino et al., 1971) but only 250–1000 per T cell (Nossal et al., 1972). Even this kind of difference cannot be accorded any unusual significance, assuming a reasonable uniformity of the cell population, because T and B cells have different functions. Indeed, the density of the surface receptors on T and B cells may well be related to, or required for, such functions. Finally, it is noteworthy that the principal protagonists in the polemic as to whether T cells possess surface immunoglobulins have recently reiterated their original positions (Marchalonis, 1975; Vitetta and Uhr, 1975).

5.4   MOBILITY OF SURFACE IMMUNOGLOBULINS

Examination of individual cells for the distribution of surface immunoglobulins, as detected with fluorescein-labeled anti-immunoglobulin antiserum, has revealed several patterns. The stain was found to be diffuse on a few cells and patchy on most cells (Pernis, Forni and Amante, 1970) while, in other studies, the fluorescent indicator gave a cap-like appearance at one pole of the cell (Raff, 1970; Raff, Sternberg and Taylor, 1970). However, all of these patterns can be observed if the temperature is carefully controlled (Taylor et al., 1971); at 0 °C both the diffuse pattern and the patches were observed whereas at room temperature, or at 37 °C, most cells showed cap fluorescence. The pattern of diffuse and patch staining which was observed at 0 °C was rapidly converted to cap staining when the suspension was warmed to 37 °C, 50 percent of the stained cells forming caps in 1.5 minutes at 37°C and in 4 minutes at 24°C.

The fact that cap formation does not occur at 37 °C in the presence of sodium azide or dinitrophenol suggests that the process is dependent upon the metabolic activity of the cells. In addition to the dependence on temperature and cellular metabolic activity, there also is a requirement for the divalency of the anti-immunoglobulin. Thus, monovalent anti-immunoglobulin after being labeled with fluorescein does not induce either cap or patch formation at 37 °C, but the percentage of cells showing a diffuse staining pattern is equal to the percentage of cells that stain with divalent anti-immunoglobulin antisera. The absence of patch formation using monovalent anti-immunoglobulins suggests that the lymphocyte has a diffuse distribution of immunoglobulins before testing and that the patch pattern observed is a consequence of divalency.

A further indication that cap formation is dependent on multiple valences was obtained using an indirect labeling procedure. Mouse cells were reacted first with monovalent anti-mouse immunoglobulin antisera prepared in rabbits, washed and then incubated at 37 °C with divalent anti-rabbit immunoglobulin antisera labeled with fluorescein. The resultant cap formation can best be explained as being due to the association between surface immunoglobulins and monovalent anti-mouse immunoglobulin antisera, and that this is followed by cross linking of the latter with divalent antibodies.

Not only are the surface immunoglobulins mobile within the plane of the membrane but the immunofluorescence technique has shown that they are internalized as well. Pinocytosis was not observed at 0 °C or when monovalent anti-immunoglobulin antisera were used but, subsequent to cap formation at 37 °C, pinocytosis was detectable within 5 minutes, was marked within 10 minutes and was extensive within 30 minutes.

Several hours after the addition of divalent anti-immunoglobulin the surface immunoglobulins were no longer detectable. On the other hand, cells which had been similarly incubated with monovalent anti-immunoglobulin did not show any loss of surface immunoglobulins, as judged by the aforementioned indirect staining procedure. Moreover, the loss of surface immunoglobulins was selective since other surface-associated antigens showed a normal staining intensity when tested by indirect immunofluorescence (Taylor et al., 1971).

Subsequent to their disappearance under the influence of anti-immunoglobulins, surface immunoglobulins reappear after about 6 hours provided that the cells are washed and then incubated in tissue culture medium (Loor, Forni and Pernis, 1972). If the anti-immunoglobulins are not removed, cells incubated in tissue culture medium do not recover their surface immunoglobulins.

The results of ultrastructural studies are entirely consistent with immunofluorescence studies and, in addition, provide greater resolution. Monovalent anti-immunoglobulin which had been linked to ferritin was used by de Petris and Raff (1973) to show that the surface immunoglobulin was distributed, both diffusely and randomly, in single molecules or small clusters of two to four molecules. The gaps between molecules were between 50 and 200 nm. The diffuse staining which was observed with ferritin-conjugated monovalent anti-immunoglobulins was converted to patches at 0 °C by the addition of divalent antibodies which were specific for the monovalent anti-immunoglobulin. Under these circumstances, the ferritin label was in clusters separated by large unstained areas. At 37 °C, capping occurred over that pole of the cell containing the Golgi apparatus and other organelles (Figure 5.1a). Details of both the cap and a pinocytotic vesicle are shown in Figure 5.1b. Pinocytosis was not dependent upon the cross linking of surface immunoglobulins because ferritin-labeled monovalent anti-immunoglobulins were, to some extent, internalized within vesicles, when the cells were incubated at 37 °C. Thus, endocytosis was not dependent on either patch or cap formation. A

measure of the spacing of surface immunoglobulins can be obtained by determining the distance between adjacent ferritin particles and between ferritin
particles and the cell membrane. For cells showing a diffuse pattern of distribution the ferritin particles and membrane were separated by about 150–
200 nm (Karnovsky, Unanue and Leventhal, 1972; de Petris and Raff, 1973)
and the distance between the centers of adjacent particles was *ca.* 200 nm
(Karnovsky, Unanue and Leventhal, 1972). It is important to note that before
capping the ferritin particles appeared most often in a single layer whereas
after capping the marker was stacked in many layers with the upper layer
several thousand nanometers distant from the membrane.

The use of metabolic inhibitors has shown that cap formation is dependent
on cellular metabolism (Loor, Forni and Pernis, 1972; de Petris and Raff,
1973; Unanue, Karnovsky and Engers, 1973). Furthermore, cells which have

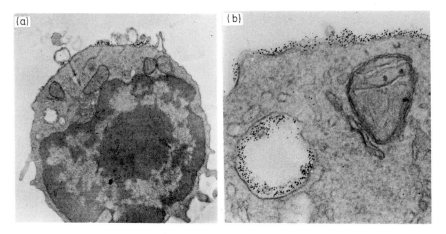

*Figure 5.1   Capping of ferritin-labeled anti-immunoglobulin. (a) Electron micrograph of cap show-
ing its relationship to cellular organelles. × 18 000. (b) Pinocytotic vesicle and details of the cap
× 76 000. (From de Petris and Raff, 1973. Dr de Petris kindly furnished the photographs. Reproduced
courtesy of the authors and Macmillan)*

been inactivated by prolonged incubation in phosphate-buffered saline solution and are dead as assessed by the dye exclusion test also do not show
the capping phenomenon (Unanue, Karnovsky and Engers, 1973). On the
other hand, patch formation is unaffected by metabolic inhibitors.

From these data a number of conclusions can be drawn with regard to
the mobility of surface immunoglobulins. First, there is no evidence either
for or against any substantial movement of surface immunoglobulins in the
absence of anti-immunoglobulin antisera. The fact that fluorescein-labeled
monovalent anti-immunoglobulin antisera yielded a diffuse pattern even at
37 °C does not provide information on the presence or absence of movement
of surface immunoglobulins because even substantial movement would remain unnoticed by the usual procedures if the diffuse pattern were maintained.
The very detection of specific binding of monovalent anti-immunoglobulin
to surface immunoglobulins indicates that the surface immunoglobulins
remain linked to the membrane while bound to the monovalent antibody.
This must be so because a complex composed of a single immunoglobulin

molecule and a single monovalent anti-immunoglobulin molecule would be soluble.

Patch formation appears to be solely a consequence of the ability of each divalent anti-immunoglobulin molecule to bridge two surface immunoglobulin molecules. This can occur if the individual surface immunoglobulin molecules of the resting lymphocyte are already close together, or can move close enough together to be linked by an antibody molecule; that is, within the 140 nm length of a linking antibody molecule (Green, 1969). Since the formation of patches is not affected by metabolic inhibitors and occurs readily at 4 °C, the simple explanation that patching reflects largely an antigen–antibody interaction seems likely. That ferritin particles still appear mainly in a single layer at the time patch formation is completed permits, but does not compel, the view that the surface immunoglobulins are still embedded in the membrane. On the other hand, newly formed caps have many layers of ferritin particles and it is difficult to envision, in terms of the distance involved, how surface immunoglobulins in the upper layers could still be linked to the membrane. Indeed, the capped materials, at least in part, are quite loosely linked because some can be dislodged by gentle agitation of the cells (Karnovsky, Unanue and Leventhal, 1972). For these reasons it remains a point of conjecture as to whether the extensive movement involved in cap formation occurs while the bulk of the surface immunoglobulins are embedded in the membrane or while such immunoglobulins are merely in close proximity to the cell surface.

Although the surface immunoglobulins are randomly distributed initially, cap formation takes place at only one pole of the cell. This has been interpreted as being related to cell motility (de Petris and Raff, 1972; Karnovsky, Unanue and Leventhal, 1972), the cell essentially moving out of the antigen–antibody lattice with which it is surrounded. Capping does not appear to involve new membrane synthesis because caps are formed very rapidly and also because capping is not affected by inhibitors of protein synthesis. Moreover, the amount of labeled membrane that undergoes pinocytosis during capping corresponds to only a fraction of the total that would have to be ingested in order to account for cap formation (de Petris and Raff, 1972).

There is no correlation between pinocytosis and capping since interiorization occurs even without complex formation; that is, in the presence of monovalent anti-immunoglobulins. Furthermore, the loss of immunoglobulins from the cell surface is not totally accounted for by endocytosis since a substantial part of the capped materials are known to be shed into the extracellular medium (Wilson, Nossal and Lewis, 1972).

Although most capping studies have been performed using anti-immunoglobulin as the inducing agent, capping has also been demonstrated in response to antigens *in vitro* (Dunham, Unanue and Benacerraf, 1972; de Petris and Raff, 1973), but whether this can occur *in vivo* is not known.

Antibody which is associated with the particulate matter derived from cell homogenates (Kern, Helmreich and Eisen, 1959) is noncovalently bound through sites on the carboxyl-terminal half of the heavy polypeptide chains (Kern, Helmreich and Eisen, 1961; Swenson et al., 1969). Surface antibody and the antibody associated with the particulate matter of cell homogenates as defined by the assay procedure for the latter (Kern, Helmreich and Eisen, 1959) appear to be the same because brief treatment of intact cells with the appropriate antigen, followed by removal of soluble antigen and by homogenization, results in particulate preparations in which antibody is no longer detectable with ^{125}I-labeled antigen (D. H. Zimmerman, M. B. Hayes, I. Smith and M. Kern, unpublished results). Finally, by testing anti-immunoglobulin antibodies specific for certain portions of the carboxyl-terminal half of heavy chains, and assessing their capacity to induce capping, Fröland and Natvig (1972) concluded that the surface immunoglobulin is bound at sites relatively close to the carboxyl terminus.

5.5   SYNTHESIS AND TURNOVER OF SURFACE IMMUNOGLOBULINS

The reappearance of surface immunoglobulins following removal by treatment of cells with anti-immunoglobulin antisera (Loor, Forni and Pernis, 1972) or trypsin (Pernis *et al.*, 1971) provides the basic evidence regarding the synthesis and turnover of surface immunoglobulins. However, whether these observations reflect the total synthesis of the surface immunoglobulin molecule or its insertion into the membrane from a pool of preexisting immunoglobulin is not shown by such studies. In addition, the term 'turnover' must also be used with caution in this context because treatment with anti-immunoglobulin antisera or with trypsin is not a normal physiological event.

The appearance of iodinated cell-surface proteins in the extracellular fluid has been employed to demonstrate that surface proteins of spleen and thymus cells are released at a rapid rate (Cone, Marchalonis and Rolley, 1971). Since inhibitors of cellular respiration and inhibitors of protein synthesis markedly reduce the rate of release of surface proteins into the medium, it was concluded that the release reflects a dynamic state for surface proteins. Vitetta and Uhr (1972a, b) also used the surface iodination technique and reported that a small proportion (2–4 percent) of the released cell-surface material was accounted for as immunoglobulins. Analysis of the product revealed that it was primarily IgM monomer with a minute amount of IgG and possibly a trace of pentameric IgM; based on initial rates they calculated that 5–7 hours were required for 50 percent release. The released radioiodinated immunoglobulins were shown to be noncovalently linked to membrane fragments since the gel-electrophoretic pattern of their immune precipitates was markedly affected by detergent treatment of a sample before specific precipitation. An alteration in the pattern of density gradient centrifugation, after treatment with detergent, of the radioiodinated immunoglobulins also indicated that they were bound to the membrane.

Released surface immunoglobulins can be readily distinguished from those ordinarily secreted by cells because the latter are not associated with membrane fragments (Vitetta and Uhr, 1972b). Thus, the electrophoretic pattern of extracellular proteins derived from cells incubated with [^{3}H]tyrosine was found not to be altered by prior detergent treatment and 70–80 percent of the tritiated protein was accounted for as immunoglobulin rather than the small fraction of the total released proteins represented by surface immunoglobulins. Secreted protein also differed in that about 50 percent was found as IgG whereas there was virtually no IgG detected in the released surface immunoglobulins. These results indicate that surface immunoglobulins are not direct precursors of the bulk of the secreted immunoglobulins. Indeed, the predominant released surface product was in the form of IgM monomer which, as pointed out earlier, is present in only trace amounts in serum compared with pentameric IgM. IgG, the most abundant serum immunoglobulin, was essentially not present among the released surface immunoglobulins.

Elucidation of the mechanism by which surface immunoglobulins arise is hampered by the fact that relatively little is known about their structure. For example, it would be useful to know whether the IgM monomer found on the surface is identical to the IgM monomer that can be derived from serum IgM. Differences in the galactose content of IgM molecules derived from chick embryos have been reported (Choi and Good, 1972) and simple structural differences could, of course, determine whether a molecule will be embedded into the surface membrane or will be secreted. On the other hand, if it is assumed that the synthetic process is identical both for surface immunoglobulins and molecules destined to be secreted, it may be that such molecules are synthesized at different times in the life of the cell or that their site of localization is affected by other cellular constituents. In this regard it has been suggested that membrane-associated and secreted immunoglobulins may be separately controlled (Lerner et al., 1972) on the basis of the half-times for disappearance of membrane-associated and secreted immunoglobulins.

5.6   THE ROLE OF SURFACE IMMUNOGLOBULINS

The available evidence is consistent with the view that some surface immunoglobulins are antigen receptor sites which are vital for the expression of a variety of immune phenomena. It is most likely that the specific interaction of antigen molecules and surface antibody is one of the events that leads to both proliferation and differentiation of antibody-forming cells. As studied in tissue culture, the addition of antigen to spleen cells from normal mice results in a relative abundance of cells which produce antibodies specific for the inducing agent (Dutton and Mishell, 1967). While the induction of antibody production is also dependent on other events such as T and B cell interactions, recognition of an antigen by an antibody on the cell surface appears to function as part, at least, of the triggering mechanism.

IgD-like molecules on murine lymphocytes have been detected using surface iodination techniques (Melcher et al., 1974; Abney and Parkhouse, 1974). Such molecules represent a second major class of immunoglobulin, that is, in addition to 7-S IgM. Primarily on the basis of the time of appearance of IgD and IgM as surface immunoglobulins, Vitetta and Uhr (1975) have proposed a model for differentiation of B cells in which IgD is considered the receptor for antigen triggering and IgM the receptor for the induction of tolerance.

According to the clonal selection theory (Burnet, 1959), cells capable of reacting with a given antigen possess this quality before they ever experience that antigen. In this context, IgM-containing cells have been observed in the bursa of the 14-day-old chick embryo (Kincade and Cooper, 1971) and their appearance was unaffected by exogenous antigen. Some of the cells showed immunofluorescent caps with anti-IgM antisera, suggesting that surface immunoglobulins were also present at the time, and by the 22nd day (1 day after hatching) essentially all of the bursal lymphocytes clearly bore surface immunoglobulins (Kincade, Lawton and Cooper, 1971).

Another indication that surface immunoglobulins are involved in the differentiation of lymphoid cells was provided by injecting anti-IgM antisera into chick embryos (Kincade et al., 1970). When these antisera were injected at day 14, and bursectomy was performed at hatching, antibody-producing cells were no longer detected and neither IgM nor IgG was produced by 10 weeks of age. If anti-IgM antiserum was introduced at hatching, but following bursectomy, IgM production was inhibited but IgG production was unaffected. These results have been reasonably interpreted to mean that IgM-producing cells are progenitors of IgG-producing cells. Moreover, it is most likely that the effects produced are a result of anti-IgM antibodies reacting with surface immunoglobulins.

The differentiation of lymphoid cells in tissue culture can also be inhibited with anti-IgM antisera. Rabbit popliteal lymph node cells incubated in tissue culture differentiate within 48 hours to produce cells that secrete solely IgM (Zimmerman and Kern, 1973a, b), but this does not occur in the presence of anti-IgM antisera (D. H. Zimmerman and M. Kern, unpublished results). Anti-IgG and anti-IgA were without effect in this test system and monovalent anti-IgM antiserum was as effective as divalent in preventing the appearance of cells which secreted IgM. The efficacy of monovalent anti-IgM indicated that capping was not required.

A study of the effect of the delayed addition of anti-IgM revealed that the production was not inhibited if the antiserum was added after 48 hours in tissue culture but that maximal inhibition was observed if anti-IgM was added either at the beginning of incubation of the tissue culture or after 24 hours; that is, in advance of the differentiation process but not after it had occurred.

The means by which anti-immunoglobulin antisera inhibit the differentiation of lymphoid cells is not known. However, the fact that anti-immunoglobulins can induce blast cell formation (Sell and Gell, 1965) may be indicative of a general mechanism. Thus rather than kill cells, the anti-immunoglobulins may cause a redirection of differentiation so that the cells somehow do not differentiate to form antibody-secreting cells.

The bulk of evidence indicates that, with the exception of IgE, each cell synthesizes the immunoglobulin that becomes associated with its own surface. However, the binding of serum IgE to the plasma membrane of human basophil cells (Ishizaka, Tomioka and Ishizaka, 1970; Sullivan, Grimley and Metzger, 1971) provides a well-documented exception to the general rule. When cells are treated with anti-IgE antiserum, this IgE exhibits the surface mobility noted for other surface immunoglobulins (Sullivan, Grimley and Metzger, 1971; Becker et al., 1973) and, as well as patch and cap formation, pinocytosis also occurs when cap formation is pronounced.

Basophil cells which bear IgE can be induced to release histamine if incubated with anti-IgE antiserum (Ishizaka et al., 1969) or, alternatively, with the antigen for which the surface IgE is specific (Osler, Lichtenstein and Levy, 1968). No correlation has been observed between histamine release and cap formation although both processes are dependent on divalent anti-IgE antibody (Becker *et al.*, 1973). In addition, specific antigen causes maximal histamine release under circumstances where surface immunoglobulins remain diffuse.

These findings regarding the release of histamine suggest that cell-surface immunoglobulins may play a role in allergic phenomena. Thus, leukocytes from allergic individuals release histamine in the presence of the antigen responsible for the allergy (Lichtenstein and Osler, 1964). Further, cells from a nonallergic individual were also found to release histamine when incubated with serum from an allergic individual and then reacted with the appropriate antigen (Levy and Osler, 1966). The serum component involved has been established to be IgE (Ishizaka *et al.*, 1969).

5.7   CONCLUDING COMMENTS

A precise knowledge of the role that surface immunoglobulins play in the various immune phenomena obviously requires more study. It would be advantageous to have techniques to isolate homogeneous cell types so that their activation, modification or differentiation could be scrutinized in the absence of the encumbrances or uncertainties that exist when heterogeneous populations are studied.

Quite apart from its significance in immune phenomena, surface immunoglobulin is an example of a plasma-membrane protein. The suitability of surface immunoglobulins for further study as model plasma-membrane proteins is indicated by the knowledge available concerning the detailed structure of immunoglobulins, their digestion products and the extensive methodology which permits their modification with various reagents. In addition, the precise specificity of antisera for known markers along the length of the heavy chains can be employed as a sensitive and revealing tool. Thus, surface immunoglubulins should continue to be useful in studying the binding, orientation and mobility of plasma-membrane proteins.

REFERENCES

ABNEY, E. and PARKHOUSE, R. M. E. (1974). *Nature, Lond.*, **252**:600.
ADA, G. L. and BYRT, P. (1969). *Nature, Lond.*, **222**:1291.
BASTEN, A., MILLER, J. F. A. P., WARNER, N. L. and PYE, J. (1972). *Nature, New Biol.*, **231**:104.
BAUR, S., VITETTA, E. S., SHERR, C. J., SCHENKEIN, I. and UHR, J. W. (1971). *J. Immun.*, **106**:1133.
BECKER, K. E., ISHIZAKA, T., METZGER, H., ISHIZAKA, K. and GRIMLEY, P. M. (1973). *J. exp. Med.*, **138**:394.
BIOZZI, G., BINAGHI, R. A., STIFFEL, C. and MOUTON, D. (1969). *Immunology*, **16**:349.
BURNET, M. (1959). *The Clonal Selection Theory of Acquired Immunity.* Cambridge, England; Cambridge University Press.
BYRT, P. and ADA, G. L. (1969). *Immunology*, **17**:503.
CHOI, Y. S. and GOOD, R. A. (1972). *J. exp. Med.*, **136**:8.
CLAMEN, H. N., CAPERSON, E. A. and TRIPLETT, R. F. (1967). *Proc. Soc. exp. Biol. Med.*, **122**:1167.
CONE, R. E., MARCHALONIS, J. J. and ROLLEY, R. T. (1971). *J. exp. Med.*, **134**:1373.
DE PETRIS, S. and RAFF, M. C. (1972). *Eur. J. Immun.*, **2**:523.
DE PETRIS, S. and RAFF, M. C. (1973). *Nature, New Biol.*, **241**:257.
DUNHAM, E. K., UNANUE, E. R. and BENACERRAF, B. (1972). *J. exp. Med.*, **136**:403.
DUTTON, R. W. and MISHELL, R. I. (1967). *Cold Spring Harb. Symp. quant. Biol.*, **32**:407.
FARAH, F. S., KERN, M. and EISEN, H. N. (1960). *J. exp. Med.*, **112**:1195.

FRÖLAND, S. S. and NATVIG, J. B. (1972), *J. exp. Med.*, **136**:409.

GOLDSCHNEIDER, I. and COGEN, R. B. (1973). *J. exp. Med.*, **138**:163.

GREAVES, M. F. (1970). *Transplantn Rev.*, **5**:45.

GREAVES, M. F., TORRIGIANI, G. and ROITT, I. M. (1969). *Nature, Lond.*, **222**:885.

GREEN, N. M. (1969). *Adv. Immun.*, **11**:1.

GREY, H. W., KUBO, R. T. and CEROTTINI, J. (1972) *J. exp. Med.*, **136**:1323.

HAMMOND, E. (1970). *Expl Cell Res.*, **59**:359.

ISHIZAKA, K., TOMIOKA, H. and ISHIZAKA, T. (1970). *J. Immun.*, **105**:1459.

ISHIZAKA, T., ISHIZAKA, K., JOHNSON, S. G. O. and BENNICH, H. (1969). *J. Immun.*, **102**:884.

JONES, G., MARCUSON, E. C. and ROITT, I. M. (1970). *Nature, Lond.*, **227**:1051.

KABAT, E. A. (1956). *J. Immun.*, **77**:377.

KARNOVSKY, M. J., UNANUE, E. R. and LEVENTHAL, M. (1972). *J. exp. Med.*, **136**:907.

KERN, M., HELMREICH, E. and EISEN, H. N. (1959). *Proc. natn. Acad. Sci. U.S.A.*, **45**:862.

KERN, M., HELMREICH, E. and EISEN, H. N. (1961). *Proc. natn. Acad. Sci. U.S.A.*, **47**:767.

KINCADE, P. W. and COOPER, M. D. (1971). *J. Immun.*, **106**:371.

KINCADE, P. W., LAWTON, A. R. and COOPER, M. D. (1971). *J. Immun.*, **106**:1421.

KINCADE, P. W., LAWTON, A. R., BOCKMAN, D. E. and COOPER, M. D. (1970). *Proc. natn. Acad. Sci. U.S.A.*, **67**:1918.

KONDA, S., NAKAO, Y. and SMITH, R. T. (1972). *J. exp. Med.*, **136**:1461.

LEE, S. T., PARASKERAS, F. and ISRAELS, L. G. (1971). *J. Immun.*, **107**:1583.

LERNER, R. A., McCONAHEY, P. J., JANSEN, I. and DIXON, F. J. (1972). *J. exp. Med.*, **135**:136.

LEVY, D. A. and OSLER, A. G. (1966). *J. Immun.*, **97**:203.

LICHTENSTEIN, L. M. and OSLER, A. G. (1964). *J. exp. Med.*, **120**:507.

LOOR, F., FORNI, L. and PERNIS, B. (1972). *Eur. J. Immun.*, **2**:203.

MÄKELÄ, O. and NOSSAL, G. J. V. (1961). *J. Immun.*, **87**:447.

MANDEL, T., BYRT, P. and ADA, G. L. (1969). *Expl Cell. Res.*, **58**:179.

MARCHALONIS, J. J. (1969). *Biochem. J.*, **113**:299.

MARCHALONIS, J. J. (1975). *Science, N.Y.*, **190**:20.

MARCHALONIS, J. J., ATWELL, J. L. and CONE, R. E. (1972a). *Nature, New Biol.*, **235**:240.

MARCHALONIS, J. J., ATWELL, J. L. and CONE, R. E. (1972b). *J. exp. Med.*, **135**:456.

MARCHALONIS, J. J., CONE, R. E. and SANTER, V. (1971). *Biochem. J.*, **124**:921.

MCCONNELL, I., MUNRO, A., GURNER, B. W. and COOMBS, R. R. A. (1969). *Int. Archs Allergy appl. Immun.*, **35**:209.

MELCHER, U., VITETTA, E. S., MCWILLIAMS, M., PHILIPS-QUAGLIATA, J., LAMM, M. E. and UHR, J. W. (1974). *J. exp. Med.*, **140**:1427.

MILLER, J. F. A. P. and MITCHELL, G. F. (1968). *J. exp. Med.*, **128**:801.

MOSIER, D. and CANTOR, H. (1971). *Eur. J. Immun.*, **1**:459.

MUNRO, A. J., TAUSSIG, M. J., CAMPBELL, R., WILLIAMS, H. and LAWSON, Y. (1974). *J. exp. Med.*, **140**:1579.

NAOR, D. and SULITZNEAU, D. (1967). *Nature, Lond.*, **214**:687.

NOSSAL, G. J. V., WARNER, N. L., LEWIS, H. and SPRENT, J. (1972). *J. exp. Med.*, **135**:405.

OSLER, A. G., LICHTENSTEIN, L. M. and LEVY, O. A. (1968). *Adv. Immun.*, **8**:183.

PERNIS, B., FORNI, L. and AMANTE, L. (1970). *J. exp. Med.*, **132**:1001.

PERNIS, B., FERRANINI, M., FORNI, L. and AMANTE, L. (1971). *Progress in Immunology*, pp. 95–106. Ed. B. AMOS. New York; Academic Press.

PHILLIPS, D. R. and MORRISON, M. (1970). *Biochem. biophys. Res. Commun.*, **40**:284.

RABELLINO, E., COLON, S., GREY, H. M. and UNANUE, E. R. (1971). *J. exp. Med.*, **133**:156.

RAFF, M. C. (1970). *Immunology*, **19**:637.

RAFF, M. C., STERNBERG, M. and TAYLOR, R. B. (1970). *Nature, Lond.*, **225**:553.

REISS, E., MERTENS, E. and EHRICH, W. E. (1950). *Proc. Soc. exp. Biol. Med.*, **74**:732.

SELL, S. and GELL, P. G. H. J. (1965). *J. exp. Med.*, **122**:427.

STOBO, J. D., ROSENTHAL, A. S. and PAUL, W. E. (1973). *J. exp. Med.*, **138**:71.

SULLIVAN, A., GRIMLEY, P. M. and METZGER, H. (1971). *J. exp. Med.* **134**:1403.

SWENSON, R. M., COHEN, H. A., D'AMICO, R. P. and KERN, M. (1969). *J. biol. Chem.*, **244**:440.

TAYLOR, R. B., DUFFUS, P. H., RAFF, M. C. and DE PETRIS, S. (1972). *Nature, New Biol.*, **233**:225.

UNANUE, E. R., KARNOVSKY, M. J. and ENGERS, H. D. (1973). *J. exp. Med.*, **137**:675.

VAN BOXEL, J. A., PAUL, W. E., TERRY, W. D. and GREEN, I. (1972). *J. Immun.*, **109**:648.

VITETTA, E. S., BAUR, S. and UHR, J. W. (1971). *J. exp. Med.*, **134**:242.

VITETTA, E. S. and UHR, J. W. (1972a). *J. Immun.*, **108**:577.

VITETTA, E. S. and UHR, J. W. (1972b). *J. exp. Med.*, **136**:676.

VITETTA, E. S. and UHR, J.W. (1975). *Science, N.Y.*, **189**:964.

VITETTA, E. S., BIANCO, C., NUZZENZWEIG, V. and UHR, J. W. (1972). *J. exp. Med.*, **136**:81.
WALTERS, C. A. and WIGZELL, H. (1970). *J. exp. Med.*, **132**:1233.
WARNER, N. L., BYRT, P. and ADA, G. L. (1970). *Nature, Lond.*, **226**:942.
WIGZELL, H. and ANDERSON, B. (1969). *J. exp. Med.*, **129**:23.
WILSON, J. D., NOSSAL, G. J. V. and LEWIS, H. (1972). *Eur. J. Immun.*, **2**:225.
ZIMMERMAN, D. H. and KERN, M. (1973a). *J. Immun.*, **111**:761.
ZIMMERMAN, D. H. and KERN, M. (1973b). *J. Immun.*, **111**:1326.

6

Permeability of membranes

J. J. H. H. M. de Pont and S. L. Bonting
Department of Biochemistry, University of Nijmegen, Nijmegen, The

6.1   INTRODUCTION

One of the most striking and important properties of the plasma membrane is the large variety of ways in which substances can move through it. The membrane is readily permeable to some of them, whereas it is nearly impermeable towards others. Some substances move along their (electro)chemical gradient, whereas others are transported against this gradient. In this chapter we shall deal primarily with the first type of transport, *passive transport* or diffusion, whereas *active transport* is discussed in Chapter 7. First we shall define these two types of transport in terms of both equilibrium thermodynamics and irreversible thermodynamics. Next we shall describe two types of passive transport of non-electrolytes, simple diffusion and facilitated diffusion. The special phenomena associated with the permeability of ions will be discussed separately. This is followed by a discussion of water transport.

Given their variety, it is impossible to offer a comprehensive treatment of these processes. We shall attempt to give an introduction to the present state of knowledge in this field with selected references to reviews and original papers. For a more extensive description of these processes we refer the reader to books by Troshin (1966), Schoffeniels (1967), Stein (1967), Lakshminarayanaiah (1969), Cereijido and Rotunno (1970), Kotyk and Janáček (1970) and Harris (1972).

6.2   DEFINITIONS OF ACTIVE AND PASSIVE TRANSPORT

Criteria to distinguish between active and passive transport have been developed both from equilibrium thermodynamics and from irreversible thermodynamics. The first approach is based on the Nernst–Planck equation. In this theory it is assumed that the driving forces acting upon the membranes are independent and that the total force is the algebraic sum of these forces. The forces are: (a) the chemical potential gradient $d\mu/dx$, which for dilute solutions can be replaced by the concentration gradient dC/dx; (b) the electrical potential gradient dE/dx, and (c) the hydrostatic pressure gradient dP/dx. The resulting total force is represented by the negative gradient of the electrochemical potential $-d\bar{\mu}/dx$, for which the Nernst–Planck equation holds.

$$-\frac{d\bar{\mu}}{dx} = -RT\left[\frac{1}{C}\frac{dC}{dx} + \frac{\bar{V}}{RT}\frac{dP}{dx} + \frac{zF}{RT}\frac{dE}{dx}\right] \tag{6.1}$$

Since the flux J through a membrane can be expressed as the product of mobility, concentration and force (Teorell, 1953), equation 6.1 can be converted to

$$J = -\omega RTC\left[\frac{1}{C}\frac{dC}{dx} + \frac{\bar{V}}{RT}\frac{dP}{dx} + \frac{zF}{RT}\frac{dE}{dx}\right] \tag{6.2}$$

J is the flux expressed in $mol\,cm^{-2}\,s^{-1}$, ω is the mobility expressed in $cm\,mol\,s^{-1}\,dyn^{-1}$ and C is the concentration difference in $mol\,cm^{-3}$. Equation 6.2 cannot be applied in this form, but has to be integrated first. This has been done using different simplifying boundary conditions (Teorell, 1953; Conti and Eisenman, 1965). When no pressure gradient is present and we are dealing with an uncharged substance, equation 6.2 can be simplified to

$$J = -\omega RT\frac{dC}{dx} \tag{6.3}$$

Since ωRT is equal to the diffusion coefficient D (Einstein, 1905), equation 6.3 can be written

$$J = -D\frac{dC}{dx} \tag{6.4}$$

which represents Fick's law of diffusion.

When we assume that the concentration gradient across the membrane is homogeneous and that the diffusion path is equal to the thickness, d, of the membrane, equation 6.4 can be integrated to

$$J_{I\rightarrow II} = -D\frac{C_I - C_{II}}{d} \tag{6.5}$$

where C_I and C_{II} are the concentrations of the substance at each side of the membrane. Since the thickness of the membrane is not generally known, D/d is replaced by a new constant P, which is called the *permeability coefficient*. Equation 6.5 can now be written

$$J_{I\rightarrow II} = -P(C_I - C_{II}) \tag{6.6}$$

This equation represents the integrated form of Fick's law of diffusion.

For an uncharged substance, the diffusional flux will be directly proportional to the concentration gradient and will be directed from the high to the low concentration. When the flux is directed against this gradient, we are generally dealing with energy-requiring or active transport.

The flux of an ion will be determined by the concentration gradient ($C_I - C_{II}$) and by the potential gradient ($E_I - E_{II}$), assuming that no pressure gradient is present. For the passive transport of an ion, Ussing (1949a, b) has derived from equation 6.2 the following relationship for the ratio of the fluxes in both directions.

$$\frac{J_{I \to II}}{J_{II \to I}} = \frac{C_I}{C_{II}} \exp \left| \frac{zF(E_I - E_{II})}{RT} \right| \tag{6.7}$$

When the observed flux ratio $J_{I \to II}/J_{II \to I}$ deviates from the Ussing ratio, the ion is transported against its electrochemical gradient in one direction or the other. This transport, which requires energy, is again called 'active transport'.

The approach based on the Nernst–Planck equation is readily understandable, but has several disadvantages. First, it has been assumed that the forces acting on the membrane behave independently. Secondly, this theory does not take into account that only rarely is a single substance transported, since several substances (e.g. water, ions, metabolites) are normally transported simultaneously and these fluxes may influence each other. Finally, the application of equilibrium thermodynamics presupposes the existence of an equilibrium, which is not the case in systems involving active transport.

Another approach, which does not have these limitations, is based on the thermodynamics of irreversible processes. It has as its starting point the assumption that all fluxes in the system are related to all the driving forces present.

According to Onsager (1931a, b), each flux (J) has a linear relationship to the driving forces, X, viz. $J = LX$. We can thus write the following set of equations for an n-component system.

$$
\begin{aligned}
J_1 &= L_{11}X_1 + L_{12}X_2 + \ldots L_{1n}X_n \\
J_2 &= L_{21}X_1 + L_{22}X_2 + \ldots L_{2n}X_n \\
& \cdot \qquad \cdot \qquad \cdot \\
& \cdot \qquad \cdot \qquad \cdot \\
J_n &= L_{n1}X_1 + L_{n2}X_2 + \ldots L_{nn}X_n
\end{aligned}
\tag{6.8}
$$

Or in a more condensed form,

$$J_i = \sum_{j=i}^{n} L_{ij} X_j \tag{6.9}$$

This means that the flux of each substance can be considered as the sum of a flux due to its own (conjugated) driving force and a flux arising from the operation of the other driving forces present. Equation 6.9 can be written

$$J_i = L_{ii}\Delta\bar{\mu}_i + \sum_{ij}^{n} L_{ij}\Delta\bar{\mu}_j + L_{ir}\Delta\bar{\mu}_r \tag{6.10}$$

The first term is due to the coupling of the flux with its own conjugated force X, which is in this case $\Delta\bar{\mu}_i$, the electrochemical potential gradient. The other coefficients (L_{ij} and L_{ir}) are the *coupling coefficients* determining how the non-conjugated driving forces influence the flux J_i. The non-conjugated driving forces may be related to other fluxes (j) or to chemical reactions in the membrane (r). A very simple definition for active transport based on irreversible thermodynamics has been given by Kedem (1961), who said that when the coupling coefficient between the flux and an energy-consuming reaction (L_{ir}) is not equal to zero, then there is active transport.

The two approaches can lead to different results, as shown by the following example. As we shall see later, in several tissues there is transport of carbohydrates against their concentration gradient, which process seems to require the existence of a Na^+ gradient. Since this carbohydrate transport does not require metabolic energy directly, the coupling coefficient between the transport and an energy-consuming reaction is zero in this case, which means that there is no active transport. However, this form of transport is active in so far as the carbohydrate moves against its concentration gradient. For this reason Wilbrandt (1961) suggested the use of more descriptive terms such as 'uphill' and 'downhill' transport, so avoiding the term 'active' transport.

6.3   SIMPLE DIFFUSION OF NON-ELECTROLYTES

6.3.1   Relation to membrane lipids

Increasing interest in the cell in the second half of the nineteenth century led to the discovery of osmotic behaviour (Nägeli and Cramer, 1855; Pfeffer, 1877; de Vries, 1884). Cells behave as osmometers in the sense that their volume is regulated by the osmotic concentration of the medium. This observation led to the conclusion that cells have a plasma membrane which is easily permeable to water but impermeable or almost so to solutes (Pfeffer, 1877).

The permeability of the plasma membrane for a large series of solutes was investigated by Overton, who measured the speed with which they penetrated plant (Overton, 1895, 1896) and muscle cells (Overton, 1902). He compared the rate of penetration with the hydrophobicity (measured as the distribution coefficient between oil and water) for each substance. He noticed that the more hydrophobic the substance, the faster was its permeability rate. This offered the first indication that lipids play an important role in the permeability characteristics of membranes. Ultimately this observation led Davson and Danielli (1943) to their well-known lipid bilayer model for biological membranes. The studies of Overton were extended by other workers in the first half of this century. For example, Collander and Barlund (1933) found a nearly linear relationship between the oil/water distribution coefficient and the permeability of a great number of substances in large single cells of *Chara ceratophylla*.

The observation that large molecules permeate more slowly than expected on the basis of their lipid solubility indicated that another factor plays a role in permeation, namely the molecular volume of the molecule. This explains why a plot of the product of permeability and square root of molecular weight against the oil/water distribution (*Figure 6.1*) gives a better linear relationship than when permeability alone is plotted.

Further studies of Collander (*see* Wartiovaara and Collander, 1960) were carried out on a large series of plant, fungal, bacterial and yeast cells and the same correlation was generally found for substances with a molecular weight of > 75. Many permeation studies have also been carried out with red blood cells (*see* Sha'afi and Gary-Bobo, 1973). In a recent study Naccache and Sha'afi (1973) tested 90 different substances on human erythrocytes and found a similar dependence on lipid solubility and molecular weight, at least for the larger molecules. In epithelia such as rabbit, goldfish and bullfrog gall bladder (Wright and Diamond, 1969; Hingson and Diamond, 1972), bullfrog choroid plexus (Wright and Prather, 1970) and bullfrog and guinea-pig intestine (Hingson and Diamond, 1972) a similar relationship appears to exist. Another factor which can influence the penetration of larger molecules is their

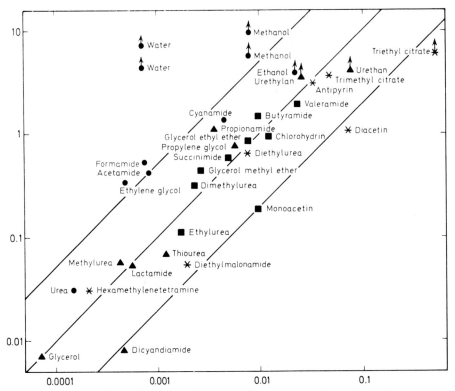

Figure 6.1   Permeability of non-electrolytes across Chara ceratophylla *membrane. The product of the permeability and the square root of the molecular weight is plotted as a function of the oil/ water partition ratio. MR$_D$ is the molar refraction of the molecules depicted, a parameter proportional to the molecular volume. The central line is calculated to be the best fitted one, the other lines represent values which are five times smaller and larger, respectively, (●) MR$_D$ < 15; (▲) MR$_D$ = 15–22; (■) MR$_D$ = 22–30; (*) MR$_D$ > 30. (From Collander, 1949, courtesy of the Scandinavian Society for Plant Physiology)*

degree of chain branching, which alters the rate of penetration through human red cell plasma membranes (Naccache and Sha'afi, 1973) and through gall bladders of several species (Hingson and Diamond, 1972). In the intestine, however, branching has only a minor influence (Hingson and Diamond, 1972), whereas in choroid plexus it has no influence at all (Wright and Prather, 1970). This difference between various tissues is probably a consequence of differing closeness of packing of the membrane lipids.

6.3.2   Pore theory

Another deviation from the Overton rule is offered by polar substances of low molecular weight (*Figure 6.1*). Early this century Ruhland and Hoffman

(1925), in a study of the penetration of the membrane of the sulphur alga *Beggiatoa*, had already shown a direct relation between the size, rather than the hydrophobicity, of the penetrating molecule and its permeability rate. They suggested the presence of pores in the membrane having a diameter large enough for the smaller molecules to penetrate, but too small for the larger molecules.

It was logical that the theories mentioned above should later be combined in the so-called 'lipid–pore theory' (Collander and Barlund, 1933: Höber, 1936, 1945; Wilbrandt, 1938). This theory postulates that the plasma membrane consists of a lipid sheet with pores, so that the permeation of lipid-soluble molecules occurs through the lipid phase and is therefore governed primarily by hydrophobicity, while the penetration of polar molecules occurs through the pores and is therefore primarily regulated by molecular size. The ratio of polar and apolar areas in the membrane would differ for different cells, being low in *Chara* and high in *Beggiatoa*, for example.

Evidence for the existence of aqueous pores has been obtained for many tissues, most extensively for erythrocytes (Solomon, 1968; Sha'afi and Gary-Bobo, 1973). In rabbit gall bladder there is also clear evidence for the presence of such pores (Diamond and Wright, 1969a). However, a comparative study by Hingson and Diamond (1972) showed that penetration through aqueous pores is much less marked in guinea-pig gall bladder and intestine, and in bullfrog choroid plexus, and is nearly absent in goldfish and bullfrog gall bladder. This suggests that there is a large difference in the abundance of pores between various tissues. The nature of these pores will be discussed in more detail in Section 6.6.

6.3.3   Lattice model

As an alternative to the lipid–pore theory, Stein (1962, 1967) has postulated a lattice model. In this model the membrane has a quasi-crystalline structure and the diffusible substances pass through vacant positions in the lattice. According to this theory the permeability P for large molecules is inversely proportional to the radius of the molecule, and thus to the cube root of the molecular weight M. This means that $P.M^{1/3}$ should be a constant. For small molecules, where the molecular radius is of the order of magnitude of a lattice constant, the product of permeability and the square root of the molecular weight $(P.M^{1/2})$ should be constant (*Figure 6.2*). Stein suggested that penetration through a membrane requires first that all hydrogen bonds between the permeant molecule and water be broken. Using the data given in *Figure 6.1*, he plotted the logarithm of $P.M^{1/2}$ as a linear function of the number of hydrogen bonds to be broken. From the slope of this curve he calculated enthalpy changes of 4–5 kcal mol^{-1} per hydrogen bond. This value agrees with the value previously calculated in other ways for the breaking of single hydrogen bonds, suggesting that penetration does indeed require previous breaking of hydrogen bonds. For the penetration of polyhydroxy compounds through artificial lipid membranes (liposomes), de Gier *et al.* (1971) have also calculated enthalpy changes of about 5 kcal per hydrogen bond. In similar experiments Stein (1967) found that there is an enthalpy decrease of 1.5 kcal mol^{-1} per additional CH_2 group for homologous series of alcohols and amides, suggesting that the rate of penetration is enhanced upon lengthening of the carbon chain. After correction for hydrogen bond breaking and for chain length, the value for the permeability coefficient is found to be 100–1000 times smaller than that for a layer of water of the same thickness. This indicates that the permeability of all these substances is primarily determined by the lipid character of the membrane. Although the suggestion that hydrogen bond

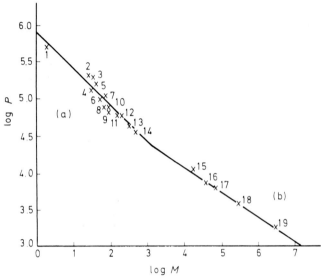

Figure 6.2   The diffusion coefficient D as a function of molecular weight M. Diffusing molecules are as follows: 1, hydrogen; 2, nitrogen; 3, oxygen; 4, methanol; 5, carbon dioxide; 6, acetamide; 7, urea; 8, n-butanol; 9, n-amyl alcohol; 10, glycerol; 11, chloral hydrate; 12, glucose; 13, lactose; 14, raffinose; 15, myoglobin; 16, lactoglobulin; 17, haemoglobin; 18, edestin; 19, erythrocruorin. All at or near 20 °C. (a) The relation $P.M^{1/2}$ constant, (b) the relation $P.M^{1/3}$ constant. (From Stein, 1962, courtesy of Associated Scientific Publishers)

formation plays a role in permeation has been supported (Diamond and Wright, 1969b; Naccache and Sha'afi, 1973), Stein's lattice theory has not met with wide acceptance. Diamond and Wright (1969a) pointed out that the thermodynamic parameters ΔF, ΔH and ΔS, which Stein (1967) interpreted as referring to the transition state for interfacial passage, can also be interpreted as the parameters of partition between solute and membrane, if one assumes that the membrane interior is rate-limiting for most solutes.

6.4   FACILITATED DIFFUSION

6.4.1   Definition

Some substances such as amino acids and carbohydrates are transported much faster through a plasma membrane than would be expected from their structure. These substances have a very weakly lipophilic character and are also too large to be transported through the water-filled pores in the membrane. Moreover, their diffusion does not obey Fick's law in the sense that at increasing concentration gradient the rate of penetration becomes constant (saturation), and that their penetration is not independent of that of chemically similar compounds (competition). In view of the faster rate of penetration than expected for simple diffusion through a lipid barrier or through water-filled pores the term *facilitated diffusion* has been coined for this type of transport (Davson and Reiner, 1942; Stein and Danielli, 1956).

A possible explanation for this type of diffusion is to assume the presence of 'carriers' in the membrane, as was suggested in 1933 by Osterhout. Such a 'carrier' would bind the substrate present on one side of the membrane and release it on the opposite side. The carrier–substrate complex would be more lipid-soluble than the free substrate, and thus would be able to traverse the lipid bilayer. So far no evidence for the existence of movable lipid-soluble carriers has been obtained (Stein, 1967, pp. 295–303). More recently it has been suggested that carriers are membrane proteins which have a binding site for permeant that can be exposed alternately to each membrane face (*see* Kotyk, 1973). LeFevre (1973) recently formulated a more detailed model of this kind for erythrocyte sugar transport. In this model the substrate-binding sites can shift from an introverted to an extraverted state under the influence of substrate. In the latter state substrate binding would be possible, whereas transport of the sugar can occur only when the substrate sites at both sides are in the introverted state.

6.4.2   Kinetics

Although the chemical nature of the 'carriers' is still unknown, the saturation of facilitated diffusion at high concentration gradients has led to kinetic models of facilitated diffusion along the lines of the Michaelis–Menten theory for enzyme activity. This kinetic treatment has been developed by LeFevre (1948) and Widdas (1952) (*see* Wilbrandt and Rosenberg, 1961; Stein, 1967; Kotyk, 1973).

Generally the model shown in *Figure 6.3* is presented. A substrate molecule

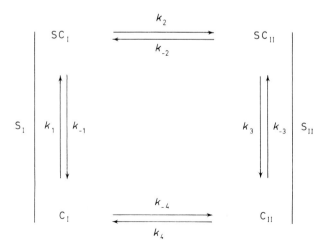

Figure 6.3   General model for facilitated diffusion

S on side I reacts with the carrier C (reaction 1), the carrier–substrate complex SC moves from side I to side II (reaction 2), whereupon the substrate is released (reaction 3) and the empty carrier moves back to side I again (reaction 4). The net flux of substrate S across the membrane will be given by

$$J_s = k_2[SC]_I - k_{-2}[SC]_{II} \tag{6.11}$$

Since the concentrations of the carrier–substrate complexes at each side of the membrane are not known, equation 6.11 does not have any direct practical use. Although it is possible to work out an equation using the eight rate constants, the total concentration of carrier, and the substrate concentrations on each side, this equation is so complicated that it cannot easily be applied and verified (Britton, 1966).

When certain simplifying assumptions are made, it is possible to derive a simple equation, capable of verification. The first assumption is that the dissociation constant K_m of the carrier–substrate complex on side I is equal to that on side II.

$$K_m = \frac{k_{-1}}{k_{+1}} = \frac{k_{+3}}{k_{-3}} = \frac{[S].[C]}{[SC]} \tag{6.12}$$

Secondly, it is assumed that the rate of transport of the loaded carrier is equal to that of the free carrier, and, thirdly, that the rate-limiting step of the transport process is the transport of the carrier–substrate complex across the membrane.

With these assumptions equation 6.11 can be converted into

$$J = \frac{D}{d}([SC]_I - [SC]_{II}) \tag{6.13}$$

in which D is the diffusion coefficient of the carrier–substrate complex and d is the thickness of the membrane. Rearrangement of equation 6.12 and replacement of $[C]$ by $([C]_t - [SC])$, where $[C]_t$ is the total carrier concentration, gives

$$[SC] = \frac{[C]_t.[S]}{K_m + [S]} \tag{6.14}$$

Combination of equations 6.13 and 6.14 gives

$$J = \frac{D.[C]_t}{d} \frac{K_m([S]_I - [S]_{II})}{(K_m + [S]_I)(K_m + [S]_{II})} \tag{6.15}$$

At infinitely high $[S]_I$ and with $[S]_{II}$ being zero,

$$J = J_{max} = \frac{D[C]_t}{d} \tag{6.16}$$

Substitution of equation 6.16 in equation 6.15 gives

$$J = J_{max} \frac{K_m([S]_I - [S]_{II})}{(K_m + [S]_I)(K_m + [S]_{II})} \tag{6.17}$$

which can be rearranged to give

$$J = J_{max} \left(\frac{[S]_I}{[S]_I + K_m} - \frac{[S]_{II}}{[S]_{II} + K_m} \right) \tag{6.18}$$

In experiments with a radioactive substrate, present only at side I, the second term in equation 6.18 can be neglected and equation 6.18 changes to

$$J = J_{max} \frac{[S]_I}{[S]_I + K_m} \qquad (6.19)$$

Since equation 6.19 is of the Michaelis–Menten type, which is valid for enzyme kinetics, it can easily be converted into forms more suitable for quantitative use, such as those given by Lineweaver and Burk (1934) and Hofstee (1952).

The value of K_m, the dissociation constant between carriers and substrate, is a measure for the affinity of the substrate and the carrier in the sense that a high K_m value represents a low affinity. Two extreme conditions can be distinguished: where the substrate concentration is low relative to K_m, and where the reverse is the case. If the substrate concentration is low compared with the K_m value, equation 6.17 approaches the form

$$J = \frac{J_{max}([S]_I - [S]_{II})}{K_m} \qquad (6.20)$$

In this case the flux is proportional to the substrate concentration gradient and thus obeys Fick's law for simple diffusion. In addition, the flux is inversely proportional to K_m, and is high when the affinity of the substrate for the carrier is high.

In the other case ($[S]_I$, $[S]_{II} \gg K_m$), equation 6.17 simplifies to

$$J = J_{max} \frac{K_m([S]_I - [S]_{II})}{[S]_I \cdot [S]_{II}} = J_{max} K_m \left(\frac{1}{S_{II}} - \frac{1}{S_I} \right) \qquad (6.21)$$

In actual measurements of the flux rate initial rates will be determined, where $[S]_I \gg [S]_{II}$. Then the term $1/[S]_I$ becomes negligible and the flux is independent of the substrate concentration offered to the membrane. This is the saturation phenomenon characteristic of facilitated diffusion. While in the previous case the transport rate is inversely proportional to K_m, in this case the rate is directly proportional to K_m, i.e. the flux is low when the affinity of the substrate to the carrier is high. Wilbrandt (1956) has experimentally verified this in human erythrocytes. At 0.03 M concentrations the order of penetration rates is glucose > mannose > galactose > arabinose > sorbose. When high concentrations (1.5 M) of the sugars are used the order of penetration rates is exactly reversed, which is in agreement with the theoretical prediction.

6.4.3   Competition and inhibition

Next to the phenomenon of saturation, another characteristic of facilitated diffusion is the phenomenon of competition, that is, inhibition of the transport of the substrate by a structural analogue. As in enzyme kinetics, there is, in principle, an effect only on the apparent K_m and none on J_{max}. The apparent K_m obtained by means of Lineweaver–Burk or Hofstee plots is related to the value of K_m measured without inhibitor, the inhibitor concentration [I] and the inhibitor–carrier dissociation constant K_i by the formula

$$K_m(\text{apparent}) = K_m\left(1 + \frac{[I]}{K_i}\right) \qquad (6.22)$$

At increasing concentration of inhibitor, the apparent K_m increases. Examples of this type of inhibition are the inhibition by L-phenylalanine of the transport of L-monoiodotyrosine (Nathans, Tapley and Ross, 1960) and the inhibitory action of mannose on transport of galactose (Riklis, Haber and Quastel, 1958) in rat intestine.

The phenomenon of competition between a physiological substrate and its structural analogues suggests that the 'carrier' is multifunctional, being able to transport several structurally related substances. The difference in specificity towards various substances is expressed in different K_m values. For sugar transport in human erythrocytes the sugars have been ordered according to decreasing activity by LeFevre and Marshall (1958). In this sequence 2-deoxy-D-glucose, immediately followed by D-glucose, is the sugar with the highest affinity towards the carrier, whereas L-glucose has the lowest affinity. From this order of affinity LeFevre and Marshall formulated the sugar configuration most effective for binding to the carrier. For other tissues the specificity pattern may be quite different. For intestine Crane (1960) found that a pyranose ring with a free OH group at position 2 is necessary and sufficient; fructose and 2-deoxy-D-glucose are not transported, in contrast to the situation for the erythrocyte.

Non-competitive inhibition is also possible in facilitated diffusion. As in the case of an enzyme, it results in a lowering of J_{max} without changing K_m. The maximal rate obtainable is related to the maximal rate without inhibitor by the equation

$$J_{max}(\text{apparent}) = \frac{J_{max}}{\left(1 + \frac{[I]}{K_i}\right)} \qquad (6.23)$$

Examples of non-competitive inhibitors are sulphydryl reagents, such as N-ethylmaleimide, which blocks glucose transport in human erythrocytes (Dawson and Widdas, 1963), and Hg^{2+} ion, which blocks sugar transport in several tissues (Passow, Rothstein and Clarkson, 1961). This type of inhibition may be either reversible or irreversible.

6.4.4.   Exchange diffusion and counter-transport

There are some special forms of facilitated diffusion which are not covered by the foregoing kinetic treatment, such as exchange diffusion, counter-transport and Na^+-coupled solute transport. *Exchange diffusion* is a form of facilitated diffusion, in which the flux of a substance or a structural analogue is in the opposite direction. Lacko and Burger (1961) have reported an example in the case of the efflux of glucose at $0\,°C$ from human erythrocytes which had previously been loaded with glucose at $37\,°C$. The efflux is markedly enhanced when D-galactose, D-mannose or D-xylose are present in the medium. In the presence of L-sorbose, D-fructose, D-arabinose and D-ribose, however, no increase in the efflux is observed. The extra efflux obtained with D-galactose in the medium is compensated by an equimolar influx of the latter substance. This phenomenon can be explained by assuming that the back-transport of the loaded carrier (k_{-2}) is faster than that of the free carrier (k_4) (*Figure 6.3*). The rate-limiting step is then the back-transport of the loaded carrier. In some cases the value of k_4 may even approach zero. This also explains the finding of LeFevre and McGinnis (1960), that in red blood cells the time for equilibrium of unlabelled glucose is much larger than that for [^{14}C]glucose (*Figure 6.4*). The only difference is that in the experiment with

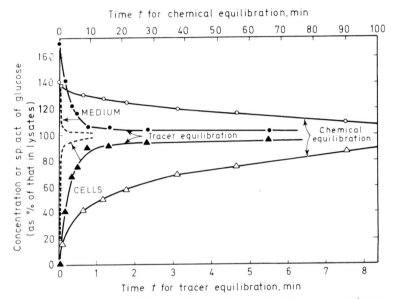

Figure 6.4   Comparative equilibration of tracer glucose and net glucose entry in human erythrocytes. Circles denote medium concentration, triangles cell concentration; open symbols refer to chemical equilibration experiment (time scale at top); solid symbols refer to tracer equilibration experiment (time scale at bottom). Broken curves represent the tracer experiments with the time scale above. (From LeFevre and McGinnis, 1960, courtesy of The Rockefeller University Press)

labelled glucose the cells had been pre-equilibrated with non-radioactive glucose. Thus the presence of glucose at both sides of the membrane enables the carrier to transport [^{14}C]glucose much faster.

Another phenomenon which can be explained by the carrier concept is *counter-transport*. Park *et al.* (1956) observed in rabbit red cells, equilibrated with xylose, a xylose efflux as soon as glucose was added to the medium. Since, prior to the addition of glucose, xylose was in equilibrium across the membrane, the xylose efflux occurs against its own developing concentration gradient. This can be explained without invoking an active-transport system by assuming that glucose and xylose compete for the same sugar carrier at the outside of the membrane, whereas on the inside only xylose is present. This leads to a faster efflux than influx of xylose, the larger glucose gradient driving the xylose against the small xylose gradient.

6.4.5   Sodium-coupled transport

The transport of sugars and of amino acids in a variety of systems (e.g. small intestine, Ehrlich ascites carcinoma, pigeon erythrocytes) has been found to be dependent on the presence of Na^+. The first report of cation-dependent amino acid uptake by erythrocytes is that of Christensen *et al.* (1952), who noticed a reduction in the accumulation of glycine and alanine when half of the external Na^+ was replaced by K^+. Although they first attributed this to the increase in the K^+ concentration rather than to the decreased Na^+ level, this view had to be reversed in the light of later findings. The first report of cation-dependent glucose transport derives from Riklis and Quastel (1958), who studied the uptake of glucose from the guinea-pig intestinal lumen. The

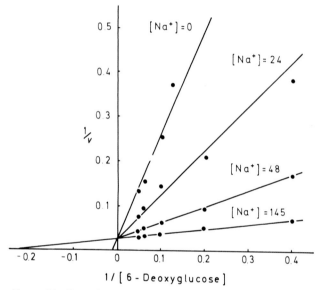

Figure 6.5   Dependence of the rate of uptake of 6-deoxyglucose in hamster intestine on the Na^+ concentration (mM) in the medium. (From Crane, Forstner and Eichholz, 1965, courtesy of Associated Scientific Publishers)

further development in this field has been reviewed extensively by Schultz and Curran (1970).

It is now clear that in all these cases the presence of Na^+ ions, which are actively transported, is essential. On the other hand, the Na^+ dependence of these transport processes is noticeable only when the transported substance moves against its own concentration gradient. In order to explain the Na^+ dependence of these transport processes, a carrier model has been postulated in which, besides the substrate-binding site, a cation-binding site is present on the carrier leading to a ternary complex of carrier, substrate and cation. Evidence for the formation of such a complex has been obtained by Crane, Forstner and Eichholz (1965). The uptake of 6-deoxyglucose in intestinal strips is measured at different concentrations of this carbohydrate and of Na^+ and the K_m is determined (*Figure 6.5*). In the presence of 145 mM Na^+ and 5 mM K^+ a value of 4 mM is found. When the Na^+ is replaced by K^+, the value of K_m increases to 100 mM, whereas the maximal rate of uptake is not affected. This suggests that the presence of a high K^+ concentration, such as that which exists inside the epithelial cells, favours the dissociation of the carrier–substrate complex (*Figure 6.6*).

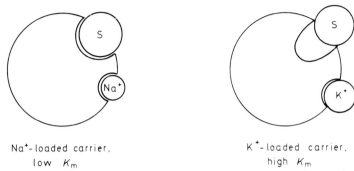

Na⁺-loaded carrier,   K⁺-loaded carrier,

low K_m high K_m

Figure 6.6   A model of the ternary carrier and the influence of Na^+ *and* K^+ *on the affinity for the substrate S. (From Crane, Forstner and Eichholz, 1965, courtesy of Associated Scientific Publishers)*

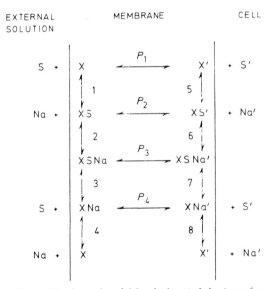

Figure 6.7   General model for the kinetic behaviour of a ternary carrier between Na, solute (S) and binding site (X). (From Schultz and Curran, 1970, courtesy of the American Physiological Society)

The kinetics of this type of carrier-mediated transport can generally be described in Michaelis–Menten terms, where both the permeant and the sodium ion can be considered as substrates. A general model for the behaviour of such a ternary carrier has been postulated by Schultz and Curran (1970), and is shown in *Figure 6.7*. Several possible variants of this model can be distinguished. In the most general case the transport of all forms of the carrier is possible except that the translocation of these forms is rate-limiting. A further condition which can be distinguished is that the formation of either the carrier–substrate complex or the carrier–Na$^+$ complex must occur before formation of the ternary complex is possible. In other variants of the model the rates of translocation of the various carrier–substrate–Na$^+$ complexes are assumed to be different and some translocations, in particular of the free carrier, are taken to be forbidden. In the equations obtained by Schultz and Curran (1970), both the maximal transport rate and the K_m constant are functions of the various rate constants and the concentrations of carrier, substrate and Na$^+$.

The carrier system for glucose transport through the intestinal epithelium is inhibited by phlorizin on the mucosal side, suggesting that the system is located in the brush border. This has been confirmed by Crane (1962): analysis of the sugar content of epithelium, lamina propria, submucosa and muscularis, separated by microdissection, shows a concentration profile with the highest level inside the cell layer, the lowest level external to the brush border and an intermediate level in the lamina propria external to the serosal side of the epithelial cells. This profile has also been demonstrated by means of autoradiography of strips of intestine which had previously been allowed to accumulate [^{14}C]galactose.

The way in which the carrier transport system for the solute is coupled to the sodium pump is discussed in Chapter 7.

6.5   DIFFUSION OF IONS

6.5.1   Diffusion potentials

As observed earlier in this chapter, the driving forces acting on a substance penetrating the membrane are, according to the Nernst–Planck theory, the chemical potential gradient, the electrical potential gradient and the osmotic or hydrostatic pressure gradient. If it is supposed that these forces operate independently of each other, then the flux for the permeating substance is a function of the sum of these forces (equation 6.2). When we are dealing with a single ion in the absence of a pressure gradient, this equation can be simplified to give

$$J = -\omega RTC \left[\frac{1}{C}\frac{dC}{dx} + \frac{zF}{RT}\frac{dE}{dx} \right] \tag{6.24}$$

When the ion is passively distributed across the membrane, influx and efflux are in equilibrium and the net flux will be zero. This will be the case if

$$\frac{1}{C}\frac{dC}{dx} = -\frac{zF}{RT}\frac{dE}{dx} \tag{6.25}$$

or

$$\frac{d\ln C}{dx} = -\frac{zF}{RT}\frac{dE}{dx} \tag{6.26}$$

Rearrangement gives

$$\frac{dE}{dx} = -\frac{RT}{zF}\frac{d\ln C}{dx} \tag{6.27}$$

Integration for C between the limits C_2 and C_1 yields the Nernst equation

$$\Delta E = E_2 - E_1 = -\frac{RT}{zF}\ln\frac{C_2}{C_1} \tag{6.28}$$

where C_1 and C_2 represent the concentrations of the ion in the compartments 1 and 2 located on either side of the membrane. Strictly we should have used activities instead of concentrations, and thus equation 6.28 is only an approximation.

In frog sartorius muscle both K^+ and Cl^- are in approximate flux equilibrium across the membrane (Hodgkin and Horowicz, 1959; Adrian, 1960). In this case the membrane potential can be calculated from the ratio of the concentrations inside and outside according to the equation

$$E \simeq \frac{RT}{F}\ln\frac{[K]_o}{[K]_i} \simeq \frac{RT}{F}\ln\frac{[Cl]_i}{[Cl]_o} \tag{6.29}$$

Thus

$$\frac{[K]_o}{[K]_i} = \frac{[Cl]_i}{[Cl]_o}$$

This equation is not valid for every tissue; for example, in human erythrocytes equation 6.29 is valid only for the Cl^- distribution. Calculation of the Nernst potential from the K^+ distribution gives a potential of $-86\,mV$ (Bernstein, 1954), whereas the measured membrane potential is only $-8\,mV$ (Jay and Burton, 1969). In smooth muscle cells of guinea-pig taenia coli neither the calculated Nernst potential for Cl^- ($-24\,mV$) nor that for K^+ ($-89\,mV$) is in agreement with the measured membrane potential of $-55\,mV$ (Casteels, 1969). Generally, Na^+ is not in flux equilibrium at all. In these cases the Nernst equation, which is valid only for equilibrium conditions, is not applicable.

6.5.2   Goldman–Hodgkin–Katz equation

An alternative relation between the concentrations of the ions and the membrane potential, which does not require flux equilibrium, was worked out by Goldman (1943). For each ion penetrating the membrane a flux equation in the form of equation 6.24 is written and these equations are integrated. In doing this the following assumptions are made: (a) there exists no net current through the membrane; (b) there is electrical neutrality in the entire system, and (c) the electrical gradient across the membrane is linear. Goldman (1943) formulated the membrane potential as a function of the mobilities and the concentrations of the ions in the membrane. This approach was later extended by Hodgkin and Katz (1949), who introduced a permeability coefficient P for each ion

$$P = \frac{\omega RT\beta}{d}$$

where ω is the mobility of the ion, β is the partition coefficient of the ion between membrane and solution and and d is the thickness of the membrane. Assuming that only K^+, Na^+ and Cl^- contribute to the total current over the membrane and that this total current, as posed by Goldman, is zero, the following relation must exist.

$$J_K + J_{Na} = J_{Cl} \tag{6.30}$$

All these assumptions lead to the equation

$$E = \frac{RT}{F} \ln \frac{P_K[K]_o + P_{Na}[Na]_o + P_{Cl}[Cl]_i}{P_K[K]_i + P_{Na}[Na]_i + P_{Cl}[Cl]_o} \tag{6.31}$$

This equation is generally called the Hodgkin–Katz version of the Goldman equation, or, less accurately, the Goldman equation. In many cases Cl^- ions are passively distributed. This means that in those cases the sum of K^+ and Na^+ currents is zero and this leads to the following abbreviated form of the Goldman equation:

$$E = \frac{RT}{F} \ln \frac{P_K[K]_o + P_{Na}[Na]_o}{P_K[K]_i + P_{Na}[Na]_i} \tag{6.32}$$

The Goldman equation can be tested by two different approaches. In the first method the intracellular membrane potential (measured against the extracellular medium) is determined as a function of the extracellular ion concentrations. The intracellular concentration of each ion is determined by correcting the total ion concentration in the tissue for the contribution of the ion present in the extracellular space. By substituting the values of the ion concentrations and the related potentials in equation 6.31, relative values of P_K, P_{Na} and P_{Cl} can be obtained. In the second method radioisotopes of these ions are used. Either the influx rate of the isotope or its efflux rate, measured after a preloading with the isotope, is determined. When the tissue is in a steady state the influx and efflux rates for each ion are expected to be equal and the permeability constants can be calculated from them. In this case, however, Fick's law (equation 6.6) does not apply, since the membrane potential also affects the flux of the charged particle. The following relation, derived from the Goldman–Hodgkin–Katz equation (equation 6.31), must here be applied:

$$J = P.\Delta C \frac{EF/RT}{1 - \exp(-EF/RT)} \tag{6.33}$$

where P is the permeability coefficient and ΔC the concentration difference of the ion and E is the membrane potential.

Equation 6.33 is valid only for the passive flux of an ion. Normally the influx of Na^+ and the efflux of K^+ are passive, while the membrane potential is negative relative to the outer medium. In that case the K^+ efflux is hindered by the membrane potential, while the Na^+ influx is assisted. This means that the factor

$$\frac{EF/RT}{1 - \exp(-EF/RT)}$$

is less than 1 for the efflux of K^+ and greater than 1 for the influx of Na^+.

The Goldman–Hodgkin–Katz equation offers a reasonably accurate description of the behaviour of the membrane potential in a large number of tissues, such as squid nerve (Hodgkin, 1958), papillary muscle (Page, 1962), *Nitella* (MacRobbie and Dainty, 1958), *Chara australis* (Hope and Walker, 1961), gall bladder (Diamond, 1962), rat portal vein (Wahlström, 1973) and toad lens (Duncan, 1969). In smooth muscle cells of taenia coli (Casteels, 1969) and in rat liver (Claret and Mazet, 1972), a large difference between the measured and calculated potential has been found, and is attributed to a contribution of an electrogenic pump to the membrane potential.

6.5.3   Ion permeabilities

The fact that in most tissues the membrane potential is negative and that in some tissues K^+ seems to be passively distributed reflects the fact that the permeability coefficient for K^+ is generally much higher than that for Na^+. In the resting squid nerve, for example, the ratio P_{Na}/P_K is 0.05. This difference in permeability for Na^+ and K^+ is a striking phenomenon. How does a membrane discriminate between two monovalent cations, which seem to differ only in size? Our present knowledge in this field has resulted mainly from studies of minerals, glasses and other non-biological materials. The difference in binding characteristics for monovalent cations has remained unclear for a long time. Jaeger (1929) found that in the mineral ultramarine at 160 °C the binding of the five monovalent cations decreased in the order of increasing non-hydrated ionic radii ($Li^+ > Na^+ > K^+ > Rb^+ > Cs^+$). It appears that the smaller the cation, the closer it can approach the negatively charged centres in the mineral. In another aluminosilicate, permutite, the order of binding is, however, completely opposite (Jenny, 1932). The order which is found here is that of the so-called lyotropic series, which suggests that in this case the hydrated radius, which is largest for the small cations, determines the order of binding. In later investigations, carried out after the development of the ion exchange resins, several other patterns have been found.

Eisenman, Rudin and Casby (1957) have prepared a series of glasses composed of SiO_2, Al_2O_3 and Na_2O in different ratios and have determined the order of preference for different cations. They find that of the 5! ($=120$) theoretical possibilities only the following 11 rank orders actually occur.

I $Cs^+ > Rb^+ > K^+ > Na^+ > Li^+$ VII $K^+ > Na^+ > Rb^+ > Li^+ > Cs^+$

II $Rb^+ > Cs^+ > K^+ > Na^+ > Li^+$ VIII $Na^+ > K^+ > Rb^+ > Li^+ > Cs^+$

III $Rb^+ > K^+ > Cs^+ > Na^+ > Li^+$ IX $Na^+ > K^+ > Li^+ > Rb^+ > Cs^+$

IV $K^+ > Rb^+ > Cs^+ > Na^+ > Li^+$ X $Na^+ > Li^+ > K^+ > Rb^+ > Cs^+$

V $K^+ > Rb^+ > Na^+ > Cs^+ > Li^+$ XI $Li^+ > Na^+ > K^+ > Rb^+ > Cs^+$

VI $K^+ > Na^+ > Rb^+ > Cs^+ > Li^+$

Each rank order is derived from the previous one by a single permutation. The first series is the lyotropic series, whereas the eleventh one is the rank order of the non-hydrated radii. The same orders have been found in biological systems (*see* Diamond and Wright, 1969a). In some cases the place of Li^+ in the rank order differs, which is attributed to the fact that in biological membranes non-coulombic forces play a more important role than in non-biological systems.

6.5.4   Ionic field strength theory

Eisenman (1961, 1962) has offered a thermodynamic explanation for the restriction on the number of rank orders. He suggests that two processes play a role: the interaction of the cations with water and the affinity of the cations for a fixed anionic site. Since for a given cation the former interaction is a constant, only the latter interaction can cause the differences. Eisenman (1961) assumed, as the simplest case, that water molecules cannot approach the anionic site and that the anionic sites are completely separated from each other ($>0.5\,nm$ distance). In that case the internal energy of the coulombic

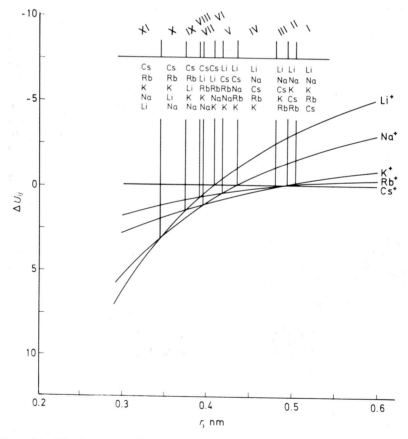

Figure 6.8 ΔU_{ij} (*in* kcal mol^{-1}) *for the cations* Li^+, Na^+, K^+, Rb^+ *and* Cs^+ (i) *relative to* Cs^+ (j) *as a function of the radius (in nm) of the fixed anionic site. Above are given the 11 different selectivity sequences. (From Eisenman, 1961, courtesy of the Publishing House of the Czechoslovak Academy of Sciences)*

force interaction between a cation i^+ and a fixed anionic site x^- is given by the equation

$$\Delta U_{i^+x^-} = -\frac{332}{r_{i^+} + r_{x^-}} \text{ kcal mol}^{-1} \qquad (6.34)$$

where r_{i^+} and r_{x^-} are the radii of the cation and the fixed anionic site, respectively. When a fixed anionic site reacts with two cations i^+ and j^+, the following equilibrium reaction can be formulated.

$$i^+x^- + j^+ \rightleftharpoons |i^+ + j^+x^- + \Delta G_{ij} \qquad (6.35)$$

In this equation ΔG_{ij} is the standard free-energy change of the reaction, which consists of the components

$$\Delta G_{ij} = (G_{ix} - G_{jx}) - (G_{i(H_2O)} - G_{j(H_2O)}) \qquad (6.36)$$

$(G_{ix} - G_{jx})$ represents the difference in free energy for the interaction of each cation with the fixed anionic site, whereas $(G_{i(H_2O)} - G_{j(H_2O)})$ represents the difference in free energy of hydration of the cations i and j. According to this theory, ΔG_{ij} is a function of the radius of the fixed anionic site, and can be obtained from ΔU_{ij} after correction for a small contribution of an entropy term, since $P\Delta V_{ij} \simeq 0$. In *Figure 6.8* ΔU_{ij} is plotted as a function of the radius of the fixed anionic site for the five alkali cations (i), with Cs^+ (j) being taken as point of reference. The various possible sequences can be read from this figure.

The ionic field strength theory is not only valid for monovalent cations but can also be applied for the permeability of the halides and the alkaline earth cations. Of the 24 permutations theoretically possible for the four halides, Eisenman (1965) predicted that only seven will actually occur. In 17 biological systems investigated only these seven sequences have been observed (Diamond and Wright, 1969a). For the alkaline earth cations, Mg^{2+}, Ca^{2+}, Ba^{2+} and Sr^{2+}, the seven sequences theoretically predicted by Sherry (1969) out of 24 permutations have been observed in 45 of 49 non-biological and biological systems (Diamond and Wright, 1969a). This finding offers strong support for the Eisenman theory.

6.5.5   Ion channels

The permeation of ions across a plasma membrane is thought to occur through the aqueous pores of the membrane. Relatively little is known so far about the number and the dimensional and chemical characteristics of these channels. However, for nerve membranes in particular some information is available. In the resting state the membrane is most permeable to K^+. For squid axon the relative permeability ratios $P_K : P_{Na} : P_{Cl}$ are $1 : 0.04 : 0.45$ (Hodgkin and Katz, 1949). Application of the Goldman–Hodgkin–Katz equation immediately explains why the membrane potential of the resting nerve is negative inside. When the nerve is stimulated, a fast and transient depolarization of the plasma membrane occurs, leading within about 1 ms to a positive potential of up to 40 mV. At the same time there is a fortyfold increase in membrane conductance (Cole and Curtis, 1939). The peak value of the positive potential approaches the Nernst potential for sodium, suggesting that an increase in Na^+ permeability has occurred. The Na^+ dependence of the height of the action potential can be confirmed by changing the extra-cellular (Hodgkin and Katz, 1949) and intracellular (Hodgkin, 1964) Na^+ concentrations. Calculations by means of the Goldman–Hodgkin–Katz equation show that at the peak of the action potential the Na^+ permeability must be 20 times as high as the K^+ permeability.

Hodgkin, Huxley and Katz (see Hodgkin, 1964; Noble, 1966) have further analysed the events taking place during nerve stimulation by means of the so-called voltage-clamp technique. In this technique the membrane potential is 'clamped' at any desired value by means of an electronic feed-back circuit, which stops the propagation of the impulse and permits determination of direction and size of the membrane current at the 'clamped' potential. In this way they demonstrated that upon stimulation there is an inward transient current carried by Na^+, followed by a lasting outward current carried by K^+. They advanced strong arguments for the existence of two independent cation channels, a sodium channel and a potassium channel. This assumption has subsequently been confirmed by means of the compounds tetrodotoxin and tetraethylammonium ion. Tetrodotoxin is a toxin from the Japanese puffer fish, which specifically blocks the sodium channels without affecting the potassium channels (Narahashi, Moore and Scott, 1964). The outward current during nerve excitation is specifically abolished by tetraethylam-monium ion (Tasaki and Hagiwara, 1957; Armstrong and Binstock, 1965), which indicates that this substance blocks the potassium channels.

Since tetrodotoxin is so specific for sodium channels, Moore, Narahashi and Shaw (1967) have used it as a quantitative marker for the sodium channels in lobster nerve membranes. Assuming that one molecule of tetrodotoxin binds per channel, they came to the surprisingly low number of 13 sodium channels per μm^2. This indicates that the sodium channels are widely spaced in the nerve membrane, roughly 1 per 300 000 membrane lipid molecules. This latter fact illustrates that the channels must have very specific configura-tions.

Sodium channels also exist in non-excitatory membranes, like that of frog skin epithelium. These sodium channels are not blocked by tetrodotoxin, but in this case the drug amiloride inhibits Na^+ diffusion. Cuthbert (1973) has determined with this drug that there are 400 sodium channels per μm^2 of frog skin, each able to pass 8000 Na^+ ions per second. This shows the tremen-dous capacity of the nerve sodium channels, which pass about 2000 Na^+ ions per impulse, the impulse lasting only about 1 ms.

The existence of separate sodium and potassium channels in the nerve membrane raises the question of the configuration and dimensions of these channels. This problem has been attacked by Hille (1971, 1972, 1973). In order to study the sodium channels, he blocked the potassium channels of a frog myelinated nerve by means of tetraethylammonium ion. Under those con-ditions the ionic currents during stimulation consist entirely of ions passing through the sodium channels. The ionic current for a series of organic (Hille, 1971) and metal (Hille, 1972) cations, present as the only cation, was measured. From these measurements Hille derived the relative permeability P_X/P_{Na} for each cation X. The resulting values of the relative permeabilities are given in Table 6.1. Two factors determining the permeability can be distin-guished. First, the size of the cation: from the van der Waals dimensions of

the cations Hille concluded that the channel must have an effective area of
0.31×0.51 nm. There is, however, a second factor, since some ions containing
an amino or hydroxyl group are more readily permeable than somewhat
smaller molecules containing a methyl group. This observation led Hille
(1971) to the conclusion that the channel wall contains six strategically
located oxygen atoms, which can form hydrogen bonds with amino or
hydroxyl groups, thereby favouring the penetration of cations with such
groups. The formation of hydrogen bonds would make the area of these molecules effectively smaller as compared with molecules of the same size containing a methyl group. From the effect of pH on the ionic current he concluded
that two of the oxygen atoms would represent an ionized carboxylic acid

Table 6.1 PERMEABILITY RATIOS FOR CATIONS IN CATION CHANNELS OF
FROG MYELINATED FIBRES. (From Hille, 1971, 1972, 1973)

Ion	Sodium channel, P_X/P_{Na}	Potassium channel, P_X/P_K
Na^+	1	<0.01
K^+	0.086	1
Li^+	0.93	<0.018
Tl^+	0.33	2.3
Rb^+	<0.012	0.31
Cs^+	<0.013	<0.077
Ca^{2+}	<0.11	nd
Mg^{2+}	<0.10	nd
Hydroxylamine	0.94	<0.025
Hydrazine	0.59	<0.029
Ammonium	0.16	0.13
Formamidine	0.14	<0.020
Guanidine	0.13	<0.013
Hydroxyguanidine	0.12	nd
Aminoguanidine	0.06	nd
Methylamine	<0.007	<0.021

Abbreviation—nd: not determined.

The following cations were reported to be tested only for the sodium channel and were nearly
impermeable: N-methylhydroxylamine, methylhydrazine, acetamidine, methylguanidine,
dimethylamine, tetramethylammonium, tetraethylammonium, ethanolamine, choline, tris(hy-
droxymethyl)aminomethane, imidazole, biguanidine, triaminoguanidine.

group with a pK_a of 5.5 (Hille, 1968). Of the metal cations only Li^+, Tl^+
and K^+ are permeable, the channel being impermeable towards Rb^+, Cs^+,
Ca^{2+} and Mg^{2+} (*Table 6.1*; *see* Hille, 1972). He postulated that metal cations
penetrate in the partly hydrated form. Comparing the properties of the
sodium channel with the 11 possibilities predicted by Eisenman (1961) (*see*
p. 133), the sodium channel behaves as permeability sequence X. There is
also a very good qualitative agreement between the permeability properties
of the sodium channel and that of a well-investigated kind of glass, NAS 11–
18 F (Eisenman, 1965).

The properties of the potassium channel have been studied similarly, using
tetrodotoxin in the medium to block the sodium channel (Hille, 1973). The
potassium channel is permeable to Rb^+ but less so to NH_4^+. An interesting
property of the potassium channel is its high permeability to Tl^+ (2.3 times the
permeability for K^+). This cation can also replace K^+ in the Na^+, K^+-ATPase

system with a 5–10 times lower half-maximal activating concentration (Britten and Blank, 1968). The potassium channel is nearly impermeable to the tested organic cations, which led Hille (1973) to suggest that the narrowest part of the potassium channel in nerve should be a circle with a diameter of 0.3 nm. The low permeability of this channel for Na^+ and Li^+, which have a smaller diameter than K^+, cannot easily be explained. Bezanilla and Armstrong (1972) suggested that an interaction of the cation with dipoles in the wall of the channel is necessary for permeation. The interaction of a Na^+ ion (diam. 0.19 nm) with dipoles of water molecules would be larger than that of K^+ (diam. 0.27 nm), so that the latter could better react with the dipoles in the channel wall. Since the specific blocking agent tetraethylammonium ion has a diameter of 0.8 nm, the potassium channel is probably wider at the outside than at the narrowest point.

A crucial problem, which still remains unsolved, is how the rapid increase and decrease in Na^+ permeability during nerve excitation is brought about. In the case of the photoreceptor membrane of vertebrate rods and cones and invertebrate rhabdomes, where changes in Na^+ permeability play a role in the excitation mechanism, evidence is accumulating that Ca^{2+} ions regulate the permeability to Na^+ (Hagins, 1972).

6.6   TRANSPORT OF WATER

6.6.1   Osmotic and tracer diffusion methods

The permeability of water can be determined in two different ways: the osmotic method and the tracer diffusion method. In the first method the rate of swelling or shrinkage of a cell under the influence of an osmotic gradient due to the presence of an impermeable solute is measured. In the second method a labelled form of water, such as ^3HHO, is used to measure the rate of water transfer into or out of the cell. The values for water permeability obtained with the first method are generally higher than those found using the second method. When both methods are measuring the diffusion rate, the values obtained by the two methods should be equal.

An attempt to explain the inequality of the two permeabilities was made by Hodgkin and Keynes (1955) by means of the concept of the 'long narrow pore'. They proposed that the pore is sufficiently long to contain several water molecules but so narrow that these molecules cannot pass each other. Transfer of water would occur by collisions at either side of the pore. When there is an osmotic gradient, there is a higher water concentration on one side, leading to more collisions on that side and thus to transfer of water to the opposite side. In the isotope method the number of collisions on the two sides will be the same, since there is no osmotic gradient. Thus the tracer molecule can be knocked back into the pore. This explains why the tracer transfer is slower than the net water transfer in the previous case. This model has been put in mathematical form by Lea (1963) and Heckmann (1972), but in order to explain the permeability difference in the two methods, it requires pore lengths of only 0.6–1.5 nm, whereas the membrane thickness is about 7.5 nm.

The results of the two methods can also be combined in another way, which yields information about the water pores. It is reasonable to assume that in osmotic water transport, water moves through the pores in a laminar flow (Pappenheimer, Renkin and Borrero, 1951; Koefoed-Johnsen and Ussing, 1953). In that case the flow is governed by Poisseuille's law

$$\frac{dV}{dt} = \frac{-n\pi r^4}{8\eta} \cdot \frac{\Delta P}{\Delta x}$$

(6.37)

where n is the number of pores, r the mean pore radius and Δx the mean length of the pores η the viscosity of water in the pore and ΔP the osmotic pressure gradient. In the case of the isotope method there is no osmotic pressure difference and the diffusion rate is given by Fick's equation with $n\pi r^2$ as the total pore surface area

$$\frac{dV}{dt} = -D.n\pi r^2 \cdot \frac{\Delta C}{\Delta x}$$

(6.38)

where ΔC is the isotope concentration difference and D the diffusion coefficient of water.

Combination of equations 6.37 and 6.38 yields a value for r, the pore radius. Thus dividing 6.37 by 6.38 gives the expression

$$r^2 = 8\eta D. \frac{\Delta C}{\Delta P}\alpha$$

(6.39)

where α is the ratio between the osmotic and diffusional transfer rate. However, since the dimensions of the permanent molecules are of the same magnitude as those of the pores, corrections must be made for friction between the water molecules and the cylinder wall (Pappenheimer, Renkin and Borrero, 1951; Renkin, 1954). Solomon and co-workers (Paganelli and Solomon, 1957; Sidel and Solomon, 1957) have calculated in this way a pore radius of 0.35 nm for human erythrocytes. They use the term 'equivalent pore radius' in order to indicate that these measurements and calculations do not actually prove the existence of such pores. In this respect it is interesting to note that addition of the polyene antibiotic nystatin to lipid bilayers increases not only the permeability to water but also the ratio of osmotic to diffusional permeability. This suggests that the antibiotic causes pores to form in the lipid bilayer (Holtz and Finkelstein, 1970). In some biological membranes, such as that of *Alga valonia*, the ratio of osmotic to diffusional permeability is 1, suggesting that in this membrane no aqueous pores are present (Gutknecht, 1968).

6.6.2   Correction for unstirred layers

Another, at least partial, explanation for the inequality of the two permeabilities is the occurrence of an unstirred layer on the outside surface of the membrane. Dainty (1963) has indicated that the existence of such an unstirred layer would reduce the isotope permeation during water transfer experiments, but not affect the osmotic water permeability. This was shown to be indeed the case for plant cells (Dainty and House, 1966), and also for lipid bilayers (Cass and Finkelstein, 1967). In the latter case the two permeabilities become equal after correction. A very prominent effect of the unstirred layer has been reported by van Os and Slegers (1973) for rabbit gall bladder, where the ratio between osmotic and diffusional permeability is reduced from 26.3 to 3.1 upon correcting for unstirred water layers of some 900 μm in width. In human erythrocytes the thickness of the unstirred layer does not exceed 5.5. μm, which means that it has only a small effect (6 percent) on the diffusional permeability (Sha'afi et al., 1967). With the aid of this correction and the use of more accurate data for diffusional permeability (Barton and Brown, 1964), an equivalent pore radius of 0.45 nm has been obtained for human erythrocytes.

6.6.3   Use of reflection coefficient

Another method for the measurement of the pore diameter has been developed by means of the concept of the reflection coefficient. When a membrane is perfectly semipermeable, no substance other than water is able to permeate. However, since in a biological membrane pores are present with a diameter about three times as large as that of a water molecule, molecules larger than water but with a diameter smaller than that of the pore will also permeate the membrane. Staverman (1951) has introduced the 'reflection coefficient', σ, which is defined as the ratio between the observed osmotic pressure and the theoretical osmotic pressure, assuming that the membrane is perfectly semipermeable and no volume flow is present. For a substance which cannot permeate the membrane, $\sigma = 1$, whereas for substances with the same diameter as water, $\sigma = 0$. For all other substances with a diameter smaller than that of the pore and larger than water the reflection coefficient is between 0 and 1. Durbin, Frank and Solomon (1956) derived the equation

$$1 - \sigma = A_s/A_w \tag{6.40}$$

where A_s and A_w represent the effective pore areas for filtration of solute and water respectively. Goldstein and Solomon (1960) plotted $(1 - \sigma)$ as a function of the radius of the penetrating molecules for a range of pore diameters in human erythrocytes (*Figure 6.9*). With this method they arrived at a radius of 0.42 nm for the pores in these cells.

Since the reflection coefficient is defined under zero volume flow, equation 6.40 does not hold for actual experiments. Dainty and Ginzburg (1963) have advanced the following modified form for the case in which solute passage is restricted to the pores:

$$1 - \sigma - \frac{\omega \bar{V}_s}{L_p} = \frac{A_s}{A_w} \tag{6.41}$$

where ω is the permeability coefficient for the solute, $\bar{V}_s$ the partial molar volume of the solute and L_p the osmotic permeability of the membrane. This extra factor represents the contribution of the solute flow to the total volume flow, and the correction factor is small when the solute moves only through the pores. Recalculation of the values of Goldstein and Solomon (1960) with equation 6.41 instead of 6.40 leads to a pore radius of 0.43 nm instead of 0.42 nm (Solomon, 1968). This value is very close to the value of 0.45 nm, obtained from a comparison of the osmotic and diffusional permeabilities (Sha'afi et al., 1967).

By means of these techniques the pores in various tissues have been determined. In squid axon the equivalent pore radius is increased from 0.47 nm to 0.62 nm upon stimulation (Villegas, Bruzual and Villegas, 1968). In the outer surface of toad skin the equivalent pore radius increases from 0.45 nm

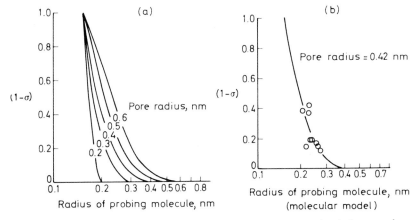

Figure 6.9 $(1 - \sigma)$ as a function of the radius of the penetration molecule for a number of pore radii (a) and a pore radius of 0.42 nm, which fits best with the measured values (b). (From Goldstein and Solomon, 1960, courtesy of The Rockefeller University Press)

to 0.65 nm upon treatment with the hormone vasopressin, which greatly increases its permeability to water (Koefoed-Johnsen and Ussing, 1953; Whittembury, 1962). However, in toad bladder there is no increase in equivalent pore radius after treatment with vasopressin, when the values are corrected for unstirred-layer effects (Hays, Harkness and Franki, 1970). Hays and co-workers postulated that the increase in the permeability of the luminal membrane is due to an increase in the number of pores rather than an increase in the diameter of the pores.

6.6.4   Reality of pores

Notwithstanding these findings and the quantitative agreement between pore radii obtained in two different ways, the existence of aqueous pores in the membrane and their function in the permeability of small molecules have been questioned (Stein, 1967; Macey and Farmer, 1970). Stein's criticism is based mainly on his finding that calculation of pore radii in human erythrocytes from the ratio of water and solute permeabilities leads for all solutes to pore radii smaller than 0.35 nm. However, this use of permeability data from two entirely different techniques (haemolysis, and time course of cell volume change in the stop-flow method) is of dubious validity (Sha'afi, Gary-Bobo and Solomon, 1971). Macey and Farmer (1970) reported that in human red blood cells the permeability of urea, methylurea and glycerol is inhibited by phloretin, while the permeability of water, 1,3-dimethylurea and other substances is not affected. Macey and Farmer (1970) proposed that urea, methylurea and glycerol are transported by a phloretin-sensitive facilitated diffusion system and that the use of these substances for the determination of the pore magnitude is not permissible. The presence of such a system in toad bladder, which is stimulated by vasopressin, has also been reported (Levine, Franki and Hays, 1973). Owen and Solomon (1972) have confirmed that the permeability of some hydrophilic substances in human red blood cells is decreased, but they found that the permeability of some lipophilic substances is increased. They interpreted their finding by assuming that phloretin reacts allosterically with some membrane protein, resulting in a conformational change leading secondarily to the observed changes in permeability. Another objection to the pore theory results from the fact that recently much higher values for osmotic permeability have been found and thus the validity of previously determined values has been questioned (House, 1974).

Although these findings indicate that the absolute values of pore diameter must be taken with reservation (Sha'afi and Gary-Bobo, 1973), the permeability behaviour of small molecules is so different from that of large lipophilic molecules that the presence of polar areas in the membrane is still the best explanation for this phenomenon.

REFERENCES

ADRIAN, R. H. (1960). *J. Physiol., Lond.*, **151**:154.
ARMSTRONG, C. M. and BINSTOCK, L. (1965). *J. gen. Physiol.*, **48**:859.
BARTON, T. C. and BROWN, D. A. J. (1964). *J. gen. Physiol.*, **47**:839.
BERNSTEIN, R. E. (1954). *Science, N. Y.*, **120**:459.
BEZANILLA, F. and ARMSTRONG, C. M. (1972). *J. gen. Physiol.*, **60**:588.
BRITTEN, J. B. and BLANK, M. (1968). *Biochim. biophys. Acta*, **159**:160.
BRITTON, H. G. (1966). *J. theor. Biol.*, **10**:28.
CASS, A. and FINKELSTEIN. A. (1967). *J. gen. Physiol.*, **50**:1765.
CASTEELS, R. (1969). *J. Physiol., Lond.*, **205**:193.
CEREIJIDO, M. and ROTUNNO, C. A. (1970). *Introduction to the Study of Biological Membranes.* New York; Gordon & Breach.
CHRISTENSEN, H. N., RIGGS, T. R., FISHER, H. and PALATINE, M. (1952). *J. biol. Chem.*, **198**:17.
CLARET, M. and MAZET, J. L. (1972). *J. Physiol., Lond.*, **223**:279.
COLE, K. S. and CURTIS, H. J. (1939). *J. gen. Physiol.*, **22**:649.
COLLANDER, R. (1949). *Physiologia Pl.*, **2**:300.
COLLANDER, R. and BARLUND, H. (1933). *Acta bot. fenn.*, **11**:1.
CONTI, F. and EISENMAN, G. (1965). *Biophys. J.*, **5**:511.
CRANE, R. K. (1960). *Physiol. Rev.*, **40**:789.
CRANE, R. K. (1962). *Fedn Proc. Fedn Am. Socs exp. Biol.*, **21**:891.
CRANE, R. K., FORSTNER, G. and EICHHOLZ, A. (1965). *Biochim. biophys. Acta*, **109**:467.
CUTHBERT, A. W. (1973). *Biochem. Soc. Trans.*, **1**:105.
DAINTY, J. (1963). *Adv. Bot. Res.*, **1**:179.
DAINTY, J. and GINZBURG, B. Z., (1963). *J. theor. Biol.*, **5**:256.
DAINTY, J. and HOUSE, C. R. (1966). *J. Physiol., Lond.*, **182**:66.
DAVSON, H. and DANIELLI, J. F. (1943). *The Permeability of Natural Membranes.* Cambridge, England; Cambridge University Press.
DAVSON, H., and REINER, J. M. (1942). *J. cell. comp. Physiol.*, **20**:325.
DAWSON, A. C. and WIDDAS, W. F. (1963). *J. Physiol., Lond.*, **168**:644.
DE GIER, J., MANDERSLOOT, J. G., HUPKES, L. V., MCELHANEY, R. N. and VAN BEEK, W. P. (1971). *Biochim. biophys. Acta*, **233**:610.
DE VRIES, H. (1884). *Jb. wiss. Bot.*, **14**:427.
DIAMOND, J. M. (1962). *J. Physiol., Lond.*, **161**:474.
DIAMOND, J. M. and WRIGHT, E. M. (1969a). *A. Rev. Physiol.*, **31**:581.
DIAMOND, J. M. and WRIGHT, E. M. (1969b). *Proc. R. Soc., Ser. B.*, **172**:273.
DUNCAN, G. (1969). *Expl Eye Res.*, **8**:315.
DURBIN, R. P., FRANK, H. and SOLOMON, A. K. (1956. *J. gen. Physiol.*, **39**:535.
EINSTEIN, A. (1905). *Annln Phys.*, **17**:549.

EISENMAN, G. (1961). *Membrane Transport and Metabolism* pp. 163–179. Ed. A. KLEINZELLER and A. KOTYK. New York; Academic Press.

EISENMAN, G. (1962). *Biophys. J.*, Part 2, **2**:259.

EISENMAN, G. (1965). *Adv. analyt. Chem. Instrumn.*, **4**:213.

EISENMAN, G., RUDIN, D. O. and CASBY, J. U. (1957). *Science, N.Y.*, **126**:831.

GOLDMAN, D. E. (1943). *J. gen. Physiol.*, **27**:37.

GOLDSTEIN, D. A. and SOLOMON, A. K. (1960). *J. gen. Physiol.*, **44**:1.

GUTKNECHT, J. (1968). *Biochim. biophys. Acta*, **163**:20.

HAGINS, W. A. (1972). *A. Rev. Biophys. Bioengng*, **1**:131.

HARRIS, E. J. (1972). *Transport and Accumulation in Biological Systems*, 3rd edn. London; Butterworths.

HAYS, R., HARKNESS, S. H. and FRANKI, N. (1970). *Urea and the Kidney*, pp. 149–156. Ed. B. SCHMIDT-NIELSEN. Amsterdam; Excerpta Medica.

HECKMANN, K. (1972). *Biomembranes*, Vol. 3, pp. 127–153. Ed F. KREUZER and J. F. G. SLEGERS. New York; Plenum Press.

HILLE, B. (1968). *J. gen. Physiol.*, **51**:221.

HILLE, B. (1971). *J. gen. Physiol.*, **58**:599.

HILLE, B. (1972). *J. gen. Physiol.*, **59**:637.

HILLE, B. (1973). *J. gen. Physiol.*, **61**:669.

HINGSON, D. J. and DIAMOND, J. M. (1972). *J. Membrane Biol.*, **10**:93.

HÖBER, R. (1936). *Physiol. Rev.*, **16**:52.

HÖBER, R. (1945). *Physical Chemistry of Cells and Tissues*. London; Churchill.

HODGKIN, A. L. (1958). *Proc. R. Soc., Ser. B*, **148**:1.

HODGKIN, A. L. (1964). *The Conduction of the Nervous Impulse*. Springfield, Illinois; Charles C. Thomas.

HODGKIN, A. L. and HOROWICZ, P. (1959). *J. Physiol., Lond.*, **145**:432.

HODGKIN, A. L. and KATZ, B. (1949). *J. Physiol., Lond.*, **108**:37.

HODGKIN, A. L. and KEYNES, R. D. (1955). *J. Physiol., Lond.*, **128**:61.

HOFSTEE, B. H. J. (1952). *Science, N.Y.*, **116**:329.

HOLTZ, R. and FINKELSTEIN, A. (1970). *J. gen. Physiol.*, **56**:125.

HOPE, A. B. and WALKER, N. A. (1961). *Aust. J. biol. Sci.*, **14**:26.

HOUSE, C. R. (1974). *Water Transport in Cells and Tissues*. London; Arnold.

JAEGER, F. M. (1929). *Trans. Faraday Soc.*, **25**:320.

JAY, A. W. L. and BURTON, A. C. (1969). *Biophys. J.*, **9**:115.

JENNY, H. (1932). *J. phys. Chem.*, **36**:2217.

KEDEM, O. (1961). *Membrane Transport and Metabolism*, pp. 87–93. Ed. A. KLEINZELLER and A. KOTYK. London; Academic Press.

KOEFOED-JOHNSEN, V. and USSING, H. H. (1953). *Acta physiol. scand.*, **28**:60.

KOTYK, A. (1973). *Biochim. biophys. Acta*, **300**:183.

KOTYK, A. and JANÁČEK, K. (1970). *Cell Membrane Transport; Principles and Techniques*. New York; Plenum Press.

LACKO, L. and BURGER, M. (1961). *Nature, Lond.*, **191**:881.

LAKSHMINARAYANAIAH, N. (1969). *Transport Phenomena in Membranes*. New York; Academic Press.

LEA, E. J. A. (1963). *J. theor. Biol.*, **5**:102.

LEFEVRE, P. G. (1948). *J. gen. Physiol.*, **31**:505.

LEFEVRE, P. G. (1973). *J. Membrane Biol.*, **11**:1.

LEFEVRE, P. G. and MARSHALL, J. K. (1958). *Am. J. Physiol.*, **194**:333.

LEFEVRE, P. G., and MCGINNIS, G. F. (1960). *J. gen. Physiol.*, **44**:87.

LEVINE, S., FRANKI, N. and HAYS, R. M. (1973). *J. clin. Invest.*, **52**:1435.

LINEWEAVER, H. and BURK, D. (1934). *J. Am. chem. Soc.*, **56**:658.

MACEY, R. I. and FARMER, R. E. L. (1970). *Biochim. biophys. Acta*, **211**:104.

MACROBBIE, E. A. C. and DAINTY, J. (1958). *J. gen. Physiol.*, **42**:335.

MOORE, J. W., NARAHASHI, T. and SHAW, T. I. (1967). *J. Physiol., Lond.*, **188**:99.

NACCACHE, P. and SHA'AFI, R. I. (1973). *J. gen. Physiol.*, **62**:714.

NÄGELI, C. and CRAMER, C. (1855). *Pflanzenphysiologische Untersuchungen*. Zürich; Schultess.

NARAHASHI, T., MOORE, J. W. and SCOTT, W. R. (1964). *J. gen. Physiol.*, **47**:965.

NATHANS, D., TAPLEY, D. F. and ROSS, J. E. (1960). *Biochim. biophys. Acta*, **41**:271.

NOBLE, D. (1966). *Physiol. Rev.*, **46**:1.

ONSAGER, L. (1931a). *Phys. Rev.*, **37**:405.

ONSAGER, L. (1931b). *Phys. Rev.*, **38**:2265.

OSTERHOUT, W. J. V. (1933). *Ergebn. Physiol.*, **35**:967.
OVERTON, E. (1895). *Vjschr. naturf. Ges. Zürich*, **40**:159.
OVERTON, E. (1896). *Vjschr. naturf. Ges. Zürich*, **41**:388.
OVERTON, E. (1902). *Pflügers Arch. ges. Physiol.*, **92**:115.
OWEN, J. D. and SOLOMON, A. K. (1972). *Biochim. biophys. Acta*, **290**:414.
PAGANELLI, C. V. and SOLOMON, A. K. (1957). *J. gen. Physiol.*, **41**:259.
PAGE, E. (1962). *J. gen. Physiol.*, **46**:189.
PAPPENHEIMER, J. R., RENKIN, E. M. and BORRERO, L. M. (1951). *Am. J. Physiol.*, **167**:13.
PARK, C. R., POST, R. L., KALHAN, C. F., WRIGHT, J. H., JOHNSON, L. H. and MORGAN, H. E. (1956). *Ciba Fdn Colloq. Endocr.*, **9**:240.
PASSOW, H., ROTHSTEIN, A. and CLARKSON, T. W. (1961). *Pharmac. Rev.*, **13**:185.
PFEFFER, W. F. P. (1877). *Osmotische Untersuchungen.* Leipzig; Engelmann.
RENKIN, E. M. (1954). *J. gen. Physiol.*, **38**:225.
RIKLIS, E. and QUASTEL, J. H. (1958). *Can. J. Biochem. Physiol.*, **36**:347.
RIKLIS, E., HABER, B. and QUASTEL, J. H. (1958). *Can. J. Biochem. Physiol.*, **36**:373.
RUHLAND, W. and HOFFMANN, C. (1925). *Planta*, **1**:1.
SCHOFFENIELS, E. (1967). *Cellular Aspects of Membrane Permeability.* Oxford; Pergamon Press.
SCHULTZ, S. G. and CURRAN, P. F. (1970). *Physiol. Rev.*, **50**:637.
SHA'AFI, R. I. and GARY-BOBO, C. M. (1973). *Prog. Biophys. molec. Biol.*, **26**:103.
SHA'AFI, R. I., GARY-BOBO, C. M., and SOLOMON, A. K. (1971). *J. gen. Physiol.*, **58**:238.
SHA'AFI, R. I., RICH, G. T., SIDEL, V. W., BOSSERT, W. and SOLOMON, A. K. (1967). *J. gen. Physiol.*, **50**:1377.
SHERRY, H. S. (1969). *Ion Exch.*, **2**:89.
SIDEL, V. W. and SOLOMON, A. K. (1957). *J. gen. Physiol.*, **41**:243.
SOLOMON, A. K. (1968). *J. gen. Physiol.*, **51**:335s.
STAVERMAN, A. J. (1951). *Recl Trav. chim. Pays-Bas Belg.*, **70**:344.
STEIN, W. D. (1962). *Comprehensive Biochemistry*, Vol. 2, pp. 283–308. Ed. M. FLORKIN and E. H. STOTZ. Amsterdam; Elsevier.
STEIN, W. D. (1967). *The Movement of Molecules across Cell Membranes.* New York; Academic Press.
STEIN, W. D. and DANIELLI, J. F. (1956). *Discuss. Faraday Soc.*, **21**:238.
TASAKI, I. and HAGIWARA, S. (1975). *J. gen. Physiol.*, **40**:859.
TEORELL, T. (1953). *Prog. Biophys. biophys. Chem.*, **3**:305.
TROSHIN, A. S. (1966). *Problems of Cell Permeability.* Oxford: Pergamon Press.
USSING, H. H. (1949a). *Acta physiol. scand.*, **17**:1.
USSING, H. H. (1949b). *Acta physiol. scand.*, **19**:43.
VAN OS, C. H. and SLEGERS, J. F. G. (1973). *Biochim. biophys. Acta*, **291**:197.
VILLEGAS, R., BRUZUAL, I. B. and VILLEGAS, G. (1968). *J. gen. Physiol.*, **51**:81s.
WAHLSTRÖM, B. A. (1973). *Acta physiol. scand.*, **89**:436.
WARTIOVAARA, V. and COLLANDER, R. (1960). *Protoplasmatologia: Handbuch der Protoplasmaforschung*, Vol. II, C 8d, pp. 1–98. Vienna; Springer-Verlag.
WHITTEMBURY, G. (1962). *J. gen. Physiol.*, **46**:117.
WIDDAS, W. F. (1952). *J. Physiol., Lond.*, **118**:23.
WILBRANDT, W. (1938). *Ergebn. Physiol.*, **40**:204.
WILBRANDT, W. (1956). *J. cell. comp. Physiol.*, **47**:137.
WILBRANDT, W. (1961). *Membrane Transport and Metabolism*, pp. 341–342. Ed. A. KLEINZELLER and A. KOTYK. London; Academic Press.
WILBRANDT, W. and ROSENBERG, T. (1961). *Pharmac. Rev.*, **13**:109.
WRIGHT, E. M. and DIAMOND, J. M. (1969). *Proc. R. Soc., Ser. B*, **172**:227.
WRIGHT, E. M. and PRATHER, J. W. (1970). *J. Membrane Biol.*, **2**:127.

7

Active transport

S. L. Bonting and J. J. H. H. M. de Pont
Department of Biochemistry, University of Nijmegen, Nijmegen, The Netherlands

7.1   INTRODUCTION: PUMPS, CARRIERS, PERMEASES

In Chapter 6 the phenomenon of passive movement of molecules and ions across membranes was discussed. Such movement follows a concentration gradient in the case of uncharged molecules or an electrochemical gradient in the case of ions. It is a 'downhill' transport. More exactly, it is transport not coupled to an input of metabolic energy.

The living organism requires, for many of its crucial processes, another type of transport, in which substances are transported 'uphill' against an electrochemical gradient. This type of transport requires energy, and is called *active transport*. The glomerular filtrate of the kidney contains glucose at approximately the plasma concentration, whereas the urine rarely contains any. This indicates that glucose must be reabsorbed in the renal tubules against its concentration gradient. In fact, micropuncture experiments on the amphibian nephron have shown that the glucose concentration falls to zero by the end of the proximal tubule. Likewise, intestinal absorption of glucose would be extremely wasteful and inefficient if it were a passive transport, since this would require the intestinal glucose concentration to be at all times greater than that of the plasma.

Another important phenomenon is the existence of cation gradients across the plasma membrane of most animal cells: the high K^+ and low Na^+ levels inside the cell against the low K^+ and high Na^+ levels of plasma and tissue fluid. The application of radioactive Na^+ showed that under equilibrium conditions there is a continuous influx of Na^+ into the red blood cell (Cohn and Cohn, 1939), suggesting that there must be an active efflux balancing a passive influx of Na^+. This idea was confirmed by the observation that the cation gradients are abolished by incubation at $0\,°C$ and are restored upon incubation at $37\,°C$ in glucose-containing saline solution, but not in the absence of glucose (Harris, 1941). These active transport processes will be

145

the subject of this chapter. The main emphasis will be on the molecular mechanism of active transport.

In reading the literature on the subject one encounters three terms: 'pump', 'carrier' and 'permease'. The term *pump* refers entirely to active transport; it denotes a system embedded in the cell membrane which is able to use cellular energy for the transport of an ion or molecule across the membrane against its electrochemical gradient. The term was coined at a time when no chemical knowledge of the system existed, and rested simply on the mechanical analogy to pumps used for transport of fluids.

The term *carrier* derives from the kinetic analysis of the so-called 'facilitated diffusion' of carbohydrates and other molecules across plasma membranes, which was described in Chapter 6. The term was introduced in order to explain the abnormally high permeability of compounds with low lipid solubility. This type of diffusion, while definitely not a form of active transport, is characterized by saturation at high substrate concentration, by a certain degree of substrate specificity and by competitive inhibition by analogous substances. It has been shown that the diffusion rate can be described by a formula derived using a Michaelis–Menten type of treatment. This kinetic behaviour would be compatible with the assumption that in the membrane a 'carrier' exists with the following properties. (a) It will form a complex with the transported substrate molecule. (b) This complex and also the carrier alone are able to traverse the membrane, while the transported molecule by itself permeates at a negligible rate. (c) The carrier–substrate complex will dissociate on the opposite side of the membrane, because of the low substrate concentration on that side. Active transport may also be described as taking place by a carrier transport system. In this case the carrier would exist in two forms: the first would be that which is released upon dissociation of the carrier–substrate complex, and this form would not be able to bind new substrate until it was transformed into the second. Transformation and/or translation to the other side of the membrane would require energy expenditure. Thus the carrier could transport the substrate only in one direction and the process would be energy-dependent.

However, the existence of carrier molecules which move back and forth through a membrane has never been convincingly demonstrated, and seems on chemical grounds rather unlikely. Stein (1967, pp. 295–303), in a useful review of the evidence, mentioned a number of cases where membrane proteins have been isolated with properties suggesting that they represent part or whole of a specific transport system, but it seems very unlikely that these molecules are carriers in the sense that they would move back and forth through the membrane. Thus, a 'carrier', in so far as it is responsible for an active transport process, appears to be a fixed membrane component able to bind the substrate to be transported, which converts chemical energy into translational energy for the movement of the substrate across the membrane, and which releases the substrate once this has reached the opposite side of the membrane. While the carrier hypothesis has been useful in formulating the kinetic analysis of facilitated diffusion, the term 'carrier' is misleading in that it seems to suggest a moving membrane component.

Similarly, the term 'permease' introduced for certain bacterial transport systems (Stein, 1967, pp. 300–303) is misleading, since it seems to suggest that it represents an enzyme which increases the permeability of the membrane in general or for a particular substance. Actually the permeases are active transport systems, pumps. Therefore, we shall use the term 'pump' rather than 'carrier' or 'permease' in this chapter to indicate an active transport system.

7.2   ELECTROGENIC AND NEUTRAL PUMPS

An ion pump can in principle operate in two different ways. It can pump two ions simultaneously, so that no net charge difference is created; such as when for every Na^+ ion pumped in one direction a K^+ ion is pumped in the opposite direction, or when a Na^+ and a Cl^- ion are pumped in the same direction. In this case the activity of the pump will not in itself cause a potential difference across the membrane, and thus we may call this a 'neutral' pump. The observed membrane potential in this situation derives entirely from the ion gradient across the membrane, and represents an ion diffusion potential.

The ion pump can also transport a single ionic species. In this case the activity of the pump itself will cause a charge imbalance and thus a potential difference across the membrane. This pump is, therefore, called an 'electrogenic' pump. The resulting potential difference will cause a passive counter-ion movement of the ionic species with the highest permeability coefficient. This ion species will build up a gradient and hence a diffusion potential of its own. The observed membrane potential in this case will approximate the algebraic sum of the electrogenic pump potential and the diffusion potential. In the previous chapter it was shown that the diffusion potential is given by the Goldman–Hodgkin–Katz equation

$$E_d = \frac{RT}{F} \ln \frac{P_K[K]_e + P_{Na}[Na]_e + P_{Cl}[Cl]_i}{P_K[K]_i + P_{Na}[Na]_i + P_{Cl}[Cl]_e}$$

where P denotes the permeability constant for a particular ionic species, $[K]_e$ the external and $[K]_i$ the intracellular K^+ concentration, etc. This equation simplifies to the Nernst equation when the contribution of only one ion (Me^+) is significant.

$$E_d = \frac{RT}{F} \ln \frac{[Me^+]_e}{[Me^+]_i}$$

The electrogenic potential E_e is equal to the product of the pumping rate (ion current, i_{Me^+}) and the membrane resistance r_m.

$$E_e = i_{Me^+} \cdot r_m$$

The ion current or pumping rate is directly dependent on ATP production (consequently on temperature and availability of substrates and oxygen), the state of activation of the ion pump, and the presence of metabolic or pump inhibitors. The temperature dependence of the electrogenic potential is about three to four times as large as that of the diffusion potential. Another difference is that upon inhibition of the pump an electrogenic potential decreases rapidly, while a diffusion potential disappears slowly.

Experimentally it is often not easy to distinguish between an electrogenic and a neutral ion pump. In both cases the membrane potential measured by means of the micropipette technique may be virtually equal to the calculated diffusion potential for the most permeable ionic species, usually K^+. Upon inhibition of the pump by cooling or by a chemical inhibitor, in both cases a nearly equivalent ion exchange will be observed as a result of the conservation of charge balance. Furthermore, the speed with which the membrane potential decreases is not a very clear-cut criterion, since this depends on such factors as the rate of penetration of the inhibitor and the rates of passive ion diffusion. A further complication is that a single ion pump system, such as the N^+, K^+-activated ATPase system, may operate in one cell type as a neutral pump and in others as an electrogenic pump. It even seems that in certain cases the type of operation may depend on the conditions to which the cells are subjected (Cross, Keynes and Rybová, 1965). This suggests that the neutral and electrogenic pumps represent modes of operation of the same mechanism in which the coupling of two ion transports may vary from $1:1$ (neutral) to $1:0$ (fully electrogenic), but may also have an intermediate value.

7.2.1   Erythrocyte

We shall now consider a few examples, in which an experimental distinction has been made between a neutral and an electrogenic cation pump. The first example is the erythrocyte. *Table 7.1* shows the K^+, Na^+ and Cl^- levels in

Table 7.1 K^+, Na^+ AND Cl^- LEVELS IN ERYTHROCYTES AND BLOOD PLASMA OF VARIOUS SPECIES (From Bernstein, 1954)

Species	Erythrocytes*			Plasma†		
	K^+	Na^+	Cl^-	K^+	Na^+	Cl^-
Man	136	19	78	5.0	155	112
Monkey	145	24	78	4.7	157	115
Rabbit	142	22	80	5.5	150	110
Rat	135	28	82	5.9	152	118
Horse	140	16	85	5.2	152	108
Sheep	46	98	78	4.8	160	116
Ox	35	104	85	5.1	150	109
Dog	10	135	87	4.8	153	112
Cat	8	142	84	4.6	158	112

* Ion levels measured in mEq per litre of cell water.
† Ion levels measured in mEq per litre of plasma.

cell water and plasma for a number of vertebrate species. There is clear evidence for the existence of ionic gradients across the erythrocyte plasma membrane. These gradients are the result of active transport, as shown by the fact that refrigeration causes a loss of K^+ and a gain of Na^+, which is reversible upon incubation at 37 °C in the presence of glucose (Harris, 1941). Incubation in the absence of glucose or in the presence of glycolytic inhibitors (iodoacetate, fluoride) also causes loss of the cation gradients. The Cl^- level is not affected by changes in metabolism (Harris and Maizels, 1952; Fitzsimons and Sendroy, 1961), and thus Cl^- appears to move passively. There was an early suggestion that only Na$^+$ is transported actively in the erythrocyte and that K$^+$ moves passively, like Cl$^-$. In that case the following relationship should exist.

$$\frac{[K^+]_i}{[K^+]_e} = \frac{[Cl^-]_e}{[Cl^-]_i}$$

Table 7.1 indicates that this is not true; the latter ratio is *ca.* 1.4 in all species but the former ratio varies between 31 and 1.7, always being larger than the chloride ratio. Calculation from the chloride ratio yields a diffusion potential of about 9 mV, which agrees with the measured potential in human erythrocytes of 8.0 ± 0.2 mV (Jay and Burton, 1969). Hence, if K$^+$ diffused freely, it would have to leak out of the cell, since the membrane potential is smaller than the K$^+$ diffusion potential. Thus we must assume that there is an active, inward transport of K$^+$. This transport is coupled to the outward transport of Na$^+$, since removal of external K$^+$ stops the active Na$^+$ efflux, while the curves for Na$^+$ efflux and K$^+$ influx versus external K$^+$ concentration are similar in shape (Glynn, 1956). It thus appears that the erythrocyte has a coupled sodium/potassium pump, which might be neutral. The coupling ratio seems to be fixed and independent of the ionic gradients (Whittam and Ager, 1965). However, the pump may not be entirely neutral, since various authors have found a slightly greater Na$^+$ efflux than K$^+$ influx (Whittam and Ager, 1965). The smallness of the membrane potential and the difficulty in impaling freely moving erythrocytes by a glass microelectrode make it unlikely that potential measurements will settle this point.

7.2.2   Neurone

As our second example we shall discuss the neurone, for which the existence of an electrogenic pump has now been proved conclusively. The first indication was the occurrence of the phenomenon of post-tetanic hyperpolarization: the return of the membrane potential to a slightly (a few millivolts) more negative value after repeated (tetanic) stimulation than existed before stimulation (Kernan, 1970, pp. 400–406; Kerkut and York, 1971, pp. 59–71). In the presence of metabolic inhibitors such as dinitrophenol, azide and cyanide, the effect disappears. Absence of glucose in the medium also reduces hyperpolarization, while addition of glucose increases it again. The hyperpolarization parallels the Na$^+$ efflux rate and the increased oxygen consumption of the nerve after tetanic stimulation. This hyperpolarization behaviour led Connelly (1959) to suggest that the phenomenon could be due to an electrogenic sodium pump, triggered by the Na$^+$ accumulation inside the axon during the tetanic stimulation.

On the other hand, Ritchie and Straub (1957) offered an explanation based on a coupled, neutral pump. The rapid active uptake of K$^+$ by the pump would deplete the extracellular K$^+$ concentration in the immediate proximity of the axonal membrane, so that the Nernst potential for potassium based on the K$^+$ level in the bathing medium underestimates the true potential.

This question has since been settled by a number of observations. Hyperpolarization is not due to a change in membrane resistance, as was shown by Nakajima and Takahashi (1966) in crayfish stretch receptor neurones. It can occur in a K$^+$-free bathing medium but decays more quickly under these conditions (Rang and Ritchie, 1968). Ouabain shortens the response within 10 minutes. Further evidence came from studies of the resting potential in snail neurones (Kerkut and Thomas, 1965). Injection of potassium acetate increased the potential by only a few millivolts but injection of sodium acetate increased it by 29 mV within 10 minutes (*Figure 7.1*). Ouabain (0.1mM and *p*-chloromercuribenzoate (2×10^{-5}M) abolished the sodium effect within 5 minutes. The membrane potential also rises and falls with oxygen pressure (Kerkut and York, 1969). The potential is strongly temperature-dependent: it rapidly falls by 29 percent for a temperature decrease from 17 to 4 °C, whereas

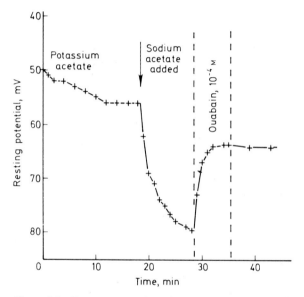

Figure 7.1   Demonstration of an electrogenic sodium pump in snail neurone. Injection of potassium acetate has little effect on the membrane potential, while injection of sodium acetate causes 24 mV hyperpolarization, which is largely abolished on addition of ouabain. (After Kerkut and York, 1971, p. 77, courtesy of Scientechnica)

for a diffusion potential a fall of only 4.5 percent would be expected (Marmor and Gorman, 1970).

The combined evidence of these experiments indicates that the membrane potential of the resting snail neurone is composed of a large potassium diffusion potential (about 60 mV) and a small electrogenic sodium pump potential (10–15 mV), both negative intracellularly. Measurements of current flowing during pump activity suggest that about one-third of the extruded sodium produces the electrogenic potential, the remaining two-thirds being coupled to the active transport of other ions, probably K$^+$ influx (Thomas, 1969). Determination of the active fluxes of sodium and potassium in the squid axon also gives a ratio of 1.5–2 (Baker *et al.*, 1969). The sigmoid shape of the curve for the K$^+$ activation of Na$^+$ efflux suggests that at least 2 K$^+$ ions co-operate in activating the Na$^+$ efflux, while 3 Na$^+$ ions are extruded per ATP molecule hydrolysed. This indicates that 3 Na$^+$ ions are extruded and 2 K$^+$ ions are pumped in for each molecule of ATP utilized.

7.2.3   Cochlea

A much more pronounced example of an electrogenic pump system has been demonstrated in the guinea-pig cochlea. In the cochlea there are two fluid compartments, the endolymph in the scala media and the perilymph in the scala vestibuli and scala tympani (*Figure 7.2*). The cationic gradients between endolymph (140–150 mM K$^+$) and perilymph (4–5 mM K$^+$) are maintained by

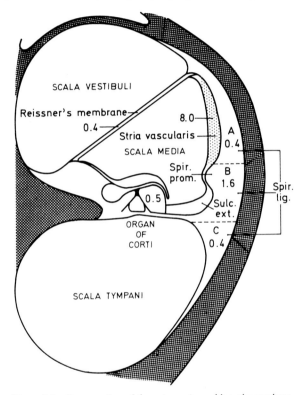

Figure 7.2   Cross section of the guinea-pig cochlea; the numbers indicate the Na$^+$,K$^+$-ATPase activities of the various structures (moles per kilogram dry wt per hour). Spir. prom., spiral prominence; sulc. ext., sulcus external; spir. lig., spiral ligament; A, B and C are three parts of the spiral ligament, analysed separately.

a high Na$^+$,K$^+$-activated ATPase activity located in the stria vascularis, an epithelial structure separating the endolymph from the blood plasma (Kuypers and Bonting, 1969). There is a steady potential difference of about 80 mV between endolymph and perilymph or blood (endolymph positive). This potential is abolished within 2 minutes following oxygen deprivation (*Figure 7.3a*), and reduced by 50 percent within 7 minutes following the start of ouabain perfusion of the perilymphatic space. After about 30 minutes the potential has reversed to about -20 mV. Perfusion of the perilymphatic space with 150 mM K$^+$-Ringer's increases the normal potential within 2 minutes up to 100 mV, whereas absence of Na$^+$ (Li$^+$- or sucrose-Ringer's) has little effect (Kuypers and Bonting, 1970a, b). The negative potential of about -20 mV after ouabain perfusion is abolished by raising the K$^+$ level in the perfusion

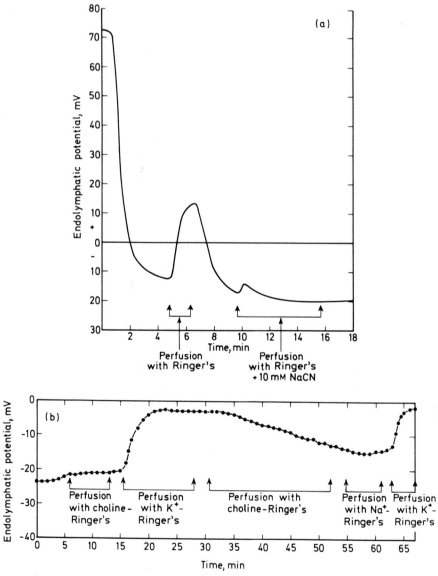

Figure 7.3   Effects of anoxia and 150 mM K$^+$-Ringer's on the endolymphatic potential in the guinea-pig cochlea. (a) Effects of anoxia and perfusion of the scala vestibuli of the anoxemic animal with normal Ringer's and subsequently with Ringer's containing 10 mM NaCN. (b) Effects on the negative endolymphatic potential (obtained by perfusion with Ringer's containing 1 mM ouabain) of perfusion of the scala vestibuli with solutions of various ionic composition. (After Kuypers and Bonting, 1970b, courtesy of Springer-Verlag)

fluid to 150 mM, but is again little affected by omitting Na^+ (*Figure 7.3b*). These and other observations indicate that the normal endolymphatic potential of about 80 mV is composed of an electrogenic potassium-pump potential of about $+100$ mV and a K^+-dominated diffusion potential of about -20 mV. Inhibition of the pump system by ouabain or anoxia abolishes the positive pump potential component, and leaves the negative K^+ diffusion potential component thus reversing the potential to about -20 mV. This means that the Na^+,K^+-activated ATPase system of the stria vascularis is an electrogenic potassium pump, which type of pump has so far not been found in any other vertebrate tissue to our knowledge.

7.3 Na^+,K^+-ACTIVATED ATPase AND CATION TRANSPORT

7.3.1   **Discovery**

The first chemical definition of a cellular transport system was made by Skou (1957) in a study of the cation pump in crab nerve. Considering that ATP has been shown to provide the energy for all sorts of cell functions, he reasoned that this would probaably also be the case for active transport. He further assumed that an ATPase would be required to make the energy of ATP available to cation transport, and that this enzyme might require Na^+ and K^+ for activity. He was able to demonstrate in a crab nerve particulate fraction a Mg^{2+}-dependent ATPase activity, which was increased considerably upon simultaneous addition of Na^+ and K^+ to the assay medium. Subsequently he found that this additional ATPase activity was completely inhibited by ouabain, one of the digitalis glycosides (Skou, 1960), which some years earlier had been shown to be powerful and specific inhibitors of cation transport in erythrocytes (Schatzmann, 1953). On the basis of these observations Skou suggested that this ouabain-sensitive, Na^+,K^+-activated ATPase (abbreviated to Na^+,K^+-ATPase) might be part of, or identical with, the cation pump in nerve cells.

7.3.2   **Properties**

For a more extensive account of the literature on the properties, occurrence and function of this important enzyme system the reader is referred to reviews by Bonting (1970) and Schwartz, Lindenmayer and Allen (1975). Only some of the more important characteristics of the Na^+,K^+-ATPase system will be discussed here.

1. The Na^+,K^+-ATPase activity is always accompanied by a Na^+,K^+-insensitive Mg^{2+}-dependent ATPase activity. The available evidence warrants the conclusion that these are two separate enzymes (Bonting, 1970, pp. 269–270). This necessitates a differential assay: the Na^+,K^+-ATPase activity is determined as the difference between the sum of the two activities (Na^+, K^+, Mg^{2+}, ATP present in the assay medium) and the Mg^{2+}-ATPase activity with Na^+ and/or K^+ omitted or ouabain present in fully inhibitory concentration (Bonting, 1970, pp. 260–264).
2. The Na^+,K^+-ATPase system requires Mg^{2+} for its activity. Optimal activity is obtained at a Mg^{2+}:ATP ratio of 1–2 (*Table 7.2*).

Table 7.2   PROPERTIES OF THE Na^+,K^+-ATPase SYSTEM IN VARIOUS SPECIES AND TISSUES

Tissue	Species	Mg^{2+}:ATP optimum*	K_m, Na^+	K_m, K^+	pH optimum	Ouabain pI_{50}§	Activity $MKH\parallel$	Reference
Nerve	Crab	2	6–8	1.8	7.2	3.9		Skou (1957, 1960)
Salt gland	Herring gull	1.5	12.5	1.5	7.2	6.3	2.7w	Bonting, Hawkins and Canady (1964)
Lens epithelium	Rabbit			0.4	7.3	5.9	0.15w	Bonting (1965)
Rectal gland	Spiny dogfish	1.5	11.7	1.0	7.0	6.8	5.7d	Bonting (1966)
Liver	Rat	1	6	0.9	7.3	3.9	0.37d	Bakkeren and Bonting (1968)
Stria vascularis	Guinea-pig	0.5–1	4.5	0.9	7.3	5.5	8.0d	Kuypers and Bonting (1969)
Pancreas	Rabbit	1.5	10	0.8	7.2	5.4	0.23d	Ridderstap and Bonting (1969b)
Kidney	Rat	0.5–1	8	0.7	7.4	3.9	6.8d	Bakkeren, van der Beek and Bonting (1971)
Gastric mucosa	Lizard	1	7.5	1.1	7.3	7.0	0.25d	Hansen et al. (1972)
Mammary gland	Guinea-pig	2	8	1.6	7.8	5.6	0.23d	Vreeswijk, de Pont and Bonting (1973)
Gall bladder	Rabbit	2	6.5	0.3	7.5	5.6	2.3d	van Os and Slegers (1970)

* Molar ratio of Mg^{2+} to ATP at which activity occurs.
† Half-maximal activation concentration (mM) for Na^+ at 5 mM K^+.
‡ Half-maximal activation concentration (mM) for K^+ at 60 mM Na^+.
§ Negative log of molar concentration of ouabain causing 50 percent inhibition.
∥ Activity in moles of ATP per kilogram per hour at $37\,°C$, d = on dry wt basis, w = on wet wt basis.

3. The enzyme is activated by the presence of both Na^+ and K^+, not by either of these alone. The half-maximal activating concentrations vary from 5 to 13 mM for Na^+ and from 0.3 to 1.8 mM for K^+ (*Table 7.2*). The cations Rb^+, Tl^+, NH_4^+, and to a lesser degree Cs^+, can replace K^+.

4. The pH optimum of Na^+,K^+-ATPase varies from 7.0 to 7.8 (*Table 7.2*), while Mg^{2+}-ATPase has a pH optimum between 8.4 and 8.9.

5. Ouabain and other cardiac glycosides inhibit the enzyme. The half-maximal inhibitory concentration for ouabain ranges from 10^{-7}M to 10^{-4} M, depending on species and tissue (*Table 7.2*). There is often a bi-

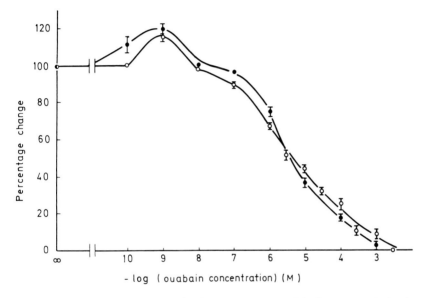

Figure 7.4   *Effect of ouabain on Na^+,K^+-ATPase activity and fluid secretion in isolated rabbit pancreas. (○) Na^+,K^+-ATPase, pI_5 5.4; (●) flow, pI_{50} 5.4. (After Ridderstap and Bonting, 1969, courtesy of the American Physiological Society)*

phasic effect, which involves a slight stimulation at a ouabain concentration of about one-threehundredth of the half-maximal inhibitory concentration (*Figure 7.4*). *Erythrophleum* alkaloids, though differing markedly in chemical structure from the cardiac glycosides (*Figure 7.5*), also specifically inhibit the Na^+,K^+-ATPase activity (Bonting, Hawkins and Canady, 1964). Both inhibitory effects are reversed in part by raising the K^+ concentration.

6. The Na^+,K^+-ATPase system is a particulate enzyme, located in the plasma membrane (Bonting, 1970, p. 270).

7. The Na^+,K^+-ATPase system is of widespread occurrence: it is present in all vertebrate tissues studied so far, and has also been detected in tissues of several invertebrate species (Bonting, 1970, p. 270–277). Its occurrence in plants is still open to question.

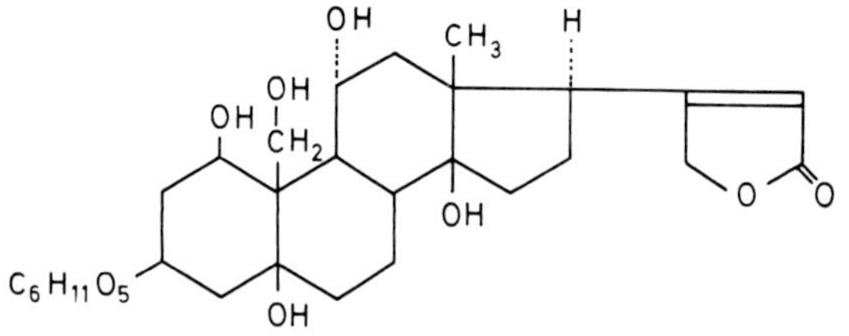

Ouabain ($C_{29}H_{44}O_{12}$)

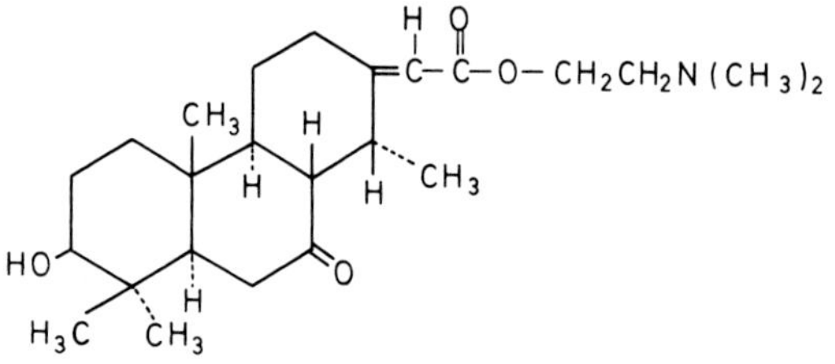

Cassaine ($C_{24}H_{39}O_4$ N)

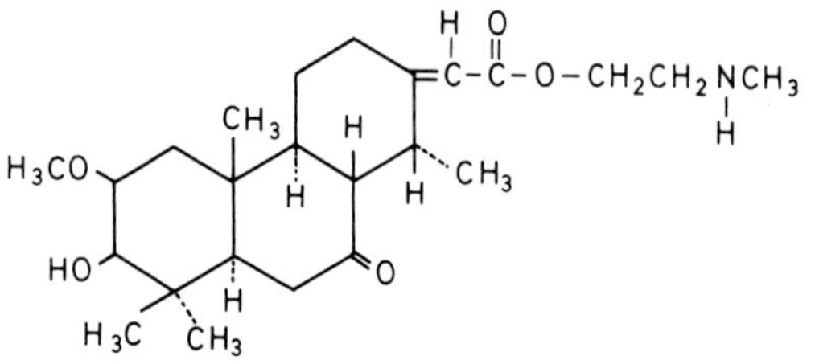

Erythrophleine ($C_{24}H_{39}O_5$ N)

Figure 7.5   Chemical structure of ouabain and the Erythrophleum alkaloids casseine and erythrophleine. The positions and configuration of the ring constituents in erythrophleine are still uncertain. (After Bonting, Hawkins and Canady, 1964, courtesy of Pergamon Press)

7.3.3   Identity with the cation pump

Early evidence for the identity of Na$^+$,K$^+$-ATPase with the cation pump system was obtained for erythrocyte membranes by Dunham and Glynn (1961) and Post et al. (1960). By using reconstituted erythrocyte ghosts it was possible to show a qualitative and quantitative agreement between the properties of the enzyme system and those of the cation transport system: both are located in the membrane, both utilize only ATP (which must be present inside), both require Na$^+$ on the inside and K$^+$ on the outside, both are inhibited by cardiac glycosides on the outside; the half-maximal activation concentrations of Na$^+$ and K$^+$ and the half-inhibitory concentration of ouabain are nearly equal for the two systems.

Further evidence comes from the observation that in all tissues known to possess a cardiac-glycoside-sensitive cation transport system Na$^+$,K$^+$-ATPase activity is present (Bonting, Caravaggio and Hawkins, 1962). Even stronger proof was derived from a comparison of the active cation fluxes and Na^+,K^+-ATPase activities, both expressed in $mol\ cm^{-2}\ s^{-1}$, for six different tissues (Bonting and Caravaggio, 1963). With the fluxes varying over a 22 000-fold range, there was a significant correlation between flux and enzyme activity, their average ratio being 2.6 (s.e. 0.19), as shown in *Table 7.3*. The Mg^{2+}-ATPase activity was not significantly correlated with the flux values. This finding clearly demonstrates the close relationship of the Na^+,K^+-ATPase system to the active cation transport involved in the maintenance of cation gradients in single cells, transport of water and electrolytes across

Table 7.3   COMPARISON OF ACTIVE CATION FLUXES AND Na^+,K^+ATPase ACTIVITIES
(After Bonting and Caravaggio, 1963)

Tissue	Temp., C	Cation flux, $10^{-14}\ mol\ cm^{-2}\ s^{-1}$	Na^+,K^+-ATPase activity, $10^{-14}\ mol\ cm^{-2}\ s^{-1}$	Ratio
Human erythrocyte	37	3.87	1.38	2.80
Frog toe muscle	17	985	530	1.86
Squid giant axon	19	1 200	400	3.00
Frog skin	20	19 700	6 640	2.97
Toad bladder	27	43 700	17 600	2.48
Electric eel, non-innervated membrane				
Sachs organ	23	86 100	38 800	2.22
		$\tau = +0.87,\ P = 0.017^*$		2.56 (s.e. 0.19)

* τ is the correlation coefficient, P indicates the significance of the correlation.

Table 7.4   RATIOS OF CATION TRANSPORT TO HYDROLYSIS OF ATP IN VARIOUS TISSUES

Tissue	Cation/ATP (Eq/mol)	References
Based on Na^+,K^+-ATPase assay		
Human erythrocytes	2.8 (Na^+)	Bonting and Caravaggio (1963)
Frog toe muscle	1.9 (Na^+)	Bonting and Caravaggio (1963)
Squid giant axon	3.0 (K^+)	Bonting and Caravaggio (1963)
Frog skin	3.0 (Na^+)	Bonting and Caravaggio (1963)
Toad bladder	2.4 (Na^+)	Bonting and Canady (1964)
Electric eel, non-innervated membrane electroplax	2.2 (Na^+)	Bonting and Caravaggio (1963)
Calf lens	3.1 (Na^+)	Bonting, Caravaggio and Hawkins (1963)
Rabbit lens	2.5 (Na^+)	Bonting, Caravaggio and Hawkins (1963)
Herring-gull nasal gland	2.0 (Na^+)	Bonting et al. (1964)
Human leukocytes	1.8 (Na^+)	Block and Bonting (1964)
Rabbit ciliary epithelium	3.0 (Na^+)	Bonting and Becker (1965)
Spiny-dogfish rectal gland	2.2 (Na^+)	Bonting (1966)
Frog sartorius muscle	2.2 (K^+)	Corrie and Bonting (1966)
Escherichia coli	1.6–3.2 (K^+)	Hafkenscheid and Bonting (1968)
Dog pancreas	1.9 (Na^+)	Ridderstap and Bonting (1969a)
Lizard gastric mucosa	2.8	Hansen et al. (1972)
Not based on Na^+,K^+-ATPase assay		
Human erythrocytes	2.7 (Na^+)	Sen and Post (1964)
	3.0 (Na^+)	Glynn (1962)
Frog muscle	2.5–3.0 (Na^+)	Harris (1967)
Frog skin	2.7–3.3 (Na^+)	Zerahn (1956)
	3.5 (Na^+)	Leaf and Renshaw (1957)
Toad bladder	2.8 (Na^+)	Leaf, Page and Anderson (1959)
Calf lens	3.0–4.5 (Na^+)	Kern, Roosa and Murray (1962)

epithelial membranes and cation gradient recovery after excitation of nerve, muscle and electric organ. A cation:ATP ratio of nearly 3 has since been found in a great number of tissues, mostly on a Na^+,K^+-ATPase basis and in some cases on the basis of extra oxygen or glucose consumption or lactate production due to cation transport (*Table 7.4*). In addition, excellent agreement between the ouabain inhibition curves for Na^+,K^+-ATPase activity and for functional activity has been demonstrated in a number of cases, including lens (Bonting, Caravaggio and Hawkins, 1963), toad bladder (Bonting and Canady, 1964), ciliary body (Bonting and Becker, 1964), choroid plexus (Vates, Bonting and Oppelt, 1964), cochlea (Kuypers and Bonting, 1970a) and pancreas (Ridderstrap and Bonting, 1969b). *Figure 7.4* illustrates this for the pancreas. Finally, it has been possible to reconstitute a stoichiometrically alkali cation transporting system by incorporating purified Na^+,K^+-ATPase into artificial phospholipid membranes (Hilden and Hokin, 1975; Sweadner and Goldin, 1975).

7.3.4   Isolation

The ouabain-insensitive Mg^{2+}-ATPase activity, which always accompanies the Na^+,K^+-ATPase system, can be abolished from the membrane fraction by treatment with $2 M$ NaI without affecting the Na^+,K^+-ATPase activity (Nakao *et al.*, 1965). After removal of the iodide 99 percent of the ATPase activity is Na^+,K^+-activated and ouabain-sensitive. Subsequently the Na^+, K^+-ATPase activity can be solubilized by detergents such as Lubrol (Nakao *et al.*, 1973) or deoxycholate (Kyte, 1971; Lane *et al.*, 1973). Further purification is then possible by means of protein separation techniques such as zonal centrifugation, ion exchange or molecular sieve chromatography and salt fractionation. Highly purified preparations with specific activities ranging from 0.8 to 4.8 moles of ATP split per gram of protein per hour have been obtained in this way from pig brain (Nakao *et al.*, 1973), canine renal medulla (Kyte, 1971; Lane *et al.*, 1973) and the rectal gland of the dogfish shark (Hokin *et al.*, 1973). A preparation of high purity has been obtained without solubilization, merely by treatment with sodium dodecyl sulphate in low concentration followed by a single zonal centrifugation, from rabbit renal outer medulla (Jørgensen, 1974a).

All these preparations show two protein bands after sodium dodecyl sulphate gel electrophoresis, only one of which is phosphorylated after previous treatment of the preparation with either ATP in the presence of Mg^{2+} and Na^+ (Collins and Albers, 1972), or with inorganic phosphate in the presence of Mg^{2+} and ouabain (Schuurmans Stekhoven *et al.*, 1976a, b). The same band also carries the ouabain binding site (Ruoho and Kyte, 1974). Its apparent molecular weight ranges from 84000 to 100000 (α band). The other, faster-moving protein band has an apparent molecular weight of 55000–57000 (β band) and represents a sialoglycoprotein (Kyte, 1972). Although it has been suggested that the native enzyme would be a lipid-embedded tetramer $\alpha_2\beta_2$ (Stein *et al.*, 1973), it now seems more likely that it is a trimer $\alpha_2\beta$, as concluded for rabbit kidney outer medulla (Jørgensen, 1974b), the molecular weight of which is in good agreement with the value of 250000 determined by radiation inactivation (Kepner and Macy, 1968) and molecular-sieve chromatography (Atkinson, Gatenby and Lowe, 1971). In any case, two α chains and one β chain are in close proximity in the native enzyme, as determined by the use of bifunctional reagents (Kyte, 1972, 1975). Further evidence is supplied by the finding that enzymatic activity is inhibited by an antibody to the β chain and by the glycoprotein-binding agent concanavalin A (Rhee and Hokin, 1975; Churchill and Hokin, 1976). The β chain appears to function as a carrier for monovalent cations with a specificity for Na^+ ions imposed upon it by the α chain (Shamoo and Meyers, 1974).

7.3.5   Phospholipids

Removal of lipids from a membrane preparation by detergents (Tanaka and Abood, 1964), phospholipase (Roelofsen, Baadenhuysen and van Deenen, 1966), as well as by organic solvents (Järnefelt, 1972), leads to complete or partial inactivation of the enzyme. Addition of various lipids to a partially delipidated preparation restores the activity, at least in part. While some authors do not observe any specificity in this process, others maintain that phosphatidylserine is the most effective of the reactivating lipids (Ohnishi and Kawamura, 1964; Wheeler and Whittam, 1970). The alleged specificity of the phosphatidylserine effect led to the suggestion that this phospholipid is essential for the Na^+,K^+-ATPase system, which would consist of a complex of the enzyme protein and phosphatidylserine (Wheeler and Whittam, 1970).

A novel attempt to settle this question has been made by de Pont, van Prooyen-van Eeden and Bonting (1973). Treatment of cattle brain membranes with phosphatidylserine decarboxylase converts phosphatidylserine quantitatively into its decarboxylation product, phosphatidylethanolamine, without any other changes in the phospholipid pattern. Assay of the Na^+, K^+-ATPase activity of the membrane preparation before and after treatment shows no significant loss in enzyme activity. This indicates that phosphatidylserine is not essential for the functioning of the Na^+,K^+-ATPase system. The role of the phospholipids appears to be to keep the enzyme in a particular conformation necessary for activity.

7.3.6   Reaction mechanism

In recent years considerable effort has been spent in elucidating the reaction mechanism of the Na^+,K^+-ATPase system (Bonting, 1970, pp. 341–350; Skou, 1971). The results of the work of Albers, Post and others have indicated that a phosphorylation, a transphosphorylation or conformational change of the phosphorylated intermediate, and a dephosphorylation occur. This sequence of events has been formulated by Siegel and Albers (1967) as

$$E_1 + ATP \underset{Na^+}{\overset{Mg^{2+}}{\rightleftharpoons}} E_1 \sim P + ADP \tag{7.1}$$

$$E_1 \sim P \rightleftharpoons E_2 \sim P \rightarrow E_2\text{–}P \tag{7.2, 7.3}$$

$$E_2\text{–}P + H_2O \overset{K^+}{\rightarrow} E_2 + P_i \tag{7.4}$$

$$E_2 \rightleftharpoons E_1 \tag{7.5}$$

The phosphorylation step 7.1 requires Mg^{2+} and Na^+ and is not inhibited by ouabain in low concentration. Evidence has been advanced for a transition of the high-energy intermediate $E_1 \sim P$ via conformational change 7.2 to another high-energy intermediate $E_2 \sim P$, which then changes (reaction 7.3) to a low-energy intermediate E_2–P, possibly with a shift of the phosphate group to another site (Robinson, 1971). Reaction 7.3 would be blocked by ouabain. The dephosphorylation step 7.4 requires K^+ and is strongly inhibited by ouabain. It is therefore possible to allow the phosphorylated intermediate $E \sim P$ to accumulate by incubating the Na^+,K^+-ATPase preparation

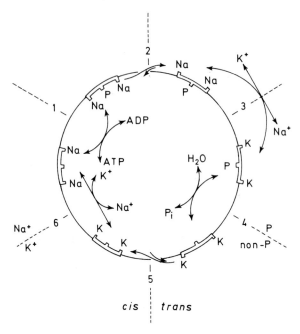

Figure 7.6   Model of the Na^+,K^+-ATPase mechanism. The circle represents the cell membrane. In the cis form the enzyme has its cation-binding sites turned inward, while in the trans form they are turned outward. Reaction 1 is the Na^+-dependent ATP–ADP exchange. Phosphorylation converts the cis form to the trans form (2). Exchange of Na^+ for K^+ takes place (3). In the presence of K^+ dephosphorylation occurs (4) and the enzyme reconverts to the cis form (5) with exchange of K^+ for Na^+ (6). (After Albers, Koval and Siegel, 1968, courtesy of Academic Press)

with terminally labelled $[^{32}P]ATP$ in the presence of Na^+, Mg^{2+} and ouabain in the absence of K^+ (Albers, Fahn and Koval, 1963). It is stable at pH 2–3, can be precipitated with trichloroacetic acid and after pepsin treatment a phosphorylated peptide can be isolated. Hydroxylamine readily dephosphorylates the peptide, which suggests that it is an acyl phosphate.

The identity of the acyl group has been investigated by Kahlenberg, Galsworthy and Hokin (1967, 1968), who converted the phosphorylated peptide into a radioactive hydroxamate derivative with tritium-labelled N-(n-propyl)-hydroxylamine, which was identified by comparison with synthetic derivatives. Although they concluded at that time that the acyl phosphate intermediate would be an L-glutamyl γ-phosphate residue, subsequent work has shown that it is an aspartyl β-phosphate. Post and Kume (1973) have isolated a phosphorylated tripeptide, in which the phosphate group is apparently bound to an aspartyl group. They based this conclusion on a comparison of the pH–stability profile, the isoelectric point and the carboxypeptidase digestibility of this tripeptide with the synthetic tripeptides prolylphosphoaspartyllysine and prolylphosphoglutamyllysine. This has been confirmed by direct chemical characterization of the bond after reduction with NaB^3H_4 to $[^3H]$-homoserine (Nishigaki, Chen and Hokin, 1974). Thus, the intermediate $E \sim P$ appears to be an aspartyl phosphate. If the phosphate binding site in E_2–P is different, it could be a serine group (Robinson, 1971). Evidence for a different chemical nature of the two phosphorylated intermediates is supplied by a comparison of the properties of their ouabain complexes (Schuurmans Stekhoven, de Pont and Bonting, 1976). There are differences in stability inhibition of dissociation by cations and antagonism between Na^+ and K^+.

The dephosphorylation step 7.4 can be measured separately as a K^+-activated ouabain-sensitive phosphatase activity with p-nitrophenyl phosphate serving as substrate. This would represent the hydrolysis of the low-energy phosphorylated intermediate E_2–P.

In this reaction sequence E_1 and $E_1 \sim P$ would be forms with inwardly oriented cation sites with high affinity for Na^+, while E_2 and E_2–P would be forms with outwardly oriented cation sites of low Na^+ affinity. The reverse would be true for the K^+ affinity of these sites. This would permit exchange of Na^+ for K^+ during step 7.3, while the reverse exchange would take place after step 7.5. A schematic representation of this reaction mechanism, proposed by Albers, Koval and Siegel (1968), is given in *Figure 7.6.*

7.4 Ca^{2+}-ACTIVATED ATPase AND CALCIUM TRANSPORT

A clear insight into active calcium transport in various tissues (erythrocytes, muscle, nerve) has arisen only in recent years. The problem has been that the greatest part of the intracellular calcium is either bound in non-ionized form by certain molecules (ATP, citrate, glutamate, phospholipids or proteins), or is sequestered in intracellular organelles (mitochondria, sarcoplasmic reticulum). The usual chemical determinations, including atomic absorption spectrophotometry, determine total calcium rather than ionized calcium. Calcium-sensitive electrodes give the concentration of ionized calcium, but are not suitable at concentrations below 0.01–0.1 mM. Our present insights have been derived mainly from ^{45}Ca flux studies, the use of the calcium-chelating agent EGTA (ethyleneglycol-bis(β-aminoethyl ether)-N,N'-tetra-acetic acid), and the use of aequorin, a protein from a jellyfish which luminesces in the presence of calcium ions.

7.4.1   **Erythrocytes**

Chemical analysis indicates an intracellular calcium concentration of less than 30 μmol per litre of cells, which implies a $[Ca^{2+}]_e/[Ca^{2+}]_i$ ratio of more than 50 (Schatzmann, 1970). The membrane potential of erythrocytes is 8–9 mV (inside negative), which would give a $[Ca^{2+}]_e/[Ca^{2+}]_i$ ratio of 0.5 provided the calcium is passively distributed across the membrane. Hence, there must be an active transport of Ca^{2+} out of the cell.

The active Ca^{2+} extrusion has been demonstrated by reconstituting erythrocyte ghosts in a medium containing Ca^{2+}, Mg^{2+}, ATP plus KCl for isotonicity, and suspending them in a physiological salt solution (Schatzmann, 1966, 1970). Ghosts reconstituted in this way extrude Ca^{2+} at a rapid rate against a considerable electrochemical gradient, provided they contain ATP

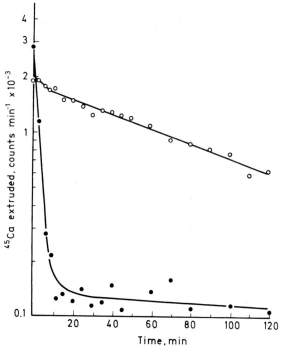

Figure 7.7   ATP-dependent extrusion of ^{45}Ca from reconstituted erythrocyte ghosts. The internal medium contained 0.75 mM Ca^{2+} 2 mM Mg^{2+} and 1 mM Tris-ATP, pH 7.2 (●) or no ATP (○). The external medium contained 130 mM Na^+, 5 mM K^+, 1 mM Ca^{2+}, 2 mM Mg^{2+}, 20 mM Tris, 161 mM Cl., pH 7.3. (After Schatzmann, 1970, courtesy of Macmillan)

and Mg^{2+} (Figure 7.7). The extrusion rate is highly temperature-dependent with a Q_{10} of 3.5. The ratio of Ca^{2+} extruded per molecule ATP hydrolysed was first estimated to be 1:1 (Schatzmann, 1973), but now appears to be 2:1 (Quist and Roufogalis, 1975), as in the case of sarcoplasmic reticulum. Strontium ion is also transported by this system, and in the presence of both Ca^{2+} and Sr^{2+} there appears to be competition between these two ions. The half-saturation concentration for intracellular Ca^{2+} is 4×10^{-6} M, while the extrusion rate is unaffected by varying the extracellular Ca^{2+} concentration from 3×10^{-7} to 5×10^{-3} M (Schatzmann, 1973). The presence of Na^+ in concentrations ranging from 0 to 130 mM has no effect on the Ca^{2+} extrusion rate.

In other experiments it could be shown that ATP does not affect $^{45}Ca^{2+}$ influx into the reconstituted ghosts. Hence the net calcium movement caused by ATP is not induced by a decrease in calcium influx, but by an increased efflux.

In analogy to the Na^+,K^+-ATPase system, which is responsible for active Na^+ (and K^+) transport, it was reasonable to look for a Ca^{2+}-activated ATPase in the erythrocyte membrane. Dunham and Glynn (1961) had already reported a Ca^{2+},Mg^{2+}-activated ATPase activity in isolated human erythrocyte membranes. Further studies by Wins and Schoffeniels (1966) and Vincenzi and Schatzmann (1967) indicate that optimal activity occurs at 0.1 mM Ca^{2+} in the presence of 2.5 mM Mg^{2+}, and that higher Ca^{2+} concentrations decrease the activity. Strontium ion can replace Ca^{2+} in activating the enzyme. The activity does not seem to be part of the Na^+,K^+-ATPase system, since it is independent of the $Na^+ : K^+$ ratio of the medium, and is not sensitive to ouabain [see, however, Bond and Green (1971) and Schatzmann and Rossi (1971)]. The enzyme is inhibited by mersalyl (half-maximal inhibiting concentration 2.3×10^{-5} M) and ethacrynic acid (half-maximal inhibiting concentration 5.5×10^{-5} M). These properties are in good agreement with those of the Ca^{2+} transport system. There is also good agreement between the half-saturation concentrations for $Ca^{2+} : 4 \times 10^{-6}$ M or less for the transport system and 2.4×10^{-6} M for the Ca^{2+},Mg^{2+}-ATPase activity (Schatzmann, 1973). In addition, the enzyme activity and the transport system have about the same pH optimum and the same temperature optimum, both require ATP, and both are inhibited by the trivalent ions of holmium and praseodymium, by temperatures of 50 °C and by preincubation with phospholipase A and C but not phospholipase D (Cha, Bak and Lee, 1971; Schatzmann, 1973). Chlorpromazine, which inhibits Ca^{2+} transport by the sarcoplasmic reticulum (Hasselbach, Makinose and Fiehn, 1970), also inhibits the erythrocyte Ca^{2+},Mg^{2+}-ATPase activity in the same concentration range (Schatzmann, 1970).

Recently, evidence for a Ca^{2+} extrusion mechanism in calf platelets, similar to that in erythrocytes, has been obtained (Robblee, Shepro and Belamarich, 1973). A subcellular fraction consisting primarily of inverted membrane vesicles accumulates 200–400 nmol calcium per milligram of protein in the presence of ATP and oxalate. The process requires ATP and releases inorganic phosphate, and Mg^{2+} is needed for both Ca^{2+} uptake and phosphate release. Salyrgan (10^{-5} M) and ADP (1 mM) inhibit both processes, but ouabain is not inhibitory. The Ca^{2+},Mg^{2+}-ATPase is stimulated by 5–10 μM Ca^{2+} and is inhibited by 2 mM EGTA. It seems reasonable to assume that this is the same type of Ca^{2+}-transporting enzyme as observed in erythrocytes.

Unfortunately, no sensitive and specific inhibitor, like ouabain for the Na^+, K^+-ATPase system and Na^+ transport, has so far been found for the Ca^{2+}, Mg^{2+}-ATPase system and Ca^{2+} transport. Possibly ruthenium red, which inhibits Ca^{2+} transport in mitochondria (Moore, 1971) and the Ca^{2+},Mg^{2+}-ATPase activities of erythrocytes (Watson, Vincenzi and Davis, 1971), may be useful in this respect. However, it does not appear to inhibit active Ca^{2+} uptake by sarcoplasmic reticulum (Vale and Carvalho, 1973).

More information is already available on the Ca^{2+},Mg^{2+}-ATPase system of sarcoplasmic reticulum, which will be discussed in the next section. Human erythrocytes contain no organelles which might accumulate calcium. Hence the calcium metabolism of these cells appears to be determined entirely by active extrusion, slow passive influx and a certain amount of binding to the membrane and to intracellular compounds. The active Ca^{2+} efflux does not seem to be coupled to the passive Na^+ influx, resulting from the Na^+ gradient generated by the Na^+,K^+-ATPase system, since the Ca^{2+} extrusion continues in the absence of a Na^+ gradient and is insensitive to ouabain. A useful role of the active Ca^{2+} efflux is that it keeps the intracellular calcium concentration below values (>0.1 mM) which would inhibit the Na^+,K^+-ATPase system (Dunham and Glynn, 1961).

7.4.2   Muscle

It appears at present that muscle contraction occurs in basically the same way in all three types of muscle: skeletal, smooth and cardiac muscle (see Inesi, 1972; Taylor, 1973; Reuter, 1973). They all contain the protein complex actomyosin, which contracts in the presence of ATP and Ca^{2+} (0.3–1.4 μM). Excitation causes influx of Na^+ and Ca^{2+} across the plasma membrane, which would release Ca^{2+} from the sarcoplasmic reticulum (SR). Calcium ion activates myosin ATPase activity, and the resulting hydrolysis of ATP leads to contraction. Relaxation occurs upon removal of Ca^{2+} by the SR, which deactivates the myosin ATPase. The resting concentration of Ca^{2+} in the myoplasm is ca. 0.1 μM (Portzehl, Caldwell and Ruegg, 1964). After excitation the concentration may rise three- to tenfold, depending on the intensity of stimulation, as observed by means of aequorin injected into a single barnacle muscle fibre (Ashley, 1970).

The entry of Ca^{2+} across the plasma membrane during excitation cannot be directly responsible for contraction, since the diffusion of this ion would be too slow. This conclusion directed attention to the SR, which closely surrounds each myofibril by means of its tubular ramifications. The work of Hasselbach and Makinose (1961, 1962, 1963) has shown that the SR can accumulate Ca^{2+} by an active, ATP-requiring process.

Although the interpretation is greatly complicated by the several compartments (interstitial space, connective tissue, SR, mitochondria, membrane-bound fraction), $^{45}Ca^{2+}$ flux studies have indicated that there can be both a parallel Na^+-Ca^{2+} influx and a Na^+-Ca^{2+} exchange (Cosmos and Harris, 1961). When there is a Na^+ influx induced by the presence of ouabain or the absence of extracellular K^+, there is an increased net Ca^{2+} uptake, which is reversed upon removing ouabain or restoring extracellular K^+. This suggests the existence of an active Ca^{2+} efflux, which is coupled to the operation of the Na^+,K^+-ATPase system.

While little is known about active Ca^{2+} transport across the outer membrane of the muscle cell, the transport into the SR has been studied in great detail after the development of an isolation technique for this system. A muscle homogenate is subjected to differential centrifugation, followed by sucrose density gradient centrifugation, yielding a vesicular preparation (Inesi, Ebashi and Watenabe, 1964). Skeletal muscle yields 2 mg of vesicles per gram of muscle, which are able to accumulate 0.2 μmol of Ca^{2+} in the presence of ATP and Mg^{2+} (van der Kloot, 1969). This is about twice the amount of Ca^{2+} combining with the myofibrillar proteins, and thus the uptake capacity is sufficient to remove all Ca^{2+} from the myofibrillar proteins during relaxation. The uptake rate is also fast enough; the vesicular preparation can lower the Ca^{2+} concentration in the medium from 10^{-6} to 10^{-7} M within 14 ms, which is well within the relaxation time of the muscle. This appears to be true also for smooth and cardiac muscle.

In the presence of oxalate and ATP a calcium oxalate precipitate is formed within the vesicles, as shown by electron microscopy (Hasselbach, 1963). This indicates that there is Ca^{2+} transport across the vesicular membrane, and not merely a binding of Ca^{2+} to the membrane. The Ca^{2+} uptake is correlated with the Ca^{2+}-induced hydrolysis of ATP. The ratio of Ca^{2+} accumulated to ATP hydrolysed is about 2 over a wide range of ATP and Ca^{2+} concentrations. Below 10^{-9} M Ca^{2+} no uptake occurs. The isolated vesicles can establish a $[Ca^{2+}]_i/[Ca^{2+}]_e$ ratio of about 100000 across their membranes. Experiments with the calcium ionophores X537 and A23187 further support

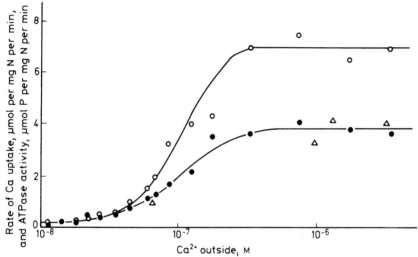

Figure 7.8   Dependence of calcium uptake (uuper curve) and Ca^{2+},Mg^{2+}-ATPase activity (lower curve) in sarcoplasmic membrane vesicles on external calcium concentration. Half-maximal activation for both activities is obtained at 2×10^{-7} M calcium. (After Hasselbach, Makinose and Fiehn, 1970, courtesy of Macmillan)

the conclusion that the major part of calcium sequestration by SR in the presence of ATP is due to transport of Ca^{2+} across the vesicular membrane, rather than to binding of Ca^{2+} to the membrane (Scarpa, Baldassare and Inesi, 1972).

The Ca^{2+},Mg^{2+}-ATPase system of the SR has been studied in some detail (Hasselbach, Makinose and Fiehn, 1970). In the presence of 5 mM ATP–Mg^{2+}, half-maximal activation of both the Ca^{2+} uptake and the activity occurs with 2×10^{-7} M Ca^{2+} (*Figure 7.8*) and maximal activation with 0.01–0.1 mM Ca^{2+} present in the medium. The system is not quite specific for ATP; other nucleoside triphosphates can also serve as substrates, but give lower rates. The enzyme is insensitive to oligomycin and ouabain (MacLennan, 1970). The reaction mechanism is similar to that for the Na^+,K^+-ATPase system: binding of ATP, transfer of the terminal phosphate under formation of a phosphorylated enzyme and hydrolysis of the latter. The formation of the phosphorylated enzyme is catalysed by Ca^{2+}, while its hydrolysis is catalysed by Mg^{2+} (Yamada and Tonomura, 1972). As in the case of Na^+,K^+-ATPase, *p*-nitrophenyl phosphate can also lead to a phosphorylated intermediate (Tenu, Ghelis and Chevalier, 1974). A tripeptide, prepared from the phosphorylated rabbit muscle enzyme by pronase treatment, appears to have the same composition (seryl- or threonylphosphoaspartyllysine) as the similarly obtained peptide from guinea-pig kidney Na^+,K^+-ATPase (Bastide *et al.*, 1973). Direct chemical identification by means of borohydride reduction, which converts the intermediate to homoserine, confirms the attachment of the phosphate group to an aspartyl residue (Degani and Boyer, 1973).

The enzyme system accounts for two-thirds of the SR protein. It can be solubilized with detergents such as Triton X-100, and then gives a band corresponding to molecular weight 90000 in sodium dodecyl sulphate gel electrophoresis (McFarland and Inesi, 1971). The solubilized enzyme reaggregates during dialysis in the presence of phospholipids with formation of membranous vesicles having full enzyme activity (Meissner, Conner and Fleischer, 1973), but addition of a proteolipid component of SR before reconstitution is required for high translocation activity (Racker and Eytan, 1975). Drugs which inhibit Ca^{2+} uptake (e.g. chlorpromazine, imipramine) also inhibit the Ca^{2+},Mg^{2+}-ATPase activity, presumably through interference with the hydrolysis of the phosphorylated enzyme (Hasselbach, Makinose and Fiehn, 1970). Phospholipid removal by treatment with phospholipase C appears to have the same effect (Martonosi *et al.*, 1971), but in this case it can be reversed by addition of unsaturated fatty acids or lysolecithin (The and Hasselbach, 1972, 1973).

7.4.3   Nerve

Most of our knowledge about Ca^{2+} movements in nerve comes from studies on squid axons (Baker, 1972). Experiments with aequorin injected into the axon have shown that 97 percent of the calcium in the normal resting nerve is located in the mitochondria, 2.5 percent is bound to organic anions (ATP, citrate, glutamate, etc.) and 0·1 percent (0.3 μM) is free ionized Ca^{2+} (Baker, Hodgkin and Ridgway, 1971). If Ca^{2+} were to be distributed passively across the axon membranes, the internal Ca^{2+} concentration would have to be 1 M. Hence, there must be an active Ca^{2+} transport system. Mitochondrial uptake is an active process (Blaustein and Hodgkin, 1969; Baker, Hodgkin and Ridgway, 1971), but cannot be solely responsible.

Evidence for active Ca^{2+} extrusion is provided by the following experiment (Blaustein and Hodgkin, 1969). A $^{45}Ca^{2+}$-loaded axon is incubated in artificial sea-water, to which 2 mM cyanide is added. After 1–2 hours a transient decrease (10–40 percent) in $^{45}Ca^{2+}$ efflux is observed, preceding the large rise in $^{45}Ca^{2+}$ efflux due to the massive release of Ca^{2+} by the mitochondria (*Figure 7.9*). The transient decrease coincides roughly with the fall in energy-rich phosphate compounds. This suggests that at least part of the calcium efflux is energy-dependent. The high Ca^{2+} efflux following the transient decrease is mainly a Ca^{2+}–Na^+ exchange, since removal of external Na^+ nearly abolishes it (Baker, 1972). This exchange mechanism has the following properties: (a) a considerable temperature dependence ($Q_{10} \simeq 3$), (b) insensitivity to ouabain, (c) inhibition by external lanthanum ions, (d) partial inhibition upon removal of external Ca^{2+}. The Ca^{2+} efflux from axons not treated with cyanide is also a Ca^{2+}–Na^+ exchange, and has the same properties. The Ca^{2+} –Na^+ exchange depends upon the existence of an electrochemical Na^+ gradient, since reversal of the Na^+ gradient by removal of external Na^+ causes Ca^{2+} influx.

There is no clear proof for an energy requirement for the Ca^{2+}–Na^+ exchange mechanism. The transient decrease in the Ca^{2+} efflux after cyanide treatment (*Figure 7.9*) is suggestive. Injection of the enzyme apyrase into the axoplasm to destroy ATP also reduces the Ca^{2+} efflux, which further supports the conclusion that energy is required. Presumably the energy is used by the Na^+,K^+-ATPase pump to maintain the Na^+ gradient, which causes

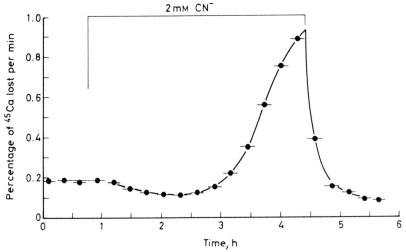

Figure 7.9   Evidence for active extrusion and active mitochondrial uptake of calcium in the squid giant axon. The transient decrease in ^{45}Ca efflux coincides with the fall in energy-rich phosphate compounds due to the effect of cyanide. The subsequent sharp increase in ^{45}Ca efflux is due to release of calcium from the mitochondria. (After Blaustein and Hodgkin, 1969, courtesy of The Physiological Society)

the passive Na^+ influx to which the Ca^{2+} efflux is coupled. Calculation shows that only 2–4 percent of the power of the Na^+,K^+-ATPase pump would be required for the Ca^{2+} efflux (Brinley, 1973). The stoichiometry of the exchange mechanism is still uncertain. If there were a neutral exchange of 1 Ca^{2+} for 2 Na^+, the gradients would be related as follows.

$$\frac{[Ca^{2+}]_i}{[Ca^{2+}]_e} = \frac{[Na^+]_i^2}{[Na^+]_e}$$

Since $[Na^+]_i/[Na^+]_e \simeq 0.1$, the calcium gradient would be only 0.01 whereas in reality it is about 0.00003. This ratio would be obtained if 1 Ca^{2+} exchanged for 3 Na^+ with the simultaneous countermovement of 1 K^+.

$$\frac{[Ca^{2+}]_i}{[Ca^{2+}]_e} = \left(\frac{[Na^+]_i}{[Na^+]_e}\right)^3 \frac{[K^+]_e}{[K^+]_i} \simeq 0.00003$$

Hence a small change in $[Na^+]_i$ could cause a large change in Ca^{2+} efflux.

Since the system rests ultimately on the activity of the Na^+,K^+-ATPase pump, ouabain should have an effect on the Ca^{2+} efflux. Baker (1972) was not able to demonstrate this in the squid axon, but it has been observed in brain slices (Tower, 1968).

Hence in the squid axon under physiological conditions the Na^+,K^+-ATPase system appears to drive a Ca^{2+} efflux, coupled to the Na^+ gradient generated by the pump system. This Ca^{2+} efflux, aided by the uptake of Ca^{2+} by the mitochondria, appears to keep the internal Ca^{2+} level low (ca. 0.3 μM), which is necessary to prevent release of transmitter. Stimulation of the nerve leads to passive Ca^{2+} entry, increase of the internal Ca^{2+} level, and thereby to release of the transmitter at the synapse (Baker, 1972). In contrast to erythrocyte and muscle, the nerve would not seem to require a Ca^{2+},Mg^{2+}-ATPase pump system to regulate its internal Ca^{2+} level. However, Ochs (1974) has recently reported the presence of a Ca^{2+},Mg^{2+}-ATPase in particulate matter of cat sciatic nerve.

7.5   ANION-ACTIVATED ATPase AND ANION TRANSPORT

We shall limit our discussion to the transport of chloride and bicarbonate, since these have been studied most extensively and only for these two anions has the existence of an ATPase type of transport been suggested.

In several tissues the movements of Cl^- and HCO_3^- appear to be passive, for example muscle (Ashley, 1970), brain (Marchbanks, 1970), erythrocytes (Garrahan, 1970) and the proximal tubule of the kidney (Burg and Grantham, 1970).

7.5.1   Tissues with active anion transport

In nerve about half of the Cl^- influx may be active, since this part is abolished by dinitrophenol (Keynes, 1963). Ouabain does not affect the Cl^- influx, hence it appears to be independent of the Na^+,K^+-ATPase cation transport system. Chloride efflux is not affected by dinitrophenol, and thus seems to be passive.

In frog skin there is evidence for an active inward Cl^- flux. Chloride moves from the mucosal to the serosal side in the absence of an electrochemical potential difference, and this movement is inhibited by dinitrophenol and iodoacetic acid in *Rana pipiens* and *R. esculenta* (Martin, 1964; Martin and Curran, 1966). A similar Cl^- influx in the frog *Leptodactylus ocellatus* is inhibited by ouabain and Cu^{2+} (Zadunaisky, Candia and Chiarandini, 1963; Zadunaisky and De Fisch, 1964).

In intestine there is evidence for active anion transport from experiments *in vivo*, which suggest the existence of active Cl^- and HCO_3^- fluxes in the ileum and active movement of HCO_3^-, but not of Cl, in the colon (Edmonds, 1970). However, experiments *in vitro* showed only passive movements for both anions in ileum and colon. No explanation for this discrepancy has been found so far.

In the gall bladder there is a parallel transport of Na^+ and Cl^- from the luminal to the serosal side. This might imply a neutral pump for NaCl or the existence of separate sodium and chloride pumps. In the light of our current knowledge neither alternative seems to be correct. Substitution of Cl^- by a variety of other monovalent anions does not block Na^+ transport. Neither is there any direct evidence for a separate chloride pump. Sodium reabsorption is inhibited by metabolic inhibitors such as dinitrophenol and cyanide, as well as by ouabain. In agreement with this latter observation, the ouabain-sensitive Na^+,K^+-ATPase system is present, and it seems to be responsible for the coupled transport of Na^+ and Cl^- (van Os and Slegers, 1970, 1971). Electrophysiological studies by Frömter (1972) indicate that the Cl^- transport does not occur through the cell as for Na^+, but between the cells through the intercellular space and the tight junction (*Figure 7.13*). It has been suggested that this separate, extracellular Cl^- transport occurs in all epithelia with low transepithelial potential difference and electrical conductance (Frömter and Diamond, 1972). Sodium would be actively extruded from the cell into the intracellular space by means of the Na^+,K^+-ATPase system, which would then be located on the lateral side of the plasma membrane. Acting as an electrogenic sodium pump, it would create a potential difference across the tight junction, which would make the Cl^- ions move passively and isotonically with the Na^+ ions.

In the ciliary epithelium of the eye, which is responsible for the formation of aqueous humour, there is evidence that in the cat at least part of the Cl^- movement to the posterior chamber is active (Holland and Gipson, 1970). Likewise HCO_3^- seems to be transported actively in this direction (Kinsey and Reddy, 1959; Kinsey, 1970).

The gastric mucosa is the most extensively studied tissue with regard to active anion transport. Its most characteristic property is its ability to secrete HCl in considerable quantity. This involves transport of H^+, which is obviously active since it can occur against a concentration gradient of more than 1 000 000 and is strongly dependent on metabolic energy. The situation for Cl^- is rather complicated, inasmuch as it appears to be transported in three different ways: by active transport, by neutral exchange, and by passive diffusion (Forte, 1970). Transport of H^+ and active Cl^- transport are rather tightly coupled, when acid secretion takes place. However, in the resting gastric mucosa there is still active Cl^- transport even in the absence of an exchangeable anion. Hence, the chloride pump appears to be electrogenic.

Recent work in our laboratory has established that in the lizard gastric mucosa the acid secretion is ouabain-sensitive and that the Na^+,K^+-ATPase system is present and operative (Hansen *et al.*, 1972). Determinations of unidirectional fluxes in this system indicate that the serosal to mucosal Cl^- flux is also sensitive to ouabain, which can virtually abolish the net Cl^- flux in this direction (Hansen, 1972). The link between the transport of H^+ and Cl^- and the Na^+,K^+-ATPase system appears to be an indirect one, the maintenance of a high intracellular K^+ concentration being essential for acid secretion.

Previously, Kasbekar and Durbin (1965) were unable to detect Na^+,K^+-ATPase activity in frog gastric mucosa, but they observed a microsomal ATPase activity which is sensitive to thiocyanate, an inhibitor of gastric secretion, and appears to be stimulated by bicarbonate. This led them to suggest a role for this enzyme activity in acid secretion (Durbin and Kasbekar, 1965).

7.5.2   **Anion-sensitive ATPase**

The gastric microsomal ATPase activity in the experiments of Kasbekar and Durbin (1965) showed stimulations of 8 percent by 5 mM NaCl, 12 percent by 20 mM NaBr and 37 percent by 20 mM NaHCO$_3$ compared with the activity in the standard medium (2 mM ATP, 1 mM Mg^{2+}, 20 mM Tris-glucuronate, pH 8.4). An additive effect of halide and bicarbonate stimulation was not found. Addition of 2 mM NaSCN inhibited the basal activity by 43 percent and the activity in the presence of 20 mM NaHCO$_3$ by 27 percent. By

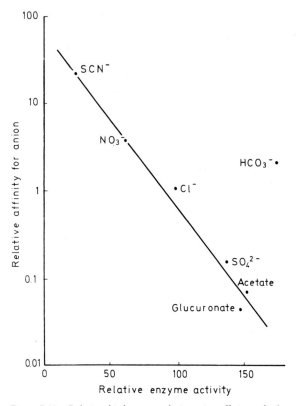

Figure 7.10   Relationship between relative anion affinity and relative enzyme activity of anion-sensitive ATPase in lizard gastric mucosa in the presence of each anion indicated. (After de Pont, Hansen and Bonting, 1972, courtesy of Associated Scientific Publishers)

analogy from what was known about the mechanism of the ouabain-sensitive Na$^+$,K$^+$-ATPase system, it was then assumed that this thiocyanate-sensitive, bicarbonate- and chloride-stimulated ATPase system would represent a system similar to the thiocyanate-sensitive transport of Cl$^-$ and HCO$_3^-$ in the gastric mucosa (Durbin and Kasbekar, 1965).

However, further studies by de Pont, Hansen and Bonting (1972) of the effects of anions on this gastric mucosal ATPase system have revealed that it behaves quite differently from the Na$^+$,K$^+$-ATPase. Unlike Kasbekar and

Durbin, they used media of constant, physiological ionic strength (100 mM Tris buffer, pH 7.5 and 8.4, 2 mM ATP, 2 mM Mg^{2+}, 60 mM Na^+) in which the anionic species (Cl^- in the control medium) is replaced by other anions without changing the cation concentrations. This excludes cation effects and also effects of ionic strength. An effect of ionic strength does indeed occur: increasing the NaCl concentration beyond 50 mM markedly decreases the activity. This effect shows that the apparent optimal activity at 25 mM $NaHCO_3$, found by Kasbekar and Durbin (1965), is only accidental, and has nothing to do with the fact that this happens to be the concentration of HCO_3^- occurring in blood plasma. Instead the activity depends very strongly on the anion present. Systematic investigation of the anion effect has led to the finding that the anion sensitivity can be expressed in a quantitative way by assuming the presence of a single anionic binding site (de Pont, Hansen and Bonting, 1972). The relative affinities of various anions, calculated from thiocyanate inhibition experiments and from anion substitution experiments, show good agreement. There is a linear relationship between the logarithm of the relative affinity and the enzyme activity with increasing activity at decreasing affinity in the order: thiocyanate > nitrate > chloride > sulphate > glucuronate and acetate (*Figure 7.10*). Only bicarbonate gives about double the activity expected from this relationship. It appears that rather than the enzyme being stimulated by anions it is inhibited, the effect increasing with increasing affinity of the anion for the enzyme.

The main differences between the Na^+,K^+-activated ATPase system and the anion-sensitive ATPase activity can be summarized as follows:

1. The Na^+,K^+-ATPase system requires the presence of both Na^+ and K^+ in order to display any activity. The anion-sensitive ATPase, on the other hand, shows activity in the presence of any single anion. The higher activity in the presence of HCO_3^- represents stimulation by such a single ion.
2. Whereas the Na^+,K^+-ATPase has a Na^+-activated phosphorylation step and a K^+-activated dephosphorylation step in its mechanism, the presence of only a single anion-binding site argues against a Cl^--activated phosphorylation and a HCO_3^--activated dephosphorylation, as suggested by Durbin and Kasbekar (1965).
3. The inhibition of Na^+,K^+-ATPase by ouabain differs strongly from that of the anion-sensitive ATPase by thiocyanate. The half-maximal inhibitory concentration of ouabain for Na^+,K^+-ATPase is about a thousand times lower than that of thiocyanate for the latter enzyme. In Na^+,K^+-ATPase there exists a specific ouabain-binding site, different from the cation-binding sites, while the apparent inhibition by thiocyanate is simply due to its high affinity for the general anion-binding site on the anion ATPase.
4. While for the Na^+,K^+-ATPase system a fixed molar ratio of 3 exists between cations transported and ATP hydrolysed for a large variety of transport processes (*Table 7.4*), the ratio between H^+ transported and ATP hydrolysed by the anion ATPase is only 0.06–0.17, as calculated from the data given by Kasbekar and Durbin (1965) and Hansen et al. (1972).

These differences indicate that the anion-sensitive ATPase cannot function in a manner similar to the way in which the Na^+,K^+-ATPase system operates in active cation transport.

Nonetheless, there are several indications for an involvement of the anion-sensitive ATPase in gastric secretion. There is reasonable, though not complete, agreement between the inhibitory concentration of sodium thiocyanate and esters of thiocyanate and isothiocyanate for the enzyme activity and the acid secretion in bullfrog gastric mucosa (Wong, Kasbekar and Forte, 1969). After stimulation of the gastric mucosa with the tetragastrin or histamine the activity of the enzyme is increased (Narumi and Kanna, 1973). Histochemical studies indicate the localization of the enzyme at the microvilli on the apical side of the oxyntic peptic cells in toad gastric mucosa (Koenig and Vial, 1970. It has also been observed in rat kidney (Katz and Epstein, 1971), in the midgut of *Hyalophora cecropia* (Turbeck, Nedergaard and Kruse, 1968), in the membrane fraction of the mammalian pancreas (Simon and Thomas, 1972), in the submandibular gland of dog (Izutsu and Siegel, 1972) and rabbit (Simon, Kinne and Knauf, 1972), in the seminiferous tubules of rodents (Setchell, Smith and Mumm, 1972) and in rat brain (Kimelberg and Bourke, 1973). Its occurrence suggests that it may be involved with HCO_3^- transport rather than with HCl secretion. The observation that its occurrence coincides with that of carbonic anhydrase (Wiebelhaus *et al.*, 1971) further strengthens this suggestion.

The main properties of the enzyme are as follows: ATP is the best substrate (50 percent activity with GTP, 10 percent with UTP), Mg^{2+} is required for activity (optimal activity at a Mg^{2+}:ATP ratio of 0.5–1.0), the pH optimum is between 8 and 9, the enzyme is SH-dependent and loses activity upon removal of phospholipid. The enzyme has been solubilized by means of the detergent Triton X-100 from gastric mucosa (Wiebelhaus *et al.*, 1971; Blum *et al.*, 1971; Sachs *et al.*, 1972; Spenny *et al.*, 1973) and from pancreas (Simon, Kinne and Sachs, 1972). Bicarbonate stimulates the solubilized enzyme up to threefold. The inhibitory effects of diisopropyl fluorophosphate and methanesulphonyl chloride suggest the presence of a serine group in the active centre. Evidence for the formation of a hydroxylamine-sensitive phosphorylated intermediate has been obtained with a microsomal fraction of gastric mucosa (Tanisawa and Forte, 1971). Its formation is increased in the presence of thiocyanate, suggesting that this substance slows down the dephosphorylation. Its stability in relation to pH and its sensitivity to hydroxylamine suggest that it is an acyl phosphate compound.

A serious problem is the fact that mitochondria of various tissues contain an anion-sensitive ATPase with similar properties (Fanestil, Baird Hastings and Mahowald, 1963; Grisolia and Mendelson, 1974; Weiner, 1975). Since microsomal fractions are usually substantially contaminated by light mitochondria or mitochondrial fragments, this raises the question whether the microsomal activity derives from the plasma membrane or is of mitochondrial origin. The latter possibility appears to be likely in the case of rat liver, where it was possible with special precautions to isolate microsomes uncontaminated by mitochondria, but in these microsomes anion-sensitive ATPase was not detectable (Izutsu and Siegel, 1975). Likewise, Soumarmon *et al.* (1974) found, after density gradient centrifugation of microsomes from rat gastric mucosa, a HCO_3^--activated ATPase coinciding with cytochrome *c* oxidase in the denser part of the gradient, while a Mg^{2+}-ATPase activity slightly inhibited by HCO_3^- coincides with 5'-nucleotidase, a plasma-membrane marker. Upon critical review of all reports on a microsomal anion-sensitive ATPase activity it appears that no clear-cut evidence that this enzyme activity is localized in plasma membranes exists, but abundant evidence for a mitochondrial localization is available. In ghosts of rabbit erythrocytes (which contain no mitochondria) a bicarbonate-sensitive ATPase has been demonstrated, but this is different in that it is not inhibited by thiocyanate or acetazolamide (Duncan, 1975).

Hence, it is probably fair to say that of the three membrane ATPases discussed in this chapter, namely Na^+, K^+-ATPase, Ca^{2+}, Mg^{2+}-ATPase and anion-sensitive ATPase, the transport role of the last of these is least certain.

7.6   COUPLED TRANSPORT OF SOLUTES AND SODIUM

Since about 1960 it has been known that the transport of several important metabolites, such as glucose and amino acids, is coupled to the operation of the sodium pump in many types of cell (Schultz and Curran, 1970). This means that these metabolites are transported by a form of facilitated diffusion. It further demonstrates another important function of the sodium pump in addition to its role in the excitation mechanism of cells such as nerve and muscle.

The first indication came from studies of amino acid uptake by erythrocytes and Ehrlich ascites tumour cells (*see* Christensen, 1962). Replacement of 50 percent of the Na^+ in the medium by K^+ reduces the level to which glycine, alanine and the non-natural amino acid α,γ-diaminobutyric acid are accumulated. Since the potassium level of the cell is reduced during amino acid accumulation, it was thought that the K^+ gradient would be responsible for the accumulation process.

However, experiments on glucose transport in the isolated intestine showed that not the K^+ gradient but the transport of Na^+ is primarily responsible. Complete replacement of Na^+ in the mucosal medium by Li^+, Mg^{2+} or choline abolishes the transport of the non-metabolized glucose analogue 3-O-methylglucose (Csáky and Thale, 1960). The same is true for glycine uptake in Ehrlich ascites tumour cells (Kromphardt et al., 1963), which is, moreover, non-competitively inhibited by ouabain (Bittner and Heinz, 1963). Further experiments on the intestine showed that the mere presence of Na^+ ions on the mucosal side is not sufficient, but that active Na^+ transport is required. Addition of cardiac glycosides such as ouabain and thevetin in the presence of mucosal Na^+ inhibits both Na^+ transport and carbohydrate transport Csáky, Hartzog and Fernald, 1961). Ouabain acts only on the serosal side Experiments with intestinal strips showed that in the presence of Na^+ in the medium an exchange occurs between previously accumulated carbohydrate and external carbohydrate; this exchange is not inhibited by ouabain, but is inhibited by phlorizin, a potent inhibitor of carbohydrate uptake (Crane, 1965).

Such observations led to the model presented in *Figure 7.11*, where E_2 is the ouabain-sensitive Na^+,K^+-ATPase sodium pump, which maintains a passive Na^+ influx across the mucosal membrane. This Na^+ influx activates the phlorizin-sensitive facilitated diffusion system E_1 for glucose. The inward mucosal glucose transport could occur by means of a 'carrier' binding both Na^+ and glucose and then losing them on the cytoplasmic side. Supporting evidence for this model comes from the good agreement between half-maximal activating concentration of the carbohydrate (or amino acid) for Na^+ transport and for glucose transport, and also from the approximately 1:1 ratio between Na^+ flux and carbohydrate flux (Schultz and Zalusky, 1964). Sodium-dependent binding of L-histidine to a mucosal brush-order preparation of hamster intestine (Faust, Burns and Misch, 1970) and Na^+-dependent, phlorizin-sensitive binding of glucose to a brush-border preparation of rat kidney (Frasch *et al.*, 1970) have been reported.

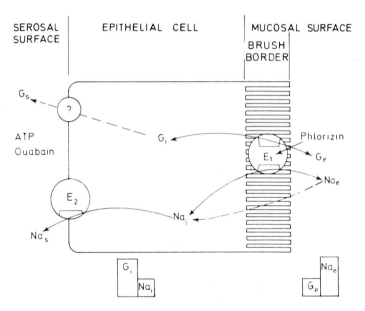

Figure 7.11   Model for the coupled transport of Na^+ *and glucose in the intestine.* E_1 *is the phlorizin-sensitive facilitated diffusion system for glucose,* E_2 *the ouabain-sensitive sodium pump. The heights of the columns at the bottom indicate the prevailing concentration of glucose and* Na^+ *mucosally and intracellularly. Subscript s refers to serosal, i to intracellular and e to extracellular. (From Stein, 1967, courtesy of Academic Press)*

Further confirmation for this 'ion-gradient' model has been derived from studies on amino acid transport with reconstituted erythrocyte ghosts, in which both the internal and external medium can be controlled (Vidaver, 1964a, b, c). Glycine accumulation was shown to be dependent on the $[Na^+]_e : [Na^+]_i$ ratio and independent of the absolute values of $[Na^+]_e$ and $[Na^+]_i$. When the ratio is reversed, the glycine influx changes to an efflux. The saturable component of the glycine influx obeys Michaelis–Menten kinetics. The K_m for glycine is dependent on the square of the Na^+ concentration, suggesting that 1 glycine molecule is transported together with 2 Na^+ ions, which has been confirmed in this case but does not hold for other cells (Christensen, 1970).

Two further findings deserve to be mentioned. The co-transport carrier itself does not appear to require energy to function, but rather it depends on the energy supplied by the Na^+ gradient. Depleting Ehrlich ascites tumour cells of ATP by incubation with cyanide does not stop glycine influx as long as a Na^+ gradient is present (Eddy, Mulcahy and Thomson, 1967). A possible explanation for the way in which the co-transport carrier loses the transported metabolite on the cytoplasmic side of the membrane is suggested by observations on the transport of 6-deoxyglucose in the everted intestinal sac (Crane, Forstner and Eichholz, 1965). The K_m value for the carbohydrate, determined at a series of mucosal Na^+ concentrations obtained by replacing Na^+ with K^+, increases from 4 mM at 145 mM Na^+ to 40 mM at 24 mM Na^+ and to 100 mM in the absence of Na^+. This would indicate that the affinity of the carrier for carbohydrate is much lower on the cytoplasmic side with a low Na^+ concentration than at the mucosal side with a high Na^+ concentration. Similar observations have been made for the uptake of glycine in rabbit ileum (Peterson, Goldner and Curran, 1970). Since the increase in K_m for 6-deoxyglucose is larger when Na^+ is replaced by K^+ than when it is replaced by the cation Tris, this suggests that the carrier affinity for the carbohydrate is regulated by Na^+ and K^+. At the external side of the membrane the high $Na^+:K^+$ ratio would favour binding of the carbohydrate, while on the cytoplasmic side the low $Na^+:K^+$ ratio would favour dissociation of carbohydrate.

Although a strong case can thus be made for Crane's model of Na^+-dependent solute transport, there are puzzling elements yet to be resolved. For example, both on energetic grounds and from observations of the effects of external K^+ concentrations, it has been suggested that not only the Na^+ gradient but also the K^+ gradient can contribute energy to the solute transport process (Eddy and Hogg, 1969; Schafer and Heinz, 1971). In Crane's model this might mean that K^+ ions move out of the cell on the returning, substrate-free carrier. Some investigators claim that even this would not supply sufficient energy in all cases (Schafer and Heinz, 1971; Heinz, 1972, pp. 116–129). An alternative model, involving direct coupling of Na^+ and solute transport at the same site on the intestinal brush-border membrane, has been proposed by Kimmich (1973). This model, however, requires a rather contrived explanation for the fact that ouabain inhibits the two transport systems only when applied to the serosal side; under these circumstances, ouabain would have to be transported actively across the serosal membrane in order to act on the interior side of the mucosal membrane. However, so far ouabain has always been found to act at the external side of membranes. The model is also in conflict with the observation of Quigley and Gotterer (1969) that the isolated brush border has only 0.3 percent as much Na^+,K^+-ATPase activity as the other plasma membranes of rat intestinal mucosa. Placing the directly coupled transport system on the serosal side would conflict with the exclusive localization of the carbohydrate transport system in the brush border (Stirling and Kinter, 1967; Stirling, 1967).

7.7   COUPLED TRANSPORT OF WATER AND SODIUM

In addition to the coupled transport of Na^+ with various solutes, the Na^+, K^+-ATPase system has another important role in coupled transport, namely in the transport of water by epithelial tissues. For many years it was believed that water in the nephron, gall bladder and intestine is transported actively. There was some evidence for this assumption; for example, resorption of plasma from the small intestine is possible in the dog, showing that water is transported against an osmotic gradient. Placing a piece of rabbit intestine between two chambers with identical Ringer's solution makes the water level on the serosal side rise, even if the latter was higher to start with. Thus water is transported against a hydrostatic pressure difference, a process that is stopped by the addition of a metabolic inhibitor.

However, in these cases there is transport not merely of water, but also of Na^+, both of which are transported isotonically. In addition, Brodsky *et*

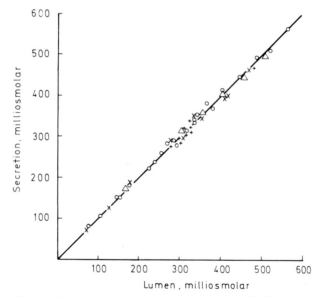

Figure 7.12 *Relationship between osmolarity of gall-bladder secretion and osmolarity of luminal bathing medium. (Δ) No HCO_3^- or glucose; ($\bigcirc$) HCO_3^- and glucose; ($\times$) HCO_3^-; ($+$) sucrose. (After Diamond, 1965, courtesy of the Society of Experimental Biology)*

al. (1955) were able to calculate that the water transport taking place in the kidney, if generated purely by an osmotic gradient, would require about a thousand times as much energy as is available from cellular energy metabolism.

These considerations led Diamond (1965) to the conclusion that there must be a coupling between water transport and salt or ion transport. He investigated the coupling mechanism in the isolated gall bladder. In a brilliant series of experiments he tested seven possibilities for this coupling mechanism: (a) classical osmosis, (b) filtration, (c) electro-osmosis, (d) pinocytosis, (e) local osmosis, (f) a double-membrane effect and (g) co-diffusion. He was able to exclude all but local osmosis, in which process active Na^+ transport generates an osmotic gradient, not *across* the epithelial layer, but *within* the layer. This process, in contrast to the last two mechanisms, requires that the secreted fluid be isotonic with the internal medium in the gall bladder over a wide concentration range, and this indeed proves to be the case (*Figure 7.12*). Thus, the primary process is active Na^+ transport, while water follows passively and isotonically.

Independently and nearly simultaneously, Kaye *et al.* (1966) succeeded in

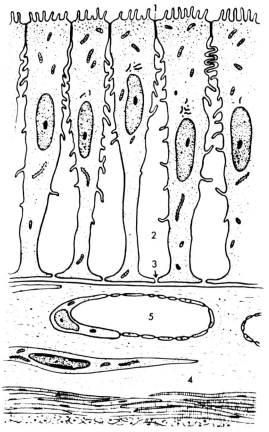

Figure 7.13   Structure of secreting gall bladder, showing tight junction (1), intercellular space (2) with tight junction (1) on the luminal side and narrow opening (3) towards the lamina propria (4), containing capillaries (5). Sodium enters the cell passively from the lumen and is transported actively across the lateral cell membranes into cellular space. The resulting local hypertonicity draws water through the cell into the space, with dilution to isotonicity of the fluid and distension of the space. The resulting hydrostatic pressure expels the fluid into the lamina propria, where it is carried away by the capillaries. (After Kaye et al., 1966, courtesy of The Rockefeller University Press)

visualizing the process of local osmosis electron microscopically. Gall-bladder epithelium, fixed in the resting state (cooling, presence of ouabain, or absence of Na^+) shows no spaces between adjacent cells, but when fixed in a state of active reabsorption there are large intercellular spaces (*Figure 7.13*). These spaces are closed off near the luminal side by a 'tight junction'. On the opposite side there is a narrow opening towards the lamina propria.

After fixation with osmium tetroxide–pyroantimonate, high Na^+ concentrations are detected in the distended intercellular spaces. Histochemically, ATPase activity is noticed along the lateral cell membranes. Assuming that this ATPase activity represents, at least in part, the Na^+,K^+-ATPase pump, this means that Na^+ ions are pumped from the cells across lateral membranes into the intercellular spaces, so causing a local hypertonicity in these spaces, which leads to passive water movement across the lateral membranes. The inflow of water gradually abolishes the hypertonicity, but at the same time distends the intercellular space. This creates a hydrostatic pressure which leads to expulsion of the isotonic solution through the narrow opening into the lamina propria, where it is carried away by the capillaries. The Na^+ and water so removed from the epithelial cells are supplemented by passive diffusion from the lumen across the apical plasma membrane.

In a further extension of this mechanism Diamond concluded that the Na^+ transport sites must be located on the lateral membrane close to the tight junction in order to obtain an isotonic secretion (Diamond and Bossert, 1967). However, Slegers has calculated that a distribution of the pump sites over the entire lateral membrane, but with decreasing activity from the tight junction to the opening, should be capable of giving an isotonic secretion (Slegers and van't Hoff-Grootenboer, 1971).

The isotonic transport of water coupled to active Na^+ transport by means of local osmosis is now recognized as a rather common phenomenon. It has been demonstrated in intestine (Parsons, 1967), kidney collecting tubules (Ganote et al., 1968), pancreas (Ridderstap, 1969), and appears to operate in frog skin and toad bladder.

7.8   CONCLUDING REMARKS

In this chapter we have been able to give only a very limited survey of active transport through mammalian cell membranes. A wealth of evidence is available at present, and much more is yet to be discovered. Our main emphasis has been on the pump systems responsible for active transport. The only pump systems which have been chemically identified, even partially, are the transport ATPases: the Na^+,K^+-ATPase, responsible for Na^+ and K^+ transport; Ca^{2+},Mg^{2+}-ATPase responsible for Ca^{2+} transport; and anion ATPase, which may function in anion transport.

The Na^+,K^+ATPase system, which has been studied most extensively, appears to have a multiple function: maintenance of cation gradients and electrical potentials across the plasma membrane, which are essential in excitatory processes such as in nerve metabolism; the coupled, isotonic transport of water essential in absorption and secretion as in intestine, gall bladder and kidney; and coupled transport of metabolites and Na^+. Much less is known about the role of the other two transport ATPases. The main function of the Ca^{2+},Mg^{2+}-ATPase seems to be to maintain an extremely low free Ca^{2+} concentration in the cytoplasm, which may be necessary in general for the maintenance of a sufficiently high permeability of the plasma membrane and to permit calcium to function as a transmitter in processes like muscle contraction and synaptic transmission.

There may well exist several other transport ATPases. Their discovery will follow the discovery of specific inhibitors, such as ouabain in the case of Na$^+$, K$^+$-ATPase, in order to identify and relate the particular ATPase to the transport function. In addition, much remains to be learned about the molecular mechanism of the known pump systems. Hence, the field of active transport across plasma membranes continues to offer a great challenge to present and future investigators.

REFERENCES

ALBERS, R. W., FAHN, S. and KOVAL, G. J. (1963). *Proc. natn. Acad. Sci. U.S.A.*, **50**:474.
ALBERS, R. W., KOVAL, G. J. and SIEGEL, G. J. (1968). *Molec. Pharmac.*, **4**:324.
ASHLEY, C. C. (1970). *Membranes and Ion Transport*, Vol. 2, pp. 1–31. Ed. E. E. BITTAR. London; Wiley–Interscience.
ATKINSON, A., GATENBY, A. D. and LOWE, A. G. (1971). *Nature, Lond.*, **233**:145.
BAKER, P. F. (1972). *Prog. Biophys. molec. Biol.*, **24**:177.
BAKER, P. F., HODGKIN, A. L. and RIDGWAY, E. B. (1971). *J. Physiol., Lond.*, **218**:709.
BAKER, P. F., BLAUSTEIN, M. P., KEYNES, R. D., MANIL, J., SHAW, T. I. and STEINHARDT, R. A. (1969). *J. Physiol., Lond.*, **200**:459.
BAKKEREN, J. A. J. M. and BONTING, S. L. (1968). *Biochim. biophys. Acta*, **150**:467.
BAKKEREN, J. A. J. M., VAN DER BEEK, J. A. C. M. and BONTING, S. L. (1971). *Pflügers Arch. ges. Physiol.* **325**:77.
BASTIDE, F., MEISSNER, G., FLEISCHER, S. and POST, R. L. (1973). *J. biol. Chem.*, **248**:8385.
BERNSTEIN, R. E. (1954). *Science, N.Y.*, **120**:459.
BITTNER, J. and HEINZ, E. (1963). *Biochim. biophys. Acta*, **74**:392.
BLAUSTEIN, M. P. and HODGKIN, A. L. (1969). *J. Physiol., Lond.*, **200**:497.
BLOCK, J. B. and BONTING, S. L. (1964). *Enzymol. biol. clin.*, **4**:183.
BLUM, A. L., SHAH, G., PIERRE, T. ST., HELANDER, H. F., SUNG, C. P., WIEBELHAUS, V. D. and SACHS, G. (1971). *Biochim. biophys. Acta*, **249**:101.
BOND, G. M. and GREEN, J. W. (1971). *Biochim. biophys. Acta*, **241**:393.
BONTING, S. L. (1965). *Invest. Ophthal.*, **4**:723.
BONTING, S. L. (1966). *Comp. Biochem. Physiol.*, **17**:953.
BONTING, S. L. (1970). *Membranes and Ion Transport*, Vol. 1, pp 257–363. Ed. E. E. BITTAR. London; Wiley–Interscience.
BONTING, S. L. and BECKER, B. (1964). *Invest. Ophthal.*, **3**:523.
BONTING, S. L. and CANADY, M. R. (1964). *Am. J. Physiol.*, **207**:1005.
BONTING, S. L. and CARAVAGGIO, L. L. (1963). *Archs Biochem. Biophys.*, **101**:37.
BONTING, S. L., CARAVAGGIO, L. L. and HAWKINS, N. M. (1962). *Archs Biochem. Biophys.*, **98**:413.
BONTING, S. L., CARAVAGGIO, L. L. and HAWKINS, N. M. (1963). *Archs Biochem. Biophys.*, **101**:37.
BONTING, S. L., HAWKINS, N. M. and CANADY, M. R. (1964). *Biochem. Pharmac.*, **13**:13.
BONTING, S. L., CARAVAGGIO, L. L., CANADY, M. R. and HAWKINS, N. M. (1964). *Archs Biochem. Biophys.*, **106**:49.
BRINLEY, F. J., JR. (1973). *Fedn Proc. Fedn Am. Socs exp. Biol.*, **32**:1735.
BRODSKY, W. A., REHM, W. S., DENNIS, W. H. and MILLER, D. G. (1955). *Science, N.Y.*, **121**:302.
BURG, M. B. and GRANTHAM, J. J. (1970). *Membranes and Ion Transport*, Vol. 3, pp. 49–77. Ed. E. E. BITTAR. London; Wiley–Interscience.
CHA, Y. N., BAK C. S. and LEE, K. S. (1971). *J. gen. Physiol.*, **57**:202.
CHRISTENSEN, H. N. (1962). *Biological Transport*. New York; Benjamin.
CHRISTENSEN, H. N. (1970). *Membranes and Ion Transport*, Vol. 1, pp. 365–394. Ed. E. E. BITTAR. London; Wiley–Interscience.
CHURCHILL, L. and HOKIN, L. E. (1976). *Biochim. biophys. Acta*, in press.
COHN, W. E. and COHN, E. R. (1939). *Proc. Soc. exp. Biol. Med.*, **41**:445.
COLLINS, R. C. and ALBERS, R. W. (1972). *J. Neurochem.*, **19**:1209.
CONNELLY, C. M. (1959). *Rev. Mod. Phys.*, **31**:475.
CORRIE, W. S. and BONTING, S. L. (1966). *Biochim. biophys. Acta*, **120**:91.
COSMOS, E. and HARRIS, E. J. (1961). *J. gen. Physiol.*, **44**:1121.
CRANE, R. K. (1965). *Fedn. Proc. Fedn. Am. Socs exp. Biol.*, **24**:1000.
CRANE, R. K., FORSTNER, G. and EICHHOLZ, A. (1965). *Biochim. biophys. Acta*, **109**:467.
CROSS, S. B., KEYNES, R. D. and RYBOVÁ, R. (1965). *J. Physiol., Lond.*, **181**:865.

CSÁKY T. Z., HARTZOG, H. G. and FERNALD, G. W. (1961). *Am. J. Physiol.*, **200**:459.

CSÁKY, T. Z. and THALE, M. (1960). *J. Physiol., Lond.*, **151**:59.

DEGANI, C. and BOYER, P. D. (1973). *J. biol. Chem.*, **248**:9222.

DE PONT, J. J. H. H. M., HANSEN, T. and BONTING, S. L. (1972). *Biochim. biophys. Acta*, **274**:189.

DE PONT, J. J. H. H. M., VAN PROOYEN-VAN EEDEN, A. and BONTING, S. L. (1973). *Biochim. biophys. Acta*, **323**:487.

DIAMOND, J. M. (1965). *Symp. Soc. exp. Biol.*, **19**:329.

DIAMOND, J. M. and BOSSERT, W. H. (1967). *J. gen. Physiol.*, **50**:2061.

DUNCAN, C. J. (1975). *Life Sci.*, **16**:955.

DUNHAM, E. T. and GLYNN, I. M. (1961). *J. Physiol., Lond.*, **156**:274.

DURBIN, R. P. and KASBEKAR, D. K. (1965). *Fedn Proc. Fedn Am. Socs exp. Biol.*, **24**:1377.

EDDY, A. A., and HOGG, M. C. (1969). *Biochem. J.*, **114**:807.

EDDY, A. A., MULCAHY, M. and THOMSON, P. J. (1967). *Biochem. J.*, **103**:863.

EDMONDS, C. J. (1970). *Membranes and Ion Transport*, Vol. 3, pp. 79–110. Ed. E. E. BITTAR. London; Wiley–Interscience.

FANESTIL, D. D., BAIRD HASTINGS, A. and MAHOWALD, T. A. (1963). *J biol. Chem.*, **238**:836.

FAUST, R. G., BURNS, M. J. and MISCH, D. W. (1970). *Biochim. biophys. Acta*, **219**:507.

FITZSIMONS, E. J. and SENDROY, J. (1961). *J. biol. Chem.*, **236**:1595.

FORTE, J. G. (1970). *Membranes and Ion Transport*, Vol. 3, pp. 111–165. Ed. E. E. BITTAR. London; Wiley-Interscience.

FRASCH, W., FROHNERT, P. P., BODE, F., BAUMONN, K. and KINNE, R. (1970). *Pflügers Arch. ges. Physiol.*, **320**:265.

FRÖMTER, E. (1972). *J. Membrane Biol.*, **8**:259.

FRÖMTER, E. and DIAMOND, J. M. (1972). *Nature, New Biol.*, **235**:9.

GANOTE, C. E., GRANTHAM, J. J., MOSES, H. L., BURG, M. B. and ORLOFF, J. (1968). *J. Cell Biol.*, **36**:355.

GARRAHAN, P. J. (1970). *Membranes and Ion Transport*, Vol. 2, pp. 185–215. Ed. E. E. BITTAR. London; Wiley-Interscience.

GLYNN, I. M. (1956). *J. Physiol., Lond.*, **134**:278.

GLYNN, I. M. (1962). *J. Physiol., Lond.*, **160**:18P.

GRISOLIA, S. and MENDELSON, J. (1974). *Biochem. biophys. Res. Commun.*, **58**:968.

HAFKENSCHIED, J. C. M. and BONTING, S. L. (1968). *Biochim. biophys. Acta*, **151**:204.

HANSEN, T. (1972). *Thesis*, University of Nijmegen.

HANSEN, T., BONTING, S. L., SLEGERS, J. F. G. and DE PONT, J. J. H. H. M. (1972). *Pflügers Arch. ges. Physiol.*, **334**:141.

HARRIS, E. J. (1967). *J. Physiol., Lond*, **193**:455.

HARRIS, E. J and MAIZELS, M. (1952). *J. Physiol., Lond.*, **118**:40.

HARRIS, J. E. (1941). *J. biol. Chem.*, **141**:579.

HASSELBACH, W. (1963). *Proc. R. Soc., Ser. B*, **160**:501.

HASSELBACH, W. and MAKINOSE, M. (1961). *Biochem. Z.*, **333**:518.

HASSELBACH, W. and MAKINOSE, M. (1962). *Biochem. biophys. Res. Commun.*, **7**:132.

HASSELBACH, W. and MAKINOSE, M. (1963). *Biochem. Z.*, **339**:94.

HASSELBACH, W., MAKINOSE, M. and FIEHN, W. (1970). *Calcium and Cellular Function*, pp. 75–84. Ed. A. W. CUTHBERT. London; Macmillan.

HEINZ, E. (1972). *Sodium-Linked Transport of Organic Solutes*. Berlin; Springer-Verlag.

HILDEN, S. and HOKIN, L. E. (1975). *J. biol. Chem.*, **250**:6296.

HOKIN, L. E., DAHL, J. L., DEUPREE, J. D., DIXON, J. F., HACKNEY, J. F. and PERDUE, J. F. (1973). *J. biol. Chem.*, **248**:2593.

HOLLAND, M. G. and GIPSON, C. C. (1970). *Invest. Ophthal.*, **9**:20.

INESI, G. (1972). *A. Rev. Biophys. Bioengng*, **1**:191.

INESI, G., EBASHI, S. and WATANABE, S. (1964). *Am. J. Physiol.*, **207**:1339.

IZUTSU, K. T. and SIEGEL, I. A. (1972). *Biochim. biophys. Acta*, **284**:478.

IZUTSU, K. T. and SIEGEL, I. A. (1975). *Biochim. biophys. Acta*, **382**:193.

JÄRNEFELT, J. (1972). *Biochim. biophys. Acta*, **266**:91.

JAY, A. W. L. and BURTON, A. C. (1969). *Biophys. J.*, **9**:115.

JØRGENSEN, P. L. (1974a). *Biochim. biophys. Acta*, **356**:36.

JØRGENSEN, P. L. (1974b). *Biochim. biophys. Acta*, **356**:53.

KAHLENBERG, A., GALSWORTHY, P. R. and HOKIN, L. E. (1967). *Science, N.Y.*, **157**:434.

KAHLENBERG, A., GALSWORTHY, P. R. and HOKIN, L. E. (1968). *Archs Biochem. Biophys.*, **126**:331.

KASBEKAR, D. K. and DURBIN, R. P. (1965). *Biochim. biophys. Acta*, **105**:472.

KATZ, A. I. and EPSTEIN, F. H. (1971). *Enzyme*, **12**:499.

KAYE, G. I., WHEELER, H. O., WHITTOCK, R. T. and LANE, N. (1966). *J. Cell Biol.*, **30**:237.

KEPNER, G. R. and MACEY, R. I. (1968). *Biochim. biophys. Acta*, 163:188.

KERKUT, G. A. and THOMAS, R. C. (1965). *Comp. Biochem. Physiol.*, 14:167.

KERKUT, G. A. and YORK, B. (1969). *Comp. Biochem. Physiol.*, 28:1125.

KERKUT, G. A. and YORK, B. (1971). *The Electrogenic Sodium Pump*. Bristol; Scientechnica.

KERN, H. L., ROOSA, P. and MURRAY, S. (1962). *Expl Eye Res.*, 1:385.

KERNAN, R. P. (1970). *Membranes and Ion Transport*, Vol. 1, pp. 395–431. Ed. E. E. BITTAR. London; Wiley-Interscience.

KEYNES, R. D. (1963). *J. Physiol., Lond.*, 169:690.

KIMELBERG, H. R. and BOURKE, R. S. (1973). *J. Neurochem.*, 20:347.

KIMMICH, G. A. (1973). *Biochim. biophys. Acta*, 300:31.

KINSEY, V. E. (1970). *Membranes and Ion Transport*, Vol. 3, pp. 185–209. Ed. E. E. BITTAR. London; Wiley–Interscience.

KINSEY, V. E. and REDDY, D. V. N. (1959). *Documenta ophth.*, 13:7.

KOENIG, C. and VIAL, J. D., (1970). *J. Histochem. Cytochem.*, 18:340.

KROMPHARDT, H., GROBECKER H., RING, K. and HEINZ, E. (1963). *Biochim. biophys. Acta*, 74:549.

KUYPERS, W. and BONTING, S. L. (1969). *Biochim. biophys. Acta*. 173:477.

KUYPERS, W. and BONTING, S. L. (1970a). *Pflügers Arch. ges. Physiol.*, 320:348.

KUYPERS, W. and BONTING, S. L. (1970b). *Pflügers Arch. ges. Physiol.*, 320:359.

KYTE, J. (1971). *J. biol. Chem.*, 246:4157.

KYTE, J. (1972). *J. biol. Chem.*, 247:7642.

KYTE, J. (1975). *J. biol. Chem.*, 250:7443.

LANE, L. K., COPENHAVER, J. H., JR., LINDENMAYER, G. E. and SCHWARTZ, A. (1973). *J. biol. Chem.*, 248:7197.

LEAF, A., PAGE, L. B. and ANDERSON, J. (1959). *J. biol. Chem.*, 234:1625.

LEAF, A. and RENSHAW, A. (1957). *Biochem. J.*, 65:90.

MACLENNAN, D. H. (1970). *J. biol. Chem.*, 245:4508.

MACLENNAN, D. H., SEEMAN, P., ILES, G. H. and YIP, C. C. (1971). *J. biol. Chem.*, 246:2702.

MARCHBANKS, R. M. (1970). *Membranes and Ion Transport*, Vol. 2, pp. 145–184. Ed. E. E. BITTAR. London: Wiley–Interscience.

MARMOR, M. and GORMAN, A. L. F. (1970). *Science, N.Y.*, 167:65.

MARTIN, D. W. (1964). *J. cell. comp. Physiol.*, 63:245.

MARTIN, D. W. and CURRAN, P. F. (1966). *J. cell. comp. Physiol.*, 67:367.

MARTONOSI, A., DONLEY, J., PUCELLI, A. G. and HALPIN, R. A. (1971). *Archs Biochem. Biophys.*, 144:529.

MCFARLAND, B. H. and INESI, G. (1971). *Archs Biochem. Biophys.*, 145:456.

MEISSNER, G., CONNER, G. E. and FLEISCHER, S. (1973). *Biochim. biophys. Acta*, 298:246.

MOORE, C. L. (1971). *Biochim. biophys. Res. Commun.*, 42:298.

NAKAJIMA, S. and TAKAHASHI, K. (1966). *J. Physiol., Lond.*, 187:105.

NAKAO, T., TASHIMA, Y., NAGANO, K. and NAKAO, M. (1965). *Biochem. biophys. Res. Commun.*, 19:755.

NAKAO, T., NAKAO, M., NAGAI, F., KAWAI, K., FUJIHIRA, Y., HARA, Y. and FUJITA, M. (1973). *J. Biochem., Tokyo*, 73:781.

NARUMI, S. and KANNA, M. (1973). *Biochim. biophys. Acta*, 311:80.

NISHIGAKI, I., CHEN, F. T. and HOKIN, L. E. (1974). *J. biol. Chem.*, 249:4911.

OCHS, S. (1974). *Brain Res.*, 81:413.

OHNISHI, T. and KAWAMURA, H. (1964). *J. Biochem., Tokyo*, 56:377.

PARSONS, D. S. (1967). *Br. med. Bull.*, 23:252.

PETERSON, S. C., GOLDNER, A. M. and CURRAN, P. F. (1970). *Am. J. Physiol.*, 219:1027.

PORTZEHL, H., CALDWELL, P. C. and RUEGG, J. C. (1964). *Biochim. biophys. Acta*, 79:581.

POST, R. L. and KUME, S. (1973). *J. biol. Chem.*, 248:6993.

POST, R. L., MERRITT, C. R., KINSOLVING, C. R. and ALBRIGHT, C. D. (1960). *J. biol. Chem.*, 235:1796.

QUIGLEY, J. R. and GOTTERER, G. S. (1969). *Biochim. biophys. Acta*, 173:456.

QUIST, E. E. and ROUFOGALIS, B. D. (1975). *FEBS Lett.*, 50:135.

RACKER, E. and EYTAN, E. (1975). *J. biol. Chem.*, 250:7533.

RANG, H. P. and RITCHIE, J. M. (1968). *J. Physiol., Lond.*, 196:183.

REUTER, H. (1973). *Prog. Biophys. molec. Biol.*, 26:1.

RHEE, H. M. and HOKIN, L. E. (1975). *Biochem. biophys. Res. Commun.*, 63:1139.

RIDDERSTAP, A. S. (1969). *Thesis*, University of Nijmegen.

RIDDERSTAP, A. S. and BONTING, S. L. (1969a). *Am. J. Physiol.*, 216:547.

RIDDERSTAP, A. S. and BONTING, S. L. (1969b). *Am. J. Physiol.*, 217:1721.

RITCHIE, J. M. and STRAUB, R. W. (1957). *J. Physiol., Lond.*, 136:80.

ROBBLEE, L. S., SHEPRO, D. and BELAMARICH, F. A. (1973). *J. gen. Physiol.*, 61:462.

ROBINSON, J. D. (1971). *Nature, Lond.*, **233**:419.

ROELOFSEN, B., BAADENHUYSEN, H. and VAN DEENEN, L. L. M. (1966). *Nature, Lond.*, **212**:1379.

RUOHO, A. and KYTE, J. (1974). *Proc. natn Acad. Sci. U.S.A.*, **71**:2352.

SACHS, G., SHAH, G., STRYCH, A., CLINE, G. and HIRSCHOWITZ, B. I. (1972). *Biochim. biophys. Acta*, **266**:625.

SCARPA, A., BALDASSARE, J. and INESI, G. (1972). *J. gen. Physiol.*, **60**:735.

SCHAFER, J. A. and HEINZ, E. (1971). *Biochim. biophys. Acta*, **249**:15.

SCHATZMANN, H. J. (1953). *Helv. physiol. pharmac. Acta*, **11**:346.

SCHATZMANN, H. J. (1966). *Experientia*, **22**:364.

SCHATZMANN, H. J. (1970). *Calcium and Cellular Function*, pp. 85–95. Ed. A. W. CUTHBERT. London; Macmillan.

SCHATZMANN, H. J. (1973). *J. Physiol., Lond.*, **235**:551.

SCHATZMANN, H. J. and ROSSI, G. L. (1971). *Biochim. biophys. Acta*, **241**:379.

SCHULTZ, S. G. and CURRAN, P. F. (1970). *Physiol. Rev.*, **50**:637.

SCHULTZ, S. G. and ZALUSKY, R. (1964). *J. gen. Physiol.*, **47**:567, 1043.

SCHUURMANS STEKHOVEN, F. M. A. H., DE PONT, J. J. H. H. M. and BONTING, S. L. (1976). *Biochim. biophys. Acta*, **419**:137.

SCHUURMANS STEKHOVEN, F. M. A. H., VAN HEESWIJK, M. P. E., DE PONT, J. J. H. H. M. and BONTING, S. L. (1976). *Biochim. biophys. Acta*, **422**:210.

SCHWARTZ, A., LINDENMAYER, G. E. and ALLEN, J. C. (1975). *Pharmac. Rev.*, **27**:1.

SEN, A. K. and POST, R. L. (1964). *J. biol. Chem.*, **239**:345.

SETCHELL, B. P., SMITH, M. W. and MUMM, E. A. (1972). *J. Reprod. Fert.*, **28**:413.

SHAMOO, A. E. and MEYERS, M. (1974). *J. Membrane Biol.*, **19**:163.

SIEGEL, G. J. and ALBERS, R. W. (1967). *J. biol. Chem.*, **242**:4972.

SIMON, B. and THOMAS, L. (1972). *Biochim. biophys. Acta*, **288**:434.

SIMON, B., KINNE, R. and KNAUF, H. (1972). *Pflügers Arch. ges. Physiol.*, **337**:177.

SIMON, B., KINNE, R. and SACHS, G. (1972). *Biochim. biophys. Acta*, **282**:293.

SKOU, J. C. (1957). *Biochim. biophys. Acta*, **23**:394.

SKOU, J. C. (1960). *Biochim. biophys. Acta*, **42**:6.

SKOU, J. C. (1971). *Curr. Topics Bioenergetics*, **4**:357.

SLEGERS, J. F. G. and VAN'T HOFF-GROOTENBOER, A. E. (1971). *Pflügers Arch. ges Physiol.*, **327**:167.

SOUMARMON, A., LEWIN, M., CHERET, A. M. and BONFILS, S. E. (1974). *Biochim. biophys. Acta*, **339**:403.

SPENNEY, J. G., STRYCH, A., PRICE, A. H., HELANDER, H. F. and SACHS, G. (1973). *Biochim. biophys. Acta*, **311**:545.

STEIN, W. D. (1967). *The Movement of Molecules across Cell Membranes*. New York; Academic Press.

STEIN, W. D., LIEB, W. R., KARLISH, S. J. D. and EILAM, Y. (1973). *Proc. natn. Acad. Sci. U.S.A.*, **70**:275.

STIRLING, C. E. (1967). *J. Cell Biol.*, **35**:605.

STIRLING, C. E. and KINTER, W. B. (1967). *J. Cell Biol.*, **35**:585.

SWEADNER, K. J. and GOLDIN, S. M. (1975). *J. biol. Chem.*, **250**:4022.

TANAKA, R. and ABOOD, L. G. (1964). *Archs Biochem. Biophys.*, **108**:47.

TANISAWA, A. S. and FORTE, J. G. (1971). *Archs Biochem. Biophys.*, **147**:165.

TAYLOR, E. W. (1973). *Curr. Topics Bioenergetics*, **5**:201.

TENU, J. P., GHELIS, C. and CHEVALLIER, J. (1974). *Biochimie*, **56**:791.

THE, R. and HASSELBACH, W. (1972). *Eur. J. Biochem.*, **28**:357.

THE, R. and HASSELBACH, W. (1973). *Eur. J. Biochem.*, **39**:63.

THOMAS, R. C. (1969). *J. Physiol., Lond.*, **201**:495.

TOWER, D. B. (1968). *Expl Brain Res.*, **6**:273.

TURBECK, B. O., NEDERGAARD, S. and KRUSE, H. (1968). *Biochim. biophys. Acta*, **163**:354.

VAN DER KLOOT, W. (1969). *Science, N.Y.*, **164**:1294.

VAN OS, C. H. and SLEGERS, J. F. G. (1970). *Pflügers Arch. ges. Physiol.*, **319**:49.

VAN OS, C. H. and SLEGERS, J. F. G. (1971). *Biochim. biophys. Acta*, **241**:89.

VALE, M. G. P. and CARVALHO, A. P. (1973). *Biochim. biophys. Acta*, **325**:29.

VATES, T. S., BONTING, S. L. and OPPELT, W. W. (1964). *Am. J. Physiol.*, **206**:1165.

VIDAVER, G. A. (1964a). *Biochemistry*, **3**:662.

VIDAVER, G. A. (1964b). *Biochemistry*, **3**:795.

VIDAVAR, G. A. (1964c). *Biochemistry*, **3**:803.

VINCENZI, F. F. and SCHATZMANN, H. J. (1967). *Helv. physiol. pharmac. Acta*, **25**:CR233.

VREESWIJK, J. H. A., DE PONT, J. J. H. H. M. and BONTING, S. L. (1973). *Biochim. biophys. Acta*, **330**:173.

WATSON, E. L., VINCENZI, F. F. and DAVIS, P. W. (1971). *Biochim. biophys. Acta*, **249**:606.

WEINER, M. W. (1975). *Am. J. Physiol.*, **228**:815.

WHEELER, K. P. and WHITTAM R. (1970). *J. Physiol., Lond.*, **207**:303.

WHITTAM, R. and AGER, M. E. (1965). *Biochem. J.*, **97**:214.

WIEBELHAUS, V. D., SUNG, C. P., HELANDER, H. F., SHAH, G., BLUM, A. L. and SACHS, G. (1971). *Biochim. biophys. Acta*, **241**:49.

WINS, P. and SCHOFFENIELS, E. (1966). *Biochim. biophys. Acta*, **120**:341.

WONG, R. T., KASBEKAR, D. K. and FORTE, J. G. (1969). *Proc. Soc. exp. Biol. Med.*, **313**:534.

YAMADA, S. and TONOMURA, Y. (1972). *J. Biochem., Tokyo*, **72**:417.

ZADUNAISKY, J. A. and DE FISCH, F. W. (1964). *Am. J. Physiol.*, **207**:1010.

ZADUNAISKY, J. A., CANDIA, O. A. and CHIARANDINI, D. J. (1963). *J. gen. Physiol.*, **47**:393.

ZERAHN, K. (1956). *Acta physiol. scand.*, **36**:300.

8

Membrane excitability

David O. Carpenter
Neurobiology Department, Armed Forces Radiobiology Research Institute, Bethesda, Maryland

8.1 INTRODUCTION

Some biological membranes are capable of producing brief electrical pulses known as *action potentials*. The best-known tissues with this property are nerve and muscle, but some other cells, including giant algae and various other plant cells, may also generate action potentials. Lucas (1909) in amphibian muscle and Adrian (1914) in nerve first demonstrated that action potentials obey an 'all-or-none' principle. Thus if an action potential is initiated it occurs at maximal and constant amplitude and with an unchanging wave form.

There are two primary functions of action potentials. The first is that they are the means whereby information is carried over long distances without decrement. This is especially important in nerves which must rapidly convey information between the periphery and the central nervous system, and is essential for pathways connecting various parts of the brain and spinal cord. The second function is that the action potential is the trigger mechanism for neurosecretion in nerves or contraction by muscles. By virtue of the rapid spread of the action potential over the entire surface of a muscle fiber, contraction is initiated nearly simultaneously throughout the fiber and a twitch results. It is true that neurosecretion and muscular contraction can be initiated by electrical depolarization independently of action potentials. In fact some muscle fibers such as the 'slow' fibers of amphibians (Kuffler and Vaughan Williams, 1953) and the giant muscle fibers of the barnacle (Hagiwara and Naka, 1964) do not normally generate action potentials but instead give a graded contraction which is proportional to the magnitude of a local depolarization of the muscle membrane. Some nerve cells which do not have processes going great distances also do not normally produce action potentials, such as the majority of cells in the retina (Werblin and Dowling, 1969) and probably some cells in the olfactory bulb (Rall and Shepherd, 1968). However, these exceptional nerve and muscle cells have many characteristics unlike those of most inexcitable membranes. Consequently we will consider in this chapter not only the basis of generation of action potentials but also the other properties of nerve and muscle cell membranes which are characteristic or unique.

Because the electrical excitability of cells can be studied only by intracellular recording (usually from one cell at a time), most of our knowledge comes from studies of large invertebrate cells. Mammalian systems are much more difficult to study rigorously owing to the small dimensions of the cells. Consequently the discussion which follows will not be limited to mammalian cell membranes. Insofar as it is possible to test, the mechanisms involved in excitability do not appear to be different in mammalian and more primitive systems.

8.2   PASSIVE ELECTRICAL PROPERTIES OF THE EXCITABLE MEMBRANE

The term 'passive properties' refers to the electrical characteristics of the membrane other than those responsible for the action potential. The most important of these properties are membrane resistance, capacitance and resting potential.

Figure 8.1 shows the simplest equivalent circuit for biological membranes. Membranes have electrical characteristics of a resistance and a capacitance

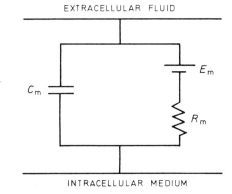

EXTRACELLULAR FLUID

Figure 8.1   Equivalent electrical circuit of biological membranes

INTRACELLULAR MEDIUM

in parallel. Although this equivalent circuit is certainly not unique to excitable membranes, in many cases the values of membrane resistance are greater for electrically excitable membranes than for those not excitable. The specific resistance R_m for red blood cell membranes is of the order of $10–15 \, \Omega cm^2$ (Johnson and Woodbury, 1964) whereas that for frog eggs (Cole and Guttman, 1942) is $170 \, \Omega cm^2$. In contrast, R_m of the squid giant axon, which is the classic preparation for the study of nerve electrical properties, is about $1000 \, \Omega cm^2$ (Cole, 1940). Many other nervous tissues apparently have even higher values of R_m, including the cell bodies which give rise to the giant axon of the squid, which have specific resistance of the order of 25 times greater than that characteristic of the axon (Carpenter, 1973a). Other molluscan nerve cell bodies have R_m values which are of the order of $10^5 \, \Omega \, cm^2$ (Carpenter, 1970; Marmor, 1971a). In contrast to the considerable variation in values of R_m in biological tissue, it appears likely that the membrane capacitance, C_m, is more or less a biological constant and has a value of approximately $1 \, \mu F \, cm^2$ (Cole, 1968).

Because the events underlying action potentials are very dependent upon potential, the very high specific membrane resistance of nerve and muscle is an important characteristic which makes the cell exquisitely sensitive to small currents. *Figure 8.2* illustrates the effects of applying small currents

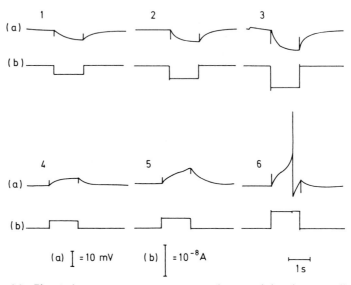

Figure 8.2   *Electrical responses to constant-current pulses recorded in the giant cell in the visceral ganglion of* Aplysia. *In each frame, record* (a) *records voltage shifts obtained following passage of varying currents* [*record* (b)] *passed through a second electrode independently inserted in the cell. Initial* RMP *is* − 55 mV. *The peak of the action potential in 6 is not shown but was 110 mV (peak to peak)*

across a nerve cell membrane, in this case a neurone from the visceral ganglion of the marine mollusc *Aplysia californica*. Two independent microelectrodes have been inserted into one nerve cell; one was used for passing current and the other for recording the resulting potential change. The amount of current passed through the current electrode is monitored and is illustrated in records (b) whereas voltage responses are illustrated in records (a) With both hyperpolarizing (1 to 3) and depolarizing (4 to 6) current administration there is a gradual change in voltage over a period of time (in this case of approximately 0.5 s) and finally the voltage trace reaches a steady value. The time required to approach this new steady-state value is determined by the time constant, $\tau = R_m C_m$. The maximal voltage shift is determined by Ohm's law, $V = IR$, where I is the applied current and R the input resistance. For hyperpolarizing pulses the increase in the voltage is nearly a linear function of the applied current. For depolarizing voltage changes, however, a point is soon reached as illustrated in number 5 where a second event begins to occur and the voltage trace does not reach a plateau. If this pulse were continued it would lead to an action potential, as shown in number 6. The events in 5 and 6 demonstrate active processes superimposed on the passive responses of the membrane.

Since the charging of C_m is an exponential process τ can be measured directly from a voltage shift resulting from a constant current pulse, and is the time required for the voltage to reach 63 percent of the maximal value. In the cell illustrated τ has a value of about 300 ms, whereas in squid axon τ is about 1 ms. The difference is probably due entirely to the very high R_m of the neuronal cell bodies, since C_m of a very similar nerve cell body is the same as squid axon (i.e. 1 μF cm^{-2}) (Marmor, 1971b). Since τ determines the speed with which voltage changes occur, its value is especially important in nerve cells receiving a high-frequency mixture of excitatory and inhibitory input.

Although not proven as yet, it appears very likely that it is a general principle that the R_m value of nerve cell bodies is one or two orders of magnitude greater than that of axons. This is very important for the whole process of synaptic transmission where chemical substances released from the terminus of one cell act on the dendrites or somal membranes of another to cause a current flow. Because of its higher R_m the cell body, where synaptic integration takes place, is much more sensitive to small synaptic currents than are other parts of the neurone.

8.3   THE ORIGIN OF RESTING MEMBRANE POTENTIAL

8.3.1   The role of concentration gradients

Although there is some potential difference between the inside and outside of most cells the potential across the membrane, or resting membrane potential (RMP), of excitable cells is much higher than that of most other tissues and is usually somewhere between 40 to 100 mV (inside negative). The origin of the RMP has been the subject of discussion since the late 1800s when Bernstein developed his hypothesis (first published in its complete form in 1902), based to a great extent upon the work of Nernst, suggesting that the RMP was a diffusion potential for K$^+$ resulting from inequalities of ionic concentrations across the plasma membrane. Bernstein viewed the living cell as a membrane permeable to K$^+$ but not to Na$^+$, which separates the aqueous interior from the exterior. He suggested that action potentials result from a transient increase in the permeability to all ions, both positive and negative, with a decay of the potential to zero.

The majority view of the origin of the RMP and action potentials derives much from the Bernstein hypothesis. Nevertheless it is now known to be inadequate in several respects. The membrane is not completely impermeable to Na$^+$, and during an action potential the depolarization overshoots zero by a sizable voltage in most excitable cells. Furthermore, recent evidence indicates that in some, but not all, cells ions in the cytoplasm exist in a state very different from that of the external solution, causing a very much lower internal conductivity than would be expected (Carpenter, Hovey and Bak, 1973).

Whereas sea-water contains approximately 450 mM Na$^+$, 10mM K$^+$ and 500 mM Cl$^-$, squid axoplasm contains about 65 mM Na$^+$, 320 mM K$^+$ and 100 mM Cl$^-$ per kilogram wet weight (*see* Brinley, 1965). The lack of negative charge on the inside is presumably caused by the presence of organic ions which cannot diffuse across the membrane.

Bernstein's suggestion that RMP is a diffusional potential for K$^+$ would indicate that

$$V = \frac{RT}{F}\ln\frac{[K^+]_o}{[K^+]_i} \tag{8.1}$$

Since the internal medium contains more than 30 times as much K$^+$ as the

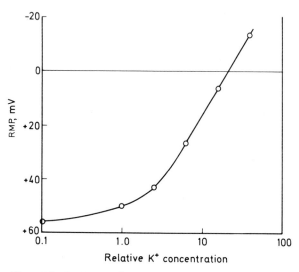

Figure 8.3   Resting membrane potential as a function of external K$^+$ concentration (logarithmic scale). The slope at high [K$^+$] approaches that of a potassium electrode, but deviates markedly at physiologic concentrations. (From Curtis and Cole, 1942, courtesy of the Wistar Press)

external, the ratio can be manipulated easily by changes in [K$^+$]$_o$. *Figure 8.3* shows the variation of the RMP of the squid giant axon as a function of [K$^+$]$_o$. At relatively high [K$^+$]$_o$ the RMP behaves as a K$^+$ electrode with a change of 58 mV per tenfold change in [K$^+$]$_o$. However, at physiological ranges of [K$^+$]$_o$ the variation of RMP is much less than expected of a K$^+$ electrode.

Although these results indicated that the K$^+$ concentration gradient alone does not explain membrane potential, later work considered the possibility that other ions in addition to K$^+$ contributed in various degrees. In 1943 Goldman studied the electrical properties of artificial membranes separated by solutions of varying ionic composition. He presented a theoretical analysis showing that the ionic movements resulting from diffusional and electrical forces were consistent with the assumption that these ions were in a constant electric field within the membrane. His analysis was applied specifically to nerve cell membranes by Hodgkin and Katz (1949). They derived an equation, commonly known as the Goldman–Hodgkin–Katz or constant field equation, which attempted to explain the resting membrane potential on the basis of permeabilities to K^+, Na^+ and Cl^- and which also became the basis for an explanation of the action potential to be discussed later. This equation had the form

$$V = \frac{RT}{F} \ln \frac{P_K+[K^+]_o + P_{Na}+[Na^+]_o + P_{Cl}-[Cl^-]_i}{P_K+[K^+]_i + P_{Na}+[Na^+]_i + P_{Cl}-[Cl^-]_o} \tag{8.2}$$

where V is the potential across the membrane, R is the universal gas constant, T is absolute temperature, F is the Faraday constant, P_{K^+}, P_{Na^+} and P_{Cl^-} are the permeabilities respectively for K^+, Na^+ and Cl^-, and the subscripts 'i' and 'o' indicate ionic concentration on the inside of the cell and in the external medium respectively.

Although the constant field equation has been a very useful tool it does not appear satisfactorily to explain the membrane potential in squid axon. Hodgkin and Katz (1949) calculated ratios of $P_{K^+} : P_{Na^+} : P_{Cl^-} = 1:0.04:0.45$. However, direct measurements of relative permeabilities by Caldwell et al. (1960) have led Caldwell and Keynes (1960) to conclude that the actual $P_{K^+} : P_{Cl^-}$ ratio is $1:0.15$. Thus it seems unlikely that either P_{Cl^-} or P_{Na^+} is large enough to explain the considerable deviation from the behavior of a K^+ electrode at physiologic $[K^+]_o$.

In contrast to squid axon, the Cl^- permeability of frog muscle fibers (Hodgkin and Horowicz, 1959) and some invertebrate nerve cell bodies (Russell and Brown, 1972) is about equal in magnitude to P_{K^+}. Furthermore, in those neurones with high R_m values the ratio of P_{Na^+} to P_{K^+} is greater than in squid axon, since the higher R_m is primarily due to a greater resistance to K^+. In neurones of the land snail this ratio is about $1:10$ (Moreton, 1968). Under these circumstances either or both Na^+ and Cl^- will contribute to diffusional potentials. A similar situation may apply at the frog node of Ranvier, where Stämpfli (1959) has shown that the RMP does not follow the K^+ concentration gradient.

The most direct test of the role of the K^+ concentration gradient in generation of the RMP has been made in experiments on the perfused squid axon. This giant axon is large enough for the internal medium to be removed, leaving a membrane capable of maintaining a normal RMP and the ability to produce action potentials, and with the possibility of controlling both internal and external media. This technique was independently developed in the laboratories of Baker and Hodgkin in England and Tasaki in America. Both groups have examined the effects of varying the K^+ concentration gradient, but have come to quite different conclusions as regards the Bernstein hypothesis. Baker, Hodgkin and Shaw (1962) replaced the axoplasm with isotonic KCl or NaCl. They showed that with KCl on the outside and NaCl on the inside (the reverse of the normal situation), the RMP became 40–60 mV inside positive. With identical solutions on both sides the RMP fell to zero. They concluded that the K^+ concentration gradient is the primary source of RMP, although they found that the RMP did not rise beyond 60 mV even when the K^+ gradient was increased. Thus when $[K^+]_i$ was increased from 150 to 600 mM with a $[K^+]_o$ of 10 mM they observed a 10 mV increase in RMP rather than the 35 mV expected. However, they explained this deviation on the basis of a decreased P_{K^+} at increased potentials, with a greater contribution from other ions.

Tasaki, Watanabe and Takenaka (1962) have shown a similar lack of sensitivity of the RMP to dramatic changes of $[K^+]_i$. They found that removal of nine-tenths of the internal K^+ (replaced by sucrose) caused only a 10 mV fall in RMP when 59 mV would be expected from the Bernstein hypothesis.

Tasaki (1968) has presented an alternative interpretation of the origin of the RMP which deserves a more serious consideration than it has received. He rejected the possibility of RMP arising as a K^+ diffusion potential on the basis of the failure of the RMP to follow the K^+ concentration gradient at physiological levels of $[K^+]_o$ and on variation of $[K^+]_i$ in perfusion experiments. He suggested that biological membranes (or at least the squid axon membrane) contain a considerable fixed negative charge on the external side of the membrane, which results in the low anionic conductance. The RMP results, in his opinion, from the sum of (a) the ionic diffusion potential in the axoplasm, (b) a phase-boundary potential arising between the inner surface of the membrane and the axoplasm, (c) a diffusion potential within the membrane (between inner and outer surfaces), (d) a phase-boundary potential between the external membrane surface and the external medium, and (e) the potential difference across the surrounding layers. He believes that the major origin of RMP is the phase-boundary potential at the external membrane surface, which contains the greatest amount of fixed charge. Thus, at low of physiologic levels of $[K^+]_o$ this fixed charge preferentially binds divalent cations, but at high $[K^+]_o$ the RMP becomes a K^+ electrode as a result of extensive binding of K^+ to this fixed charge, displacing the divalent cations.

8.3.2   The role of electrogenic pumps

Nerve and muscle cells, like other biologic tissues, are not completely impermeable to Na^+, and maintain their concentration gradients of Na^+ and K^+ by an active transport process generally referred to as the *sodium pump*. This process appears to depend upon activity of a membrane enzyme, Na^+,K^+-activated ATPase, which utilizes energy from ATP to cause removal of Na^+ and accumulation of K^+ in the cell (Albers, 1967). The activity of this enzyme system is stimulated by internal Na^+ and thus functions to regulate the Na^+ concentration gradient. Activity requires Mg^{2+} and Na^+ on the inside, and K^+ on the outside. Furthermore, activity is relatively specifically inhibited by the cardiac glycosides. The 'pump' is then the energy source which creates the concentration gradients across cell membranes.

In some cells the sodium pump has a more direct role in generation of the RMP. Whereas concentration gradients would be created effectively by a pump which exchanges 1 Na^+ for 1 K^+, in fact the pump appears to operate in such a way that more Na^+ is moved than K^+, with a ratio of 3 Na^+ : 2 K^+ reported for red blood cells (Post, Albright and Dayani, 1967) and various ratios ranging from 1 : 1 to 3 : 1 reported for nerve and muscle (*see* Thomas, 1972). The fraction of Na^+ moved which is not balanced by an opposite movement of K^+ functions as a net current across the membrane resistance, and may generate a potential (hence the term *'electrogenic pump'*). Electrogenic pumps have been shown to generate potential in a great variety of nerve and muscle cells (*see* Thomas, 1972). Although little potential is produced directly by the pump in squid axon, as a result of the relatively low R_m (Hodgkin and Keynes, 1955) there can be up to 50 mV generated in this fashion in cells

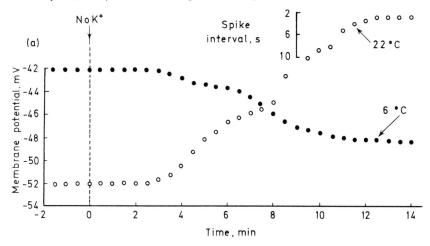

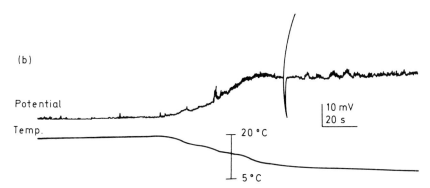

Figure 8.4   *Resting membrane potential from the giant neurone of the visceral ganglion of* Aplysia *as a function of time after exposure to* K^+*-free sea-water at 6 °C and 22 °C (a) and during cooling from 20 °C to 5 °C (b). In (a) the preparation was perfused with normal artificial sea-water containing 10 mM* K^+ *until the arrow, at which time the solution was changed to one identical except that it lacked* K^+. *At 22 °C the neurone depolarized so much 9 minutes after exposure to the* K^+*-free sea-water that it began to discharge. This makes* RMP *values unclear, and so the frequency of discharge is indicated*

with very high R_m values, such as *Aplysia* neurones (Carpenter, 1970). In these cells it can be demonstrated that the sodium pump is responsible for maintenance of normal resting potential at its hyperpolarized levels. This is shown in *Figure 8.4* from an experiment on an *Aplysia* neurone, examined at two different temperatures. Because Na^+ transport is an enzymatic process with a high temperature coefficient it can be greatly reduced by cooling. *Figure 8.4a* shows the effect of removal of external potassium at 6 °C and 22 °C. If membrane potential is determined by the Goldman–Hodgkin–Katz equation, it follows that when one removes external potassium, membrane potential should become more negative as a result of the increase in the K^+ concentration gradient. At 6 °C this is what happens. However, at 22 °C the initial membrane potential is greater by 10 mV and on removal of external potassium the membrane potential depolarizes, ultimately leading to discharge in spite of the fact that the K^+ concentration gradient is increased. This results because the pump cannot operate in absence of external K^+. *Figure 8.4b* shows the effect on the membrane potential of this cell of a sudden change in temperature. In other experiments it has been demonstrated that this temperature dependence is abolished by cardiac glycosides and lack of Na^+ (Carpenter, 1970).

These results indicate that an electrogenic Na^+ pump contributes to the RMP under normal resting conditions and is responsible for its marked temperature dependence. The total potential generated by the pump is, however, even greater than the 10 mV difference in RMP at 6 °C and 22 °C as a result of the fact that the $P_{Na^+} : P_{K^+}$ ratio is relatively large and P_{Na^+} has a higher temperature coefficient than does P_{K^+} (Carpenter, 1970). In this case the results at 6 °C indicate that a 6 mV hyperpolarization results from the change in K^+ concentration gradient when the pump is not operating. At 22 °C the potential change is of the order of a 15 mV depolarization. The electrogenic pump was thus directly responsible for generation of over 20 mV of RMP. This is much greater than the contribution of the RMP in many other tissues because of the very high R_m in these cells. But this neurone is unable to maintain RMP in absence of the electrogenic pump.

Electrogenic sodium pumps have been demonstrated in a wide variety of mammalian tissues, including spinal motor neurones, blood vessels, smooth muscle, sympathetic ganglia, intestinal smooth muscle and skeletal muscle (Thomas, 1972). The presence of a potential generated by the pump is directly correlated to membrane resistance (Carpenter, 1973a). The extent of the functional significance of electrogenic pumps is still unclear. Early reports of synaptic activation of sodium pumps have been pretty much discounted (*see* Thomas, 1972). Recently we have shown that an electrogenic Na^+ pump is the generator mechanism underlying activity of specific cold-sensing afferent fibers from rat skin (Pierau, Torrey and Carpenter, 1974, 1975). Other roles for electrogenic sodium pumps and perhaps electrogenic pumps for other ions may become apparent.

8.4   RECTIFICATION

The single most important distinguishing characteristic of electrically excitable membranes is that the ionic permeabilities of these cells vary with voltage. This property is illustrated in *Figure 8.5*, which shows a current–voltage $(I–V)$ relationship obtained from an *Aplysia* neurone in normal sea-water, K^+-free sea-water and Na^+-free sea-water. In this experiment two independent electrodes were inserted into one large nerve cell body. One was used to pass constant-current pulses of various sizes and the other to record the voltage (as in *Figure 8.2*). The voltage shift (measured after the charging of C_m) is plotted against current. In an electrically nonexcitable cell the voltage shift should obey Ohm's law over the full range of currents, and a straight line would result. However, in excitable cells R_m changes with potential resulting in the deviations from linearity. In *Figure 8.5* there is a reasonably linear portion of the $I–V$ curve near the origin (RMP) but a decrease in resistance both on depolarization (delayed rectification) and hyperpolarization (anomalous rectification).

Another important feature of excitable cells is that the rectifying properties are specific for some ions and not for others. The rectification in *Figure 8.5*

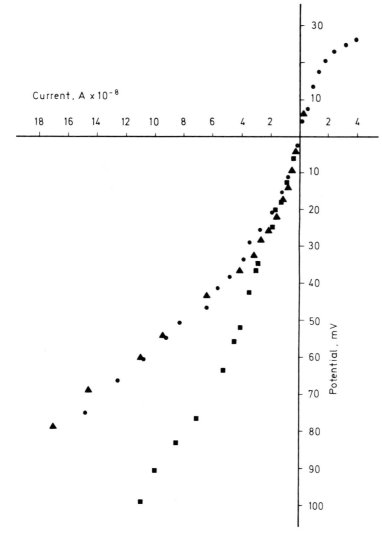

Figure 8.5   Current–voltage relations in the giant neurone of Aplysia at 20 °C in normal sea-water (▲), sea-water with no K⁺ (■) and sea-water where Tris cation has replaced all Na⁺ and blocks action potentials (●). Two electrodes were independently inserted into the cell for recording and current passage, respectively. The input resistance at RMP is $2.5 \times 10^{-6}\ \Omega$. Currents to the right of the origin are depolarizing and those to the left hyperpolarizing

on depolarizing pulses results from an increased K^+ conductance (Na^+ has been removed from the medium). The anomalous rectification also is probably due to a voltage-dependent K^+ conductance (Adrian and Freygang, 1962). This conductance has somewhat different properties from the delayed rectification in that it is reduced dramatically by removal of external K^+ and by cooling (Marmor, 1971b).

Grundfest has for many years stressed the significance of rectification in distinguishing excitable membranes (Grundfest, 1961). He has promoted the terms *activation* for an increase in a particular ionic conductance and *inactivation* for a decrease, and has shown that at least Na^+, K^+ and Cl^- conductances can be independently increased or decreased. Grundfest suggested that receptor and synaptic parts of neurones and muscle cells are not electrically excitable; that is, they do not show rectification. Although the concept is not universally accepted it is an important suggestion indicating that electrical excitability may not be a uniform property over all the membrane of a cell. Support for the view that a neurone may be a mosaic of electrically excitable and inexcitable membrane comes also from experiments by B. O. Alving (unpublished, owing to her premature death) demonstrating that spike electrogenesis in *Aplysia* neurones occurs in patches of active membrane, with many other patches of membrane functioning purely passively.

An examination *in vivo* of the basic properties of mammalian excitable tissue is extraordinarily difficult owing to the heterogeneity of the tissue, the small size of the cells and the difficulty in manipulation of the external medium. All of these problems except size can be avoided using tissue culture techniques. The use of neuroblastoma cells and the application of somatic cell hybridization of normal neurones has opened the possibility for enormous advances with mammalian tissues. These cell lines make it possible to study populations of genetically homogeneous cells.

Nelson, Ruffner and Nirenberg (1969) have studied the electrical properties of mouse neuroblastoma cells. They found a variety of degrees of electrical excitability and proposed an important hypothesis suggesting that there is a sequence of development of electrical responses as neuroblasts develop into neurones. They suggested that all cells begin with passive membranes, then develop delayed rectification (an increase in g_{K^+} on depolarization), followed by partial responses and finally full action potentials. All four types of responses were found in the neuroblastoma cells they studied.

These techniques of tissue culture allow, for the first time, attempts at a genetic analysis of electrical excitability. Minna *et al.* (1971) made hybrids of neuroblastoma cells and L cells (a mouse fibroblast). The hybrids were dividing cells which showed independent inheritance of electrical responses from both parent cells.

It is important to distinguish rectification from a simple change in conductance. In electrically excitable cells the conductance changes as a result of the change in potential: this is rectification. Many electrically nonexcitable cells can undergo conductance changes as a result of chemical or mechanical stimuli, but this is a different matter and unrelated to the ability to generate action potentials. For instance L cells show a hyperpolarizing response on stimulation resulting from a large conductance increase, but have linear I–V curves and do not generate action potentials (Minna *et al.*, 1971). Cone receptor cells in the turtle retina respond to light with a change in conductance, but also have a linear $I-V$ curve and do not generate action potentials (Baylor and Fuortes, 1970). On this basis these cells must be classed as electrically inexcitable. The conductance changes in these cases may be to the same ions as are involved in the action potential, but clearly are mediated by a different mechanism.

8.5   ACTION POTENTIALS

When Bernstein originally presented his theory of the functioning of nerve, he postulated that during an action potential, membrane potential in the fiber decreases to zero owing to a sudden increase in the permeability of the membrane to K^+. With the advent of modern electronics and microelectrode techniques it became apparent that the peak of the action potential overshoots zero and reaches a positive value. This result was inexplicable until the suggestion was made that the underlying mechanism of the action potential was a brief increase in the permeability of the membrane to Na^+ followed by an increase in permeability of the membrane to K^+. Thus the increase in permeability to Na^+ caused the membrane to depolarize and to reach a value approaching equilibrium potential (or the Nernst potential) for Na^+, after which the membrane became very permeable to K^+ and membrane potential was restored to its original value (Hodgkin and Katz, 1949). Building on this information, a theory of nerve conduction was developed by Hodgkin and Huxley in a series of papers in 1952, which led to the award of the Nobel Prize to them in 1963. Their experimental observations and model are one of the major contributions in the study of nerve cells in the twentieth century.

Cole and Curtis (1939) first demonstrated that during passage of an action potential, R_m in the squid giant axon fell from about 1000 to 25 Ω cm^2. Thus the permeability of the membrane increased during spikes. Cole (1949) introduced the use of the voltage-clamp technique, which has since been widely used and which made possible the experimental observations leading to the Hodgkin–Huxley model of nerve action potentials. Under voltage clamp, current is suddenly applied across the membrane in adequate quantity and polarity to hold membrane potential at a predetermined level. The current necessary to accomplish this is measured by a separate amplifier.

Using the voltage-clamp technique in squid axons Hodgkin, Huxley and Katz (1952) studied the time dependence and direction of the currents resulting from clamps to various potentials. After a brief capacitative current, they found that hyperpolarizing clamps always gave a small, sustained inward current. Depolarizing clamps of less than 15 mV always gave an outward current which was small initially but increased with time. Depolarizations of between 15 and 110 mV, however, showed an early inward current which decayed rapidly into a larger and prolonged outward current. At depolarized clamps of about 110 mV this early inward current disappeared and at more depolarized potentials it reversed. The inversion potential of the early current closely approximated the Nernst potential for Na^+. This early current was absent if Na^+ was replaced with choline, which will not support an action potential, although the late current was unaffected by the Na^+ replacement. On this basis Hodgkin and Huxley (1952a) suggested that the early current was carried by Na^+ whereas the late, outward current was carried by K^+.

After a detailed analysis of how these currents vary as a function of voltage and time (Hodgkin and Huxley, 1952b, c), they proposed that the time, voltage and ionic dependence of the early inward and late outward currents observed under voltage-clamp conditions corresponded respectively to the ionic currents underlying the rising and falling phases of the action potential. They proposed that the circuit diagram of the excitable membrane was as shown in *Figure 8.6*, with the conductances* to Na$^+$ and K$^+$, g_{Na^+} and g_{K^+}, independently variable in time and in parallel with a constant leak resistance, due primarily to Cl$^-$ (Hodgkin and Huxley, 1952d).

Hodgkin and Huxley also developed a series of empirical equations which

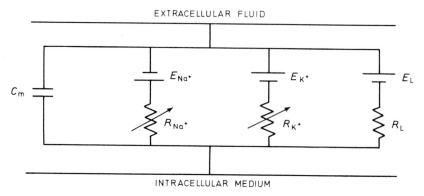

EXTRACELLULAR FLUID

C_m E_{Na^+} E_{K^+} E_L

R_{Na^+} R_{K^+} R_L

INTRACELLULAR MEDIUM

Figure 8.6   Circuit diagram for the excitable membrane as drawn from Hodgkin and Huxley (1952d), courtesy of The Physiological Society

could describe the voltage shifts observed with time during an action potential. The expression for current, I, which best fitted the action potential voltage shifts was

$$I = C\frac{\partial V}{\partial T} + (V - V_{K^+})\bar{g}_{K^+}\, n^4 + (V - V_{Na^+})\bar{g}_{Na^+}\, m^3h + (V - V_L)g_L \qquad (8.3)$$

where the first term is the capacitative current, the second that due to K$^+$, the third Na$^+$ and the fourth the leak current; C is membrane capacitance, V is potential, V_{K^+}, V_{Na^+} and V_L are equilibrium potentials for the respective ions, and g_L is the leak conductance. $\bar{g}_{K^+}$ and $\bar{g}_{Na^+}$ are the maximal possible conductance to K$^+$ and Na$^+$ with n and m respectively time-dependent amplitude factors for these conductance pathways. For the Na$^+$ pathway another term, h, was required to account for time-dependent turning off of this conductance. The terms m, n and h were assumed to be potential-dependent but independent of each other. If the probability of blockade of this sodium 'channel' is $(1 - h)$ then the Na$^+$ conductance is $g_{Na^+} = \bar{g}_{Na^+}m^3h$ and the K$^+$ conductance is $g_{K^+} = \bar{g}_{K^+}n^4$. The exponents were arbitrarily determined to give the best fit. *Figure 8.7* illustrates the computed action potential and the time dependence of Na$^+$ and K$^+$ conductances obtained by this information.

The Hodgkin–Huxley model of the action potential has received strong support from studies using tetrodotoxin (TTX), the toxin of the Japanese

---

* Although permeability (P) and conductance (G) are not actually identical they will be considered so for the purpose of this article.

puffer fish, which blocks the action potential of squid axon and selectively eliminates the early current (Moore and Narahashi, 1967). A comparable blocking agent for the K^+ current is tetraethylammonium ion (TEA), which both blocks the late current and greatly prolongs the action potential (Armstrong, 1969). The selectivity of these agents, together with the ionic dependence of the voltage-clamp currents and the action potentials, supports the assumption that the Na^+ and K^+ currents are independently variable with time.

While the Hodgkin–Huxley model was based on experiments only in squid giant axon it is widely considered to be applicable to all excitable tissues.

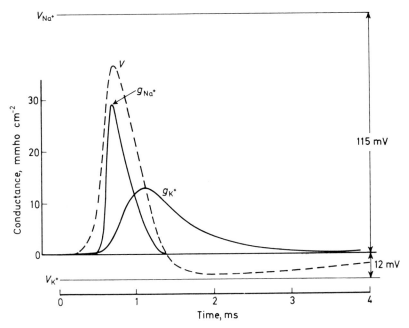

Figure 8.7   Time courses of a calculated action potential and the associated conductance changes obtained by use of the Hodgkin–Huxley model and derived from equation 8.3. (From Hodgkin, 1957, courtesy of The Royal Society)

Models have been presented attempting to modify the model to accommodate the differences in action potentials in tissues such as heart muscle (Noble, 1962) and snail neurones (Moreton, 1968).

Some excitable tissues have action potentials which are not dependent upon Na^+ but rather on Ca^{2+} (Hagiwara and Nakajima, 1966), while other tissues have action potentials to which both Na^+ and Ca^{2+} contribute (Carpenter and Gunn, 1970; Geduldig and Junge, 1968; Hagiwara and Kidokoro, 1971). In these tissues the early inward current consists of a component dependent upon external Ca^{2+}, which is blocked by Co^{2+} and Mn^{2+} (Hagiwara, Hayashi and Takahashi, 1969; Geduldig and Gruener, 1970). The Hodgkin–Huxley formulation may still apply for such cells but with a 'channel' for calcium replacing or being present in addition to a 'channel' for sodium. Evidence is now beginning to appear indicating that tissues may exhibit both

Na^+ and Ca^{2+} currents in spikes from immature cells but become one or the other with differentiation. Thus embryonic chick heart cells become progressively more sensitive to TTX with development (McDonald, Sachs and DeHaan, 1972), while embryonic tunicate muscle cells show spikes dependent on both Na^+ and Ca^{2+} whereas in maturity they are dependent only on Ca^{2+} (Miyazaki, Takahashi and Tsuda, 1972).

While most electrically excitable tissues produce action potentials only in response to some external stimulus, some are pacemakers and fire repetitively from a completely inherent mechanism (Alving, 1968). The best-known and best-studied pacemaker tissue is the heart. Dudel and Trautwein (1968) and Trautwein and Kassebaum (1961) have studied the mechanisms underlying the pacemaker discharge of Purkinje fibers of the heart and concluded that there is an appreciable resting g_{Na^+} relative to g_{K^+}. They suggested that the spontaneous discharge results from a time- and voltage-dependent decrease of g_{K^+} following the action potential in the presence of a relatively low $g_{K^+} : g_{Na^+}$ ratio. A similar situation applies for pacemaker discharge in invertebrate neurones, which also have a low $g_{K^+} : g_{Na^+}$ ratio (Carpenter, 1973b). In other respects the action potentials are not unlike those initiated by synaptic or mechanical means.

A vigorous dissent to Hodgkin and Huxley's formulation of nerve action potentials has been made by Tasaki (1968), who has used the same preparation and many of the same techniques to study the properties of nerve. Tasaki rejects their model of nerve excitation on two separate grounds. He feels that the model proposed by Hodgkin and Huxley is not possible from a consideration of the general principles of physical chemistry. In addition, he has experimental evidence of normal action potentials under circumstances incompatible with the Hodgkin–Huxley formulation. He and his collaborators have demonstrated that they can perfuse the squid giant axon by removal of all of its axoplasmic contents and maintain normal excitability in a situation where there is only the salt of a univalent cation on the inside and only a divalent cation (usually Ca^{2+}) on the outside of the membrane. The species of univalent cation inside is relatively unimportant, and can even be Na^+ if the anion is a favorable one such as F^-. Under these conditions RMP and action potentials are of normal polarity and near-normal magnitudes. In this case it is clear that the rising phase of the action potential cannot be due to movement of Na^+, since Na^+ is present only on the inside of the cell. Similar 'normal' action potentials were observed without Na^+ on either side of the membrane, with voltage-clamp currents and impedance changes during action potentials similar to those conventionally observed (see Singer and Tasaki, 1968). Furthermore these action potentials are sensitive to TTX.

On the basis of his experiments on perfused squid axons Tasaki has concluded that the basic ionic requirement for an action potential is the presence of a divalent cation on the outside and a monovalent one on the inside. He has stressed that regardless of what other ions are present in the external medium, nerve cells are not excitable in the complete absence of a divalent ion (usually Ca^{2+}). Under conditions of internal perfusion he found that the membrane at the peak of the spike overshoot does not behave as a Na^+-sensitive membrane electrode. He feels that the early and late currents observed under voltage clamp do not represent or prove a time separation of Na^+ and K^+ conductances, but are rather a result of differences in net inward and outward fluxes which occur under any condition where one can obtain a spike. As a corollary to his rejection of the notion of a time separation of specific ionic conductances he rejects also the idea of specific channels for Na^+ and K^+.

Tasaki views the excitable membrane as an asymmetric structure containing many fixed negative charges, located primarily on the external surface. He believes that the membrane is capable of two 'stable states', depending upon whether the fixed charge at the external surface is occupied primarily by mono- or divalent cations. The resting state, characterized by a low conductance, exists when divalent ions predominate, but a depolarized, high-conductance state results if monovalent ions displace the divalent ones. He has postulated that the conductance change results from a change in conformation of the membrane macromolecules, again dependent upon whether bound to mono- or divalent cations. Tasaki believes this conformational change to be the central event in nerve excitation.

The theory which Tasaki proposes is not so easily comprehended and tested as that proposed by Hodgkin and Huxley. However, the experimental observations which he has made are in several cases very difficult to explain on the basis of the more conventional model of nervous activity. There is little question that calcium is very important in regulation of membrane conductance (Shanes, 1958), and that calcium enters nerves during action potentials (Baker, Hodgkin and Ridgway, 1971). In general it is this author's opinion that Tasaki's observations and his theory have received less careful and yet critical analysis than they deserve. One reason for this is that his approach and his publications are couched very much in physicochemical terms, which most physiologists find difficult to understand. There has further been a general feeling that the accomplishment of Hodgkin and Huxley in the development of their theory was something of such significance that it could not and should not be questioned. However, even if the Hodgkin–Huxley model is correct in every detail there is much physical chemistry in the membrane underlying g_{Na^+}, g_{K^+}, m, n and h which is not understood. The different approach taken by Tasaki, based on principles of physical chemistry, should be considered from the general point of view as well as the level of specific experimental disagreements.

Further understanding of action potentials may come from the use of artificial membranes. Artificial lipid bilayer membranes can be made which have about the same thickness and capacitance as nerve cell membranes, but which have a much larger resistance. Addition of a number of substances to these membranes results in a fall in resistance, and in some cases selective permeability to ions. In a few cases, such as with alamethicin (a cyclic polypeptide), monazomycin (an antibiotic) and 'excitability-inducing material' (EIM), a bacterial extract of unknown structure, the conductance increase upon addition to the artificial membrane shows rectification (Mueller and Rudin, 1968; Ehrenstein, 1971). In the case of EIM, if protamine is also added in small concentrations the membrane shows a stable RMP and can be made to exhibit action potential like voltage shifts which may be repetitive (Mueller and Rudin, 1968). These results with artificial membranes are exciting because of the fact that relatively simple substances added to the lipid membrane produce voltage shifts similar to normal action potentials. With further study these techniques offer the possibility of identification of the nature of the channels or ionophores which support action potentials, at least in this model system.

8.6   MECHANISMS UNDERLYING MEMBRANE
         PERMEABILITY CHANGES

During action potentials the membrane becomes more permeable at least to Na^+, K^+, and Ca^{2+}. These conductance changes are voltage-dependent, unlike conductance changes induced chemically, physically or mechanically in excitable and nonexcitable tissue. However, in many respects all these conductance changes are similar, and therefore we will consider the general problem of permeability.

It is very clear that even in squid axon, ions other than Na^+ and K^+ can support action potentials. Chandler and Meves (1965) demonstrated in a voltage-clamp analysis of perfused squid axons that the sodium 'channel' would pass ions other than Na^+, with the sequence being $Li^+ > Na^+ > K^+ > Rb^+ > Cs^+$. The $P_{Na^+} : P_{K^+}$ ratio was 12:1. Tasaki and Spyropoulos (1962) and Hille (1971) have demonstrated action potentials or early currents carried by a number of small organic cations. In this regard it is abundantly clear that the sodium channel is really not specific for Na^+ but is instead permeable to a wide variety of univalent cations with a clear sequence of ease in passage.

Similar results apply to the potassium channel. Singer and Tasaki (1968) reported a sequence for 'favorable' internal cations of $Cs^+ > Rb^+ > K^+ > NH_4^+ > Na^+ > Li^+$. Hille (1973) reported $Tl^+ > K^+ > Rb^+ > NH_4^+$. An ion such as NH_4^+ is of particular interest in that it can substitute reasonably well for both Na^+ and K^+. Binstock and Lecar (1969) have shown that NH_4^+ currents are blocked by TTX when substituting for Na^+ and by TEA when substituting for K^+, and thus suggest that these agents act on the 'channel' independently of what ions pass through.

The term usually applied by the Hodgkin school to the pathway through which ions traverse the membrane is 'channel'. This term originally was meant not to have any particular physical implications, but it has come to imply a pore, or hole through the membrane. This view has been advanced particularly by Hille (1970, 1971, 1973) and by Armstrong (1971), who have calculated dimensions for the channels on the basis of the physical sizes of ions which will substitute for Na^+ and K^+. Hille believes the sodium channel to be an oxygen-lined pore 0.3×0.5 nm in cross section, while the potassium channel is a pore with a narrow region with dimensions of 0.3×0.33 nm at its narrowest point. Armstrong suggests that TEA blocks the potassium channel by physically entering the large end (on the inside) but being unable to pass through a narrow neck region.

In order to explain the opening or closing of pores such as envisioned by Hille and Armstrong there is usually thought to be a 'gate' at some position in the pore. This is presumably a molecule with two conformational states, making the pore open or closed to ionic movement. There is little information on the gating mechanism, but this is presumably the source of the voltage dependence of conductance. Armstrong, Bezanilla and Rojas (1973) have demonstrated that sodium inactivation is abolished in squid axons perfused with pronase. They conclude that the sodium inactivation gate, at least, is a readily accessible protein that is attached to the inner end of each sodium channel. The physical nature of the pore does not necessarily rule out inter-action of ions with fixed charge in the membrane, and indeed Hille's (1971) model of the sodium channel relies on fixed oxygen molecules in the pore.

There is evidence suggesting that the conductance change obtained by one channel opening to Na$^+$ is the same for the action potential, the acetylcholine receptor and alamethicin (*see* Keynes, 1972) and EIM channels of artificial membranes (Bean *et al.*, 1969). In each case the elementary conductance is about $10^{-10}\,\Omega^{-1}$. This evidence argues for a similarity of permeability mechanisms in rectifying and nonrectifying channels.

Because the term 'channel' has been used so frequently to mean 'pore' I prefer to use the term *ionophore* (cf. Changeux, Podleski and Meunier, 1969). The ionophore is any structure in the membrane which regulates the move-ment of ions across the membrane. Thus it is both the channel and the gate. This term has been used to date primarily with respect to conductance changes resulting from combination of membrane receptors with drugs such as neurotransmitters. It is preferable because it does not imply a physical structure or ionic specificity, while pores and ion-specific channels, should they exist, can still be identified as ionophores.

Indirect support for the view of the ionophores being physical pores comes from the very fast-moving field of artificial lipid membranes. Some sub-stances, for example cyclic antibiotics such as gramicidin A, not only increase conductance but impart some level of ionic selectivity when added to lipid membranes (Mueller and Rudin, 1969). These substances are thought to pile up in such a fashion that they form a continuous opening through the mem-brane. However, other substances which impart selectivity are not cyclic, and therefore these investigations, although very intriguing, do not prove a general principle. Some progress has been made in extraction of both potassium and sodium ionophores from nervous tissue and incorporation into artificial membranes (Goodall and Sachs, 1972; Shamoo and Albers, 1973). Such experiments offer hope of ultimate identification of the actual ionophores from excitable tissues.

A 'carrier' is an ionophore clearly distinct from a pore in that it is a molecule which may exhibit selective binding of ions on one side of the membrane, the ion–carrier complex then diffusing across the membrane to deposit the ion on the other side (Eigen and Winkler, 1970). Carriers are capable of carry-ing very much less current than a pore, and should also exhibit a greater dependence on temperature than a pore. It is very unlikely that action poten-tials are a result of carrier-mediated ionic movements (Keynes, 1972).

A major contribution to the understanding of selectivity patterns for ion permeability of biological membranes has been made by Eisenman (1965), who suggested that the sequence of cation binding to membrane sites is dependent upon the interaction between two forces: (a) the attraction between the ion and fixed negative charges in the membrane, and (b) the attraction of the ion for water molecules. Thus if a membrane site has a high electric field strength, the smallest cation will be able to approach most closely the fixed charge and will be subjected to the largest attractive force. That is to say, the free-energy change involved in the attraction of the ion to the fixed charge site is greater than the change in the free energy of hydration. As a result the affinity of that particular fixed charge site will decrease with increasing ionic radius. On the other hand, if the fixed charge site has a relatively low electric field strength, the attraction of the ion to the site is much weaker than the energy of hydration. Under this circumstance the affinity will decrease with decreasing ionic radius or increasing hydrated radius and the sequence of permeabilities will be the lyotropic series.

Diamond and Wright (1969) have reviewed the selectivity sequences found in both animal and mineral systems in nature, and have shown that the relatively few sequences which exist can be predicted by Eisenman's theory. Thus in addition to the series for increased nonhydrated radii for alkali cations $(Li^+ > Na^+ > K^+ > Rb^+ > Cs^+)$ and increasing hydrated radii $(Cs^+ > Rb^+ > K^+ > Na^+ > Li^+)$ only an additional nine sequences are found instead of the 120 possible combinations. Eisenman (1961, 1962) measured the changes in free energies as the fixed charge strength was varied from low to high by two methods for each of the five alkali cations, and obtained curves which repeatedly intersected each other. The intersections resulted in the same 11 sequences which are found to occur in nature.

The combination of the success of Eisenman's theory in predicting the ionic sequences observed in nature, plus the fact that there is no clear relation between ring size of cyclic antibiotics incorporated into artificial membranes with the permeability sequences resulting (Diamond and Wright, 1969), makes the view of ionophores as being only pores obsolete. Even in cases where the sequence is that of either hydrated or nonhydrated diameters, the important consideration is that of the free-energy changes in the ion–fixed charge–water attraction and not size of the ion.

8.7　OTHER PHYSICAL CHANGES IN MEMBRANES DURING ELECTRICAL ACTIVITY

Regardless of whether one views the permeability pathway as a pore with a gate or as a protein which can exist in two conformations, it is clear that the membrane has at least two states, one which generates an action potential and one which does not. This being the case one might expect to be able to detect some other physical changes in membranes during the conduction of impulses. Studies searching for other physical changes have been pursued primarily by two groups, that of Keynes at Cambridge and Tasaki at the National Institute of Mental Health, Bethesda. Recent progress in this area has been reviewed by Cohen (1973).

It has been known for some time that physical changes accompany action potentials. Flaig (1947) demonstrated that the viscosity of axoplasm increases on stimulation. Abbott, Hill and Howarth (1958) have shown that the rising phase of the action potential is accompanied by heat production whereas the falling phase is characterized by heat absorption. Recent techniques have attempted to focus directly on the membrane of the excitable cells, monitoring changes in membrane birefringence, fluorescence, light scattering and ultraviolet or infrared absorption.

Fluorescent changes during action potentials can be obtained by attachment of a fluorescent dye to the membrane, either in bundles of nerve fibers or in single axons. One of the first and most widely used dyes is 1-anilino-8-naphthalene sulfonate. This substance, and a number of other fluorescent dyes, attaches to the membrane when added to the external medium. Changes in fluorescence which parallel the time course of single action potentials have been measured (Tasaki *et al.*, 1968). The fluorescence is voltage-dependent and therefore there is a question as to the significance of the fluorescence change. Cohen, Davila and Waggoner (1971) suggested that the changes in fluorescence result simply from the potential changes associated with the action potential, rather than reflecting an elementary event underlying the potential change. However, Tasaki, Watanabe and Hallett (1972) reported that TTX selectively abolishes the early portion of the fluorescence change under voltage-clamp conditions which result in abolition of the early inward current. They suggested that the dye is attached to a membrane protein which undergoes a conformational change during action potentials. A later fluorescence change is preserved in the presence of TTX, and this they feel is the voltage-dependent fraction.

Birefringence changes have also been measured during action potentials, although these changes are voltage-dependent. Cohen, Hille and Keynes (1968), as well as Tasaki *et al.* (1968), demonstrated that the birefringence changes closely follow the time course of the action potential. These observations have been expanded by Cohen, Hille and Keynes (1970) and Cohen *et al.* (1971). Cohen (1973) concluded that all the observed changes result from the potential dependence of birefringence changes.

Light-scattering experiments have yielded evidence for clear physical changes during action potentials which are not due solely to potential changes. Tasaki *et al.* (1968) found two components in light-scattering experiments. The first corresponded in time to the action potential, whereas the later one was very prolonged and probably reflects changes of axonal volume and movements of water following excitation (Hill, 1950). Cohen, Keynes and Landowne (1972a, b) have studied these responses under voltage clamp in squid axons, and have found some changes in light scattering which correlate well with the ionic currents and not with the changes in potential. These results suggest that the changes in light scattering may reflect an event of primary importance in the origin of the action potentials.

Other changes observed during action potentials, such as in infrared spectra (Sherebrin, 1972) and changes in ultraviolet absorption, have yielded results which are relatively difficult to interpret. However, these techniques are just beginning to be used to study nerve potentials and may in the future lead to some useful information.

Protein biochemists are also beginning to study proteins which may be involved in electrical excitability. For instance, Papakostidis, Zundel and Mehl (1972) have found conformational differences in rat brain protein fractions depending upon the ionic constituents in the medium. Such an approach may ultimately lead to identification of the protein specific for the various components of action potentials.

8.8   CONCLUSION

Although the electrical responses of nerve and muscle cells have been known since the nineteenth century to distinguish these membranes from other cell membranes, it is really only since the pioneering work of Cole, Hodgkin and Huxley that the details of the electrophysiology of these membranes have been appreciated. The accomplishments of the past forty years have been primarily the description of the electrical responses of these membranes, using voltage-clamp techniques, changes in ionic media and, recently, specific toxins which block one or another electrical parameter. These investigations have been productive and have led to some general principles regarding requirements for electrical activity of nerve and muscle and to a number of models which provide some basis for understanding the mechanisms underlying electrical responses. There is, however, a poor understanding of the physicochemical changes which occur at the excitable cell membrane during the action potential. More and more electrophysiologists, however, are getting away from their traditional isolation and are applying principles long known in other disciplines to the study of electrical responses. Eisenman, in particular, has made an enormous contribution to our understanding of the mechanisms underlying membrane permeability, in both excitable and inexcitable membranes.

There are several areas of research which show special promise of being increasingly productive in the near future. Developments in protein biochemistry may soon lead to clear characterization of proteins unique to excitable membranes which can be identified with specific electrical events. The combination of protein biochemistry and use of artificial membranes is very exciting in that it allows a membrane protein to be isolated and characterized, then incorporated into an artificial membrane where its function can be directly tested. Tissue culture of mammalian neurones offers great promise, especially when used with somatic cell hybridization techniques to obtain dividing cells with neuronal characteristics. This approach allows rigorous experimentation on basic mechanisms of excitability in mammalian neurones for the first time and also holds the possibility of study of the genetics of electrical excitability. The search for other physical changes in membranes during action potentials also promises to yield new insights into the mechanisms underlying electrical excitability.

REFERENCES

ABBOTT, B. C., HILL, A. V. and HOWARTH, J. V. (1958). *Proc. R. Soc., Ser. B*, **148**:149.
ADRIAN, E. D. (1914). *J. Physiol., Lond.*, **47**:460.
ADRIAN, R. H. and FREYGANG, W. H. (1962). *J. Physiol., Lond.*, **163**:61.
ALBERS, R. W. (1967). *Rev. Biochem.*, **36**:727.
ALVING, B. O. (1968). *J. gen. Physiol.*, **51**:29.
ARMSTRONG, C. M. (1969). *J. gen. Physiol.*, **54**:553.
ARMSTRONG, C. M. (1971). *J. gen. Physiol.*, **58**:413.
ARMSTRONG, C. M., BEZANILLA, F. and ROJAS, E. (1973). *J. gen. Physiol.*, **62**:375.
BAKER, P. F., HODGKIN, A. L. and RIDGWAY, E. B. (1971). *J. Physiol., Lond.*, **218**:709.
BAKER, P. F., HODGKIN, A. L. and SHAW, T. I. (1962). *J. Physiol., Lond.*, **164**:355.

BAYLOR, D. A. and FUORTES, M. G. F. (1970). *J. Physiol., Lond.,* **207**:77.
BEAN, R. C., SHEPHERD, N. C., CHAN, H. and EICHNER, J. (1969). *J. gen. Physiol.,* **53**:741.
BERNSTEIN, J. (1902). *Pflügers Arch. ges. Physiol.,* **92**:521.
BINSTOCK, L. and LECAR, H. (1969). *J. gen. Physiol.,* **53**:342.
BRINLEY, F. J., JR. (1965). *J. Neurophysiol.,* **28**:742.
CALDWELL, P. C. and KEYNES, R. D. (1960). *J. Physiol, Lond.,* **154**:177.
CALDWELL, P. C., HODGKIN, A. L., KEYNES, R. D. and SHAW, T. I. (1960). *J. Physiol., Lond.,* **152**:561.
CARPENTER, D. O. (1970). *Comp. Biochem. Physiol.,* **35**:371.
CARPENTER, D. O. (1973a). *Science, N.Y.,* **179**:1336.
CARPENTER, D. O. (1973b). *Proceedings 2nd International Conference on Neurobiology of Invertebrates, Tihany, Hungary, 1971,* pp. 11–33. Ed. J. SALÁNKI. Budapest; Akadémiai Kiadó.
CARPENTER, D. O. and GUNN, R. (1970). *J. cell. Physiol.,* **75**:121.
CARPENTER, D. O., HOVEY, M. M. and BAK, A. F. (1973). *Ann. N.Y. Acad. Sci.,* **204**:502.
CHANDLER, W. K. and MEVES, H. (1965). *J. Physiol., Lond.,* **180**:788.
CHANGEUX, J.-P., PODLESKI, T. and MEUNIER, J.-C. (1969). *J. gen. Physiol.,* **54**:2255.
COHEN, L. B. (1973). *Physiol. Rev.,* **53**:373.
COHEN, L. B., DAVILA, H. V. and WAGGONER, A. S. (1971). *Biol. Bull.,* **141**:382.
COHEN, L. B., HILLE, B. and KEYNES, R. D. (1968). *Nature, London,* **218**:130.
COHEN, L. B., HILLE, B. and KEYNES, R. D. (1970). *J. Physiol., Lond.,* **211**:495.
COHEN, L. B., KEYNES, R. D. and LANDOWNE, D. (1972a). *J. Physiol., Lond.,* **224**:701.
COHEN, L. B., KEYNES, R. D. and LANDOWNE, D. (1972b). *J. Physiol., Lond.,* **224**:727.
COHEN, L. B., HILLE, B., KEYNES, R. D., LANDOWNE, D. and ROJAS, E. (1971). *J. Physiol, Lond.,* **218**:205.
COLE, K. S. (1940). *Cold Spring Harb. Symp. quant. Biol.,* **8**:110.
COLE, K. S. (1949). *Archs Sci. physiol.,* **3**:253.
COLE, K. S. (1968). *Membranes, Ions and Impulses.* Berkeley; University of California Press.
COLE, K. S. and CURTIS, H. J. (1939). *J. gen. Physiol.,* **22**:649.
COLE, K. S. and GUTTMAN, R. M. (1942). *J. gen. Physiol.,* **25**:765.
CURTIS, H. J. and COLE, K. S. (1942). *J. cell. comp. Physiol.,* **19**:135.
DIAMOND, J. M. and WRIGHT, W. M. (1969). *A. Rev. Physiol.,* **31**:581.
DUDEL, J. and TRAUTWEIN, W. (1968). *Pflügers Arch. ges. Physiol.,* **267**:553.
EHRENSTEIN, G. (1971). *Biophysics and Physiology of Excitable Membranes,* pp. 463–476. Ed. W. J. ADELMAN, JR. New York; Van Nostrand–Reinhold.
EIGEN, M. and WINKLER, R. (1970). *The Neurosciences: Second Study Program,* pp. 685–696. Ed. F. O. SCHMITT. New York; The Rockefeller University Press.
EISENMAN, G. (1961). *Symposium on Membrane Transport and Metabolism,* pp. 163–179. Ed. A. KLEINZELLER and A. KOTYK. New York; Academic Press.
EISENMAN, G. (1962). *Biophys. J.,* **2**, Part II:259.
EISENMAN, G. (1965). *Proceedings 23rd International Congress on Physiological Science, Tokyo,* p. 489. Amsterdam; Excerpta Medica Foundation.
FLAIG, J. V. (1947). *J. Neurophysiol.,* **10**:210.
GEDULDIG, D. and GRUENER, R. (1970). *J. Physiol., Lond.,* **211**:217.
GEDULDIG, D. and JUNGE, D. (1968). *J. Physiol., Lond.,* **199**:347.
GOLDMAN, D. E. (1943). *J. gen. Physiol.,* **27**:37.
GOODALL, M. C. and SACHS, G. (1972). *Nature, New Biol.,* **237**:252.
GRUNDFEST, H. (1961). *Ann. N.Y. Acad. Sci.,* **94**:405.
HAGIWARA, S., HAYASHI, H. and TAKAHASHI, K. (1969). *J. Physiol., Lond.,* **205**:115.
HAGIWARA, S. and KIDOKORO, Y. (1971). *J. Physiol., Lond.,* **219**:217.
HAGIWARA, S. and NAKA, K. (1964). *J. gen. Physiol.,* **48**:141.
HAGIWARA, S. and NAKAJIMA, S. (1966). *J. gen. Physiol.,* **49**:793.
HILL, D. K. (1950). *J. Physiol., Lond.,* **111**:304.
HILLE, B. (1970). *Prog. Biophys. molec. Biol.,* **21**:1.
HILLE, B. (1971). *J. gen. Physiol.,* **58**:599.
HILLE, B. (1973). *J. gen. Physiol.,* **61**:669.
HODGKIN, A. L. (1957). *Proc. R. Soc., Ser. B,* **148**:1.
HODGKIN, A. L. and HOROWICZ, P. (1959). *J. Physiol., Lond.,* **148**:127.
HODGKIN, A. L. and HUXLEY, A. F. (1952a). *J. Physiol., Lond.,* **116**:449.
HODGKIN, A. L. and HUXLEY, A. F. (1952b). *J. Physiol., Lond.,* **116**:473.
HODGKIN, A. L. and HUXLEY, A. F. (1952c). *J. Physiol., Lond.,* **116**:497.
HODGKIN, A. L. and HUXLEY, A. F. (1952d). *J. Physiol., Lond.,* **117**:500.
HODGKIN, A. L., HUXLEY, A. F. and KATZ, B. (1952). *J. Physiol., Lond.,* **116**:424.
HODGKIN, A. L. and KATZ, B. (1949). *J. Physiol., Lond.,* **108**:37.

HODGKIN, A. L. and KEYNES, R. D. (1955). *J. Physiol., Lond.,* **128**:28.
JOHNSON, S. L. and WOODBURY, J. W. (1964). *J. gen. Physiol.,* **47**:827.
KEYNES, R. D. (1972). *Nature, Lond.,* **239**:29.
KUFFLER, S. W. and VAUGHAN WILLIAMS, E. M. (1953). *J. Physiol., Lond.,* **121**:289.
LUCAS, K. (1909). *J. Physiol., Lond.,* **38**:113.
MARMOR, M. F. (1971a). *J. Physiol., Lond.,* **218**:573.
MARMOR, M. F. (1971b). *J. Physiol., Lond.,* **218**:599.
MCDONALD, T. T., SACHS, H. G. and DEHAAN, R. L. (1972). *Science, N.Y.,* **176**:1248.
MINNA, J., NELSON, P., PEACOCK, J., GLAZER, D. and NIRENBERG, M. (1971). *Proc. natn. Acad. Sci. U.S.A.,* **68**:234.
MIYAZAKI, S.-I., TAKAHASHI, K. and TSUDA, K. (1972). *Science, N.Y.,* **176**:1441.
MOORE, J. W. and NARAHASHI, T. (1967). *Fedn Proc. Fedn Am. Socs exp. Biol.,* **26**:1655.
MORETON, R. B. (1968). *J. exp. Biol.,* **48**:611.
MUELLER, P. and RUDIN, D. O. (1968). *J. theor. Biol.,* **18**:222.
MUELLER, P. and RUDIN, D. O. (1969). *Curr. Topics Bioenergetics,* **3**:157.
NELSON, P., RUFFNER, W. and NIRENBERG, M. (1969). *Proc. natn. Acad. Sci. U.S.A.,* **64**:1004.
NOBLE, D. (1962). *J. Physiol., Lond.,* **160**:317.
PAPAKOSTIDIS, G., ZUNDEL, G. and MEHL, R. (1972). *Biochim. biophys. Acta,* **288**:277.
PIERAU, FR.-K., TORREY, P. and CARPENTER, D. O. (1974). *Brain Res., Osaka,* **73**:156.
PIERAU, FR.-K., TORREY, P. and CARPENTER, D. O. (1975). *Pflügers Arch. ges. Physiol.,* **359**:349.
POST, R. L., ALBRIGHT, C. D. and DAYANI, K. (1967). *J. gen. Physiol.,* **50**:1201.
RALL, W. and SHEPHERD, G. M. (1968). *J. Neurophysiol.,* **31**:884.
RUSSELL, J. M. and BROWN, A. M. (1972). *J. gen. Physiol.,* **60**:499.
SHAMOO, A. E. and ALBERS, R. W. (1973). *Proc. natn. Acad. Sci. U.S.A.,* **70**:1191.
SHANES, A. M. (1958). *Pharmac. Rev.,* **10**:165.
SHEREBRIN, M. H. (1972). *Nature, New Biol.,* **235**:122.
SINGER, L. and TASAKI, I. (1968). *Nerve Excitability and Membrane Macromolecules,* pp. 347–410. Ed. D. CHAPMAN. London; Academic Press.
STÄMPFLI, R. (1959). *Ann. N.Y. Acad. Sci.,* **81**:264.
TASAKI, I. (1968). *Nerve Excitation: A Macromolecular Approach.* Springfield, Illinois; Charles C. Thomas.
TASAKI, I. and SPYROPOULOS, C. (1962). *Am. J. Physiol.,* **201**:413.
TASAKI, I., WATANABE, A. and HALLETT, M. (1972). *J. Membrane Biol.,* **8**:109.
TASAKI, I., WATANABE, A. and TAKENAKA, T. (1962). *Proc. natn. Acad. Sci. U.S.A.,* **48**:1177.
TASAKI, I., WATANABE, A., SANDLIN, R. and CARNEY, L. (1968). *Proc. natn. Acad. Sci. U.S.A.* **81**:883.
THOMAS, R. C. (1972). *Physiol. Rev.,* **52**:563.
TRAUTWEIN, W. and KASSEBAUM, W. G. (1961). *J. gen. Physiol.,* **45**:317.
WERBLIN, F. S. and DOWLING, J. E. (1969). *J. Neurophysiol.,* **32**:339.

9

Membranes and the control of energy

Esteban Santiago, Josefina Eugui and Natalia
López-Moratalla
Department of Biochemistry, University of Navarra, Pamplona, Spain

9.1   INTRODUCTION

All living cells must necessarily expend energy in order to perform their great
variety of physiological functions, including the biosynthesis of their own constituent molecules. In order that an adequate flow of the different metabolic
pathways be maintained by the cell, it is clear that there must be some means
of integrating reactions geared to the production of energy with other reactions which utilize the energy produced. An important point is the fact that,
in general terms, the systems responsible for the production of energy and
those which will ultimately make use of it are confined in different cellular
compartments. The bulk of the synthesis of ATP is carried out within the
mitochondria, whereas the synthesized ATP is used to a great extent in reactions which take place in the extramitochondrial part of the cell.

From all this the decisive role to be played by cell membranes in the integration and regulation of the two types of systems can be easily understood.
Of special importance is their active function, namely the specific transport
of metabolites across the membranes. It has been recognized that the distribution of metabolites, the concentration of adenine nucleotides and the redox
state of nicotinamide nucleotides in the mitochondria and in the cytoplasmic
fluid are key factors in the integration of energy-liberating metabolic pathways with other pathways which are energy-demanding.

The present discussion will be centred, first, on a general description of
the main factors which affect the control of pathways that generate energy
and those that utilize it, emphasizing the role of the 'energy charge' and distribution of adenine nucleotides in the cytoplasmic fluid and in the mitochondria. Secondly, it will deal with the transport mechanisms involved in
the translocation of ATP from the mitochondria to the cytoplasmic fluid,

and with the entry of ADP and inorganic phosphate into the mitochondria to permit the resynthesis of ATP via the process of oxidative phosphorylation.

9.2   THE CONCEPT OF 'ENERGY CHARGE'

A well-documented aspect is the interrelation between catabolic and biosynthetic pathways; much of this information has been obtained from studies carried out *in vitro*. The important regulatory role of adenine nucleotides in these pathways has been known for some years. A number of enzyme activities have been found to be modulated by the concentrations of ATP, ADP and AMP and also by the [ATP]:[ADP] and [ATP]:[AMP] ratios; these points have been discussed in excellent reviews (Atkinson, 1966, 1968). The introduction of the concept of *energy charge* (Atkinson and Walton, 1967) has been a great contribution to the understanding of the control of metabolic reactions by means of adenine nucleotide levels. The adenine nucleotide system has been compared to an electrochemical storage cell and the 'energy charge' would be analogous to the charge of the battery.

The energy charge is defined as half of the number of anhydride-bound phosphate groups per adenosine moiety and can be easily calculated for any set of concentrations of ATP, ADP and AMP by using the equation

$$\text{Energy charge} = \frac{[ATP] + \frac{1}{2}[ADP]}{[ATP] + [ADP] + [AMP]} \tag{9.1}$$

The value of the energy charge may thus vary between 0 and 1.

After the introduction of this concept, it was established that a number of enzymes are controlled by the energy charge. The activities of enzymes in pathways leading to the generation of ATP exhibit a descending hyperbolic curve with increase in energy charge, whereas the opposite happens with enzymes belonging to energy-utilizing pathways. In both cases there is a sensitive part of the curve, markedly ascending or descending, which permits a rapid response of the enzyme activity to small changes in energy charge. The energy charge seems, therefore, to be a valid regulatory factor of catabolic and anabolic pathways or, in other words, pathways which respectively generate and utilize energy. However, other factors have been recognized as affecting the control, and also have a significant physiological role in cellular activity. ATP-generating pathways, such as the glycolytic sequence and the tricarboxylic acid cycle, are amphibolic and thus also provide metabolites for biosynthetic pathways; it has been found that these are affected not only by the concentration of the initial building blocks but also by the concentration of final products, which may act through feedback inhibition.

All these factors have been proved to be operating *in vitro* both in eukaryotes and prokaryotes; however, the information regarding the latter is far more abundant and a number of reviews have appeared (Atkinson, 1969, 1971). In general terms, a high value of the energy charge causes inhibition of a series of key enzymes in catabolic processes (phosphorylase, phosphofructokinase, pyruvate kinase, pyruvate dehydrogenase, citrate synthetase, isocitrate dehydrogenase, glutamate dehydrogenase) while simultaneously enhancing the activity of enzymes which are crucial in biosynthetic processes (glycogen   synthetase,   fructose-6-phosphate   phosphatase,   pyruvate carboxykinase, aspartokinase, phosphoribosylpyrophosphate synthetase, phosphoribosyl-ATP synthetase, ATP: citrate lyase) (Atkinson, 1969, 1971; Larner, 1971). On the other hand, an increase in concentration of intermediates of the glycolytic pathway and of the tricarboxylic acid cycle usually inhibits preceding steps and stimulates subsequent reactions within the metabolic sequence. For example, the increase in concentration of acetyl-CoA inhibits pyruvate dehydrogenase and, at the same time, activates pyruvate carboxylase and phosphoenol pyruvate carboxykinase. An increase in the level of phosphoenolpyruvate inhibits phosphofructokinase, whereas it stimulates pyruvate dehydrogenase. The importance of these factors, namely the energy charge and the concentration of intermediates, in the regulation of enzymes located at branch points in a metabolic pathway is well known. From the point of view of its regulation, one of the enzymes most thoroughly studied has been phosphofructokinase, which is affected by the interaction of a variety of regulatory factors. It has been shown (Shen et al., 1968) that the inhibitory effect of a high energy charge on phosphofructokinase from rabbit muscle is more pronounced at low concentrations of its substrate, fructose-6-phosphate; variations in the energy charge have very little effect on the activity of the enzyme in the absence of citrate, which is a negative modulator, if the concentration of fructose-6-phosphate is high. At higher concentrations of citrate, phosphofructokinase is very sensitive to the inhibitory effect of a high value of the energy charge. On the other hand, the activity of the enzyme will be highest when the energy charge or the citrate concentration is low.

However, this picture seems to be further complicated by the existence of other factors which should also be kept in mind when making predictions about the regulatory function of the energy charge.

Studies have recently been carried out (Purich and Fromm, 1972, 1973) to determine the effect of the energy charge on a number of enzymes from mammalian cells and these have indicated the importance of pH, total concentration of adenine nucleotides and of other nucleotides, substrates and products, and the concentration of Mg^{2+}. Furthermore, there are indications (Lowry et al., 1971) that some enzymes seem to respond more to [ATP]:[AMP] and [ATP]:[ADP] ratios than to the energy charge itself, which is, moreover, a rather insensitive parameter; these ratios may vary considerably with very little change in the value of the energy charge; obviously, this point will require further attention.

Oversimplifications, which might be misleading, must be avoided since a great variety of effectors may be involved in the regulation of a particular enzyme. The results obtained from experiments in vitro may be considered only as indicative of what might really be happening in vivo. However, experimentation in vivo presents serious difficulties which have not yet been completely overcome; these are, for example, the independent determination of metabolite concentrations such as ATP and so forth in both the cytoplasmic fluid and the mitochondria. Besides, the interpretation of the results, even if the distribution of the different effectors is known with reasonable accuracy, might be rather hazardous because of complications introduced by the role of hormones in the regulatory processes.

A series of studies (Gumaa, McLean and Greenbaum, 1971) has been carried out on rat liver to determine the variation in vivo of levels of a variety of metabolites, of the ratio [ATP/[ADP].[P_i]], and of the redox state of nicotinamide nucleotides as regulatory factors in the reactions involved in glycolysis, the tricarboxylic acid cycle, lipogenesis, gluconeogenesis and the pentose phosphate pathway.

Experiments carried out *in vivo* (Chagoya de Sánchez, Brunner and Pinner, 1972) have shown the importance of the energy charge in regulatory mechanisms. When the value of the energy charge increases following intraperitoneal injection of adenosine, the biosynthesis of glycogen, an ATP-utilizing process, is enhanced, with a simultaneous decrease in fatty acid catabolism, which is an ATP-generating process. These results confirm previous observations from experiments *in vitro*, although the possible effect of other factors had not been considered. It may be concluded, then, that further studies will be required in order for us to understand the role of the different modulators of enzyme activity in the integration of energy-producing and energy-consuming metabolic pathways. Attention should also be focused on the mutual influence of these modulators.

The compartmentation of modulators and other metabolites within the cell is also of importance in metabolic regulation. A decisive role in the maintenance of the levels of these compounds is played by the transport systems of intracellular membranes. Knowledge of their properties and mechanisms of action has advanced considerably in recent times.

9.3   THE TRANSPORT SYSTEM FOR ADENINE NUCLEOTIDES

The existence of a transport system for adenine nucleotides was described (Pfaff, Klingenberg and Heldt, 1965; Klingenberg and Pfaff, 1965) after the observation of the high specificity of the exchange reaction of adenine nucleotides in mitochondria. The name *translocase* was later proposed for the enzyme responsible for this process (Heldt, Jacobs and Klingenberg, 1965).

In the course of their investigations, Klingenberg's group have elucidated many of the properties of translocase, especially its specificity. Large differences of exchange between ATP, ADP, AMP and also between their corresponding deoxynucleotides were found (Klingenberg and Pfaff, 1968). On the other hand, no exchange was observed with the inosine, uridine and guanosine nucleotides. It was also striking that the exchange rate for ADP was ten times greater than that of ATP, whereas AMP was not translocated at all. The specificity of the translocase was therefore limited to ADP and ATP. It was also demonstrated that the transport consists of an obligatory exchange of nucleotides; for each molecule of adenine nucleotide leaving the mitochondrion, another molecule of adenine nucleotide must necessarily pass to the inside.

Early work on the characterization of the transport system and its kinetics has been reviewed (Klingenberg and Pfaff, 1968). First-order kinetics were observed during the initial part of the exchange reaction between non-labelled exogenous adenine nucleotides and labelled endogenous adenine nucleotides, taking into account only the concentrations of endogenous ADP and ATP, since AMP does not participate in the exchange. The dependence of the exchange rate on the concentration of adenine nucleotides was determined. At concentrations of ADP as low as 50 μM, the exchange was very close to the maximal rate. The K_m for ATP exchange was 11 μM and that for ADP exchange was 4 μM, at 2 °C. It was also demonstrated that the nucleotide translocation is highly dependent on temperature, there being a twentyfold increase when the temperature was raised from 1 °C to 25 °C. A similar dependence on temperature was observed for the phosphorylation of exogenous ADP and for other reactions controlled by the phosphorylation, a fact which has been interpreted as a consequence of the limitation of these processes by the translocation of adenine nucleotides (Heldt and Klingenberg, 1968).

Studies on the specificity of the transport system revealed a particularly interesting fact. The affinity of the carrier for ADP in the direction towards the inside of the mitochondrion was larger than that towards the outside, and the opposite was true for ATP. A series of experiments using mitochondria under different metabolic conditions showed that the state of the energy transfer system of oxidative phosphorylation affects the exchange of ATP and that of ADP scarcely at all (Pfaff and Klingenberg, 1968). These results indicated that the different specificity of the translocator for ADP and ATP could be related to the functional state of the energy transfer system. It was also observed that the ATP transport towards the inside of the mitochondrion increases in the presence of uncouplers of oxidative phosphorylation and also under conditions where ion transport is induced. All this suggests that the transport of ATP towards the inside of the mitochondrion is usually hindered by some sort of inhibition, which can be eliminated by uncouplers or by valinomycin; this would be consistent with the fact that the exchange of $(ATP)^{4-}$ for $(ADP)^{3-}$ would introduce negative charges in the mitochondrion, unless accompanied by a compensatory movement of positive ions. Uncouplers are thought to facilitate the passage of H^+ across the inner mitochondrial membrane (Mitchell, 1961; Mitchell and Moyle, 1967); this would explain the increase in exogenous ATP transport towards the interior of the mitochondrion in the presence of uncouplers of oxidative phosphorylation. Some authors (Pfaff and Klingenberg, 1968; Klingenberg, Heldt and Pfaff, 1969) have thought that this different specificity of the adenine nucleotide transport towards the inside of the mitochondrion could be partly explained by the existence of a membrane potential. The asymmetry of the translocation of ADP and ATP would disappear with an increase in the membrane permeability for H^+ and K^+.

A possible control system for adenine nucleotide translocation has also been described (Klingenberg, 1970a). The existence of a negative membrane potential on the inside would limit the entry of ATP into the mitochondrion: the electrogenic repulsion would be greater than the tendency of ATP to be exchanged by the electroneutral mechanism as a consequence of a higher pH on the outside than in the interior. The membrane potential would be thus the main factor affecting the control of the translocation of ATP and ADP, a control which would lead to an [ATP]:[ADP] ratio higher on the outside than in the interior of the mitochondrion. Therefore the translocase would act mainly by translocating ADP towards the inside of the mitochondrion, there to undergo oxidative phosphorylation to give ATP which in turn would be transferred back to the cytoplasmic fluid.

The asymmetric distribution of ATP and ADP on the two sides of the membrane has been treated in theoretical terms (Klingenberg et al., 1969) and confirmed experimentally (Heldt, Klingenberg and Milovancew, 1972); it has been shown that the [ATP]:[ADP] ratio is approximately seven times as high in the cytoplasmic fluid as inside the mitochondria when the mitochondria are in the state of respiratory control. These differences in the [ATP]:[ADP] ratios between the cytoplasmic fluid and the mitochondria have also been confirmed *in vivo* (Elbers *et al.*, 1972). If the concentration of inorganic phosphate together with the [ATP]:[ADP] ratio is known, the phosphorylation potential for each compartment may be calculated. According to Heldt, Klingenberg and Milovancew (1972), the phosphorylation potential for ATP is 2.3 kcal mol^{-1} higher in the cytoplasmic fluid than in the mitochondrial matrix in the respiratory control state. Owing to this difference a large amount of energy would be necessary to transport the ATP against the higher phosphorylation potential in the cytoplasmic fluid. The energy would be generated during electron transport, and therefore the expected P/O ratio should be decreased. It is very likely that this is so in physiological conditions. Experimentally, P/O ratios are usually obtained in situations in which the [ATP]:[ADP] ratio is kept low in the incubation medium. Owing to the fact that the phosphorylation potential in the cytoplasm is different from that in the mitochondria, it may be estimated that 20–30 percent of the total energy of the phosphorylation is spent on the translocation of the ATP and the rest on the synthesis of ATP inside the mitochondrial compartment.

The use of specific inhibitors of the translocation of adenine nucleotides has permitted a great advance in the understanding of the mechanism of this process and it has helped in the formulation of several models of the translocase.

Atractyloside, a glycoside from *Atractylis gummifera*, was the first inhibitor used. Although atractyloside was first used to inhibit oxidative phosphorylation (Vignais, Vignais and Stanislas, 1962; Bruni, Contesa and Luciani, 1962; Bruni, Luciani and Bortignon, 1965) by preventing the binding of nucleotides to mitochondria through a competitive mechanism (Bruni, Luciani and Contesa, 1964; Bruni, Luciani and Bortignon, 1965), it was finally concluded that it specifically inhibited the exchange reaction of adenine nucleotides (Klingenberg and Pfaff, 1965).

With the use of atractyloside, it has been possible to determine the number of adenine nucleotide binding sites present on the translocase (Weidemann, Erdelt and Klingenberg, 1970a). In the uptake of nucleotides by mitochondria, one portion binds specifically to the carrier, and is removable by atractyloside; a second portion is non-specifically bound and is not affected by atractyloside; and a third portion exchanges with the pool of endogenous adenine nucleotides. In this way the number of binding sites for adenine nucleotides with respect to the content of cytochrome *a* has been established; in some preparations up to 1.4 moles of adenine nucleotides could be specifically bound per mole of cytochrome *a*. This ratio is very expressive, indeed, of the large number of carrier molecules existing in the mitochondrion. It has also been possible to show up the existence of two types of binding sites for ADP in beef heart mitochondria. Approximately 20–30 percent had a high affinity, whereas the rest had a much lower affinity. The existence of two types of binding sites for ATP has also been recognized. Using bongkrekic acid, another inhibitor of translocation, evidence has been obtained (Weidemann, Erdelt and Klingenberg, 1970b) in favour of the localization of binding sites of high affinity on the inner side of the membrane. Other inhibitors structurally related to atractyloside have also been discovered. Gummiferine, identified as 4-carboxylatractyloside, inhibits the translocation of adenine in a noncompetitive fashion (Stanislas and Vignais, 1964). Owing to its high specificity and the absence of a lag phase for maximal action, this inhibitor has become a useful tool in kinetic studies (Vignais, Vignais and Defaye, 1973). A new compound of the atractyloside type, probably the epi-atractyloside, has been isolated recently. It has a high affinity for translocase, similar to that of carboxylatractyloside, and higher than that of atractyloside itself (Scherer *et al.*, 1973). These two compounds of the atractyloside type have in common an equatorial carboxyl group close to groups which are negatively charged (SO_3^-). These structural features might be responsible for the high affinity they show for the translocase, since the natural substrates, ADP and ATP, also have negative charges in close vicinity.

It has been shown (Henderson and Lardy, 1970) that bongkrekic acid specifically inhibits the translocation of adenine nucleotides through a mechanism which seems to involve a large increase in the affinity of the carrier for the adenine nucleotides, preventing their dissociation (Weidemann, Erdelt and Klingenberg, 1970b). The presence of adenine nucleotides facilitates the binding of bongkrekic acid to the translocase.

Studying the interaction of the different inhibitors of translocation has also been very revealing. The kinetic data have established that these inhibitors bind to the same site of the carrier as do the adenine nucleotides. It has been shown that in the presence of ADP, the atractyloside bound to the membrane may be liberated by bongkrekic acid (Klingenberg, Grebe and Falkner, 1971). From the interactions of these two inhibitors, atractyloside and bongkrekic acid, a model for the adenine nucleotide translocase which stresses the importance of conformational modifications has been proposed (Klingenberg *et al.*, 1972). According to this proposal the translocase may exist either in a surface-oriented state with the binding sites towards the aqueous phase, permitting its interaction with hydrophilic ligands, or in another possible conformation referred to as the 'central' state, with the binding sites sequestered from the aqueous phase. When the carrier is bound to the adenine nucleotides a steady-state distribution between the two conformations takes place, permitting translocation. However, in the presence of the inhibitors one particular conformation would be predominant, resulting in the removal of ADP in the presence of atractyloside, or in its tight binding caused by bongkrekic acid. This hypothesis is in agreement with observations on mitochondrial contraction induced by ADP. Mitochondrial contraction takes place at very low concentrations of ADP and seems to be independent of oxidative phosphorylation (Weber and Blair, 1970). It probably results from a conformational change originating in the binding of ADP to the carrier (Klingenberg, Grebe and Scherer, 1971). Bongkrekic acid added, after ADP, increases mitochondrial contraction, whereas atractyloside reverses it. It has also been observed (Leblanc and Clauser, 1972) that N-ethylmaleimide increases the mitochondrial contraction caused by ADP and prevents its reversal by atractyloside. The data obtained seem to indicate that ADP unmasks a series of additional groups, perhaps SH groups, which could be related to translocation and are rendered accessible to the alkylating agent N-ethylmaleimide. ADP would produce an intermediate contraction which disappears in the presence of atractyloside and is increased by bongkrekic acid. The conformational change on the carrier induced by ADP would be transmitted to the whole membrane, thus initiating the mitochondrial contraction.

All these data support the model described above for the translocation of adenine nucleotides (Klingenberg et al., 1972). Atractyloside, which does not pass across the membrane, would stabilize the translocase in the surface-oriented conformation, whereas under the influence of bongkrekic acid the 'central' conformation would be predominant. Other models which imply conformational modifications have also been proposed from data obtained by the use of inhibitors of translocation (Vignais et al., 1972). These compounds, which are highly selective in their interaction with the carrier, will be very useful in the future for a deeper understanding of the mechanism of translocation of adenine nucleotides.

The translocase is one of the best-characterized carriers, but very little information on its regulation is yet available. So far, research on this point has been centred mainly on the study of the stimulatory and inhibitory effects of a number of ions, as well as on the action of various acyl-CoAs.

The importance of the concentration of cations to adenine nucleotide transport has come to be recognized. Although it had previously been pointed out (Pfaff and Klingenberg, 1968; Pfaff, Heldt and Klingenberg, 1969) that K^+ and Mg^{2+} affect translocation, the effect observed had been rather limited, owing perhaps to the high concentrations of ATP and ADP used. More recently (Meisner, 1971) it has been shown that protons and several cations have a stimulatory effect on translocation. Potassium ion stimulated the translocation of ADP and ATP over 100 percent at lower concentrations but raising their concentration competitively inhibited the stimulatory effect of K^+. At low concentrations, stimulation by divalent cations was more efficient than that by monovalent cations. However, at low proton concentrations translocation was inhibited, in agreement with previous observations (Weidemann, Erdelt and Klingenberg, 1970a). The data obtained (Meisner, 1971) suggest that the cations affect the binding of the adenine nucleotides to the membrane rather than the translocation process itself. The effect of Ca^{2+} at low concentrations on the exchange of adenine nucleotides indicates an important physiological role in translocation for this cation. In this respect relevant studies have been carried out on the effect of Ca^{2+} and other cations (Spencer and Bygrave, 1972a). Stimulation was greatest with Ca^{2+} and Sr^{2+}, much less with Ba^{2+} and even smaller with Mg^{2+} and Mn^{2+}. Calcium ion mainly increases the affinity of the carrier for ATP, which in the presence of K^+ and Ca^{2+} becomes as high as that for ADP. This fact could be of importance with respect to regulation, since Ca^{2+} could modify the difference of phosphorylation potential on the two sides of the mitochondrial membrane, which, as has already been mentioned, is related to the higher affinity of the carrier for ADP in its transport towards the interior of the mitochondrion. The mechanism of Ca^{2+} accumulation in the mitochondria has also been related to its effect on the translocation process; by favouring the entry of ATP into the mitochondrion, additional energy would be liberated by its hydrolysis, which in turn could be utilized for the accumulation of Ca^{2+} through its corresponding permease.

It has been proposed (Spencer and Bygrave, 1972b) that the stimulatory effect of Ca^{2+} on translocation could be due to the binding of this cation to the membrane in the vicinity of the translocase and not to the formation of a complex with ADP. Further work is necessary to clarify the influence of the concentration of cations on the translocation.

Information is becoming available on the inhibitory effect of fatty acids on the transport of adenine nucleotides, which was first demonstrated some years ago (Wojtczak and Zaluska, 1967). More recently it has been shown (Lerner et al., 1972; Vaartjes et al., 1972) that the inhibitory effect was really due to the coenzyme A derivatives of the fatty acids. The inhibition they cause may be reversed by the addition of carnitine (Shrago et al., 1972); the carnitine probably decreases the concentration of acyl-CoA by facilitating the transport of the acyl groups to the interior of the mitochondria via acylcarnitine transferase. It has been proposed that the inhibition could be explained by a displacement of the nucleotide by the ADP moiety of the acyl-CoA or by competition for the binding site of the translocase. In mitochondria isolated from livers of alloxan-diabetic rats and hibernating squirrels, a depression of adenine nucleotide translocation has been observed (Lerner et al., 1972) which seems to be due to an increased hepatic lipid content, particularly of long-chain acyl-CoA esters, in diabetes and in hibernation; the activity of the translocase may be restored to normal levels by adding carnitine to stimulate the metabolism of the acyl-CoA esters. Using mitochondria from brown adipose tissue, which contain a large amount of free fatty acids, a stimulatory effect of serum albumin on the translocation has been demonstrated (Christiansen et al., 1973); the binding of fatty acids by albumin could increase the rate of translocation to the levels observed in liver mitochondria.

A mechanism was proposed to explain the regulatory role of the acyl-CoA esters in the energy metabolism of brown adipose tissue. It is suggested that the ATP synthesized in the mitochondria is transported to the cytoplasmic fluid and utilized for the activation of fatty acids, a prior step in the synthesis of acyl-CoA esters. The acyl-CoA esters are then thought to pass to the inside of the mitochondria through the carnitine shuttle to be oxidized. A high level of ATP in the cytoplasmic fluid will cause an increase in the concentration of acyl-CoA which, in turn, will lead to an inhibition of adenine nucleotide transport; as a consequence the concentration of ADP in the mitochondria will decrease and the energy generated by the respiratory chain will be dissipated without the formation of ATP, since the mitochondria of brown adipose tissue are loosely coupled. The inhibition of the translocase will disappear with the decrease in concentration of ATP and of acyl-CoA in the cytosplasmic fluid. This mechanism would maintain adequate levels of cytoplasmic ATP for fatty acid activation, and would permit the discharge of excess energy as heat. Such a regulatory mechanism might be extended to other metabolic situations where large amounts of fatty acids are liberated, such as diabetes and starvation, but, apart from in these cases, it would not play an important role in regulating the use and generation of energy.

A decisive role for the translocase in the regulation of energy metabolism may arise from the asymmetry of its specificity with regard to the transport of ADP and ATP, which would result in an unequal [ATP]:[ADP] ratio in the cytoplasmic fluid and mitochondria, respectively. The fact that this ratio is higher in the cytoplasmic fluid, together with a much smaller volume of water in the mitochondrial compartment, would explain the rapid response to the utilization of energy in the cytoplasmic fluid. Since ADP is preferentially transported by the translocase towards the inside of the mitochondria, a slight rise in the concentration of ADP in the cytoplasmic fluid would suffice to cause a large increase in its concentration within the mitochondrial compartment; oxidative phosphorylation would be activated and consequently the ATP synthesized would be transported outside the mitochondrion (Heldt, Klingenberg and Milovancew, 1972).

9.4   THE TRANSPORT SYSTEM FOR INORGANIC PHOSPHATE AND FOR TRICARBOXYLIC ACID CYCLE INTERMEDIATES

The transport of inorganic phosphate across the mitochondrial membranes has been found to be independent from that of adenine nucleotides. However, it appears to be linked to the transport of some intermediates of the tricarboxylic acid cycle.

The proposal that inorganic phosphate requires a transport system in mitochondria was put forward some years ago (Chappell and Crofts, 1966) and soon received further support from observations (Fonyo, 1968; Fonyo and Bessman, 1968; Tyler, 1969) which demonstrated that the transport of inorganic phosphate is inhibited by blocking sulphydryl groups of the carrier, as is the case for similar carriers in other membranes.

For the transport of tricarboxylic acid cycle intermediates two carriers were first described (Chappell and Haarhoff, 1967), one for the dicarboxylic acids malate and succinate, and another for the tricarboxylates citrate, isocitrate and cis-aconitate. A third carrier, specific for α-ketoglutarate, was also proposed (De Haan and Tager, 1968). Activators are needed for maximal activity: the entry of malate or succinate requires inorganic phosphate, and that of tricarboxylates requires malate, whereas the transport of α-ketoglutarate needs malate or succinate and is not activated by inorganic phosphate (cf. Klingenberg, 1970b).

An explanation of these observations has been given (Klingenberg, 1970a) on the basis of an exchange mechanism: (a) inorganic phosphate is transported across the membrane by means of its specific carrier and may subsequently be transported outside the mitochondrion again, via the dicarboxylate carrier, in exchange for dicarboxylates; (b) the exchange between tricarboxylates and dicarboxylates may occur via the specific carrier for tricarboxylates; (c) the exchange between α-ketoglutarate and malate or succinate may take place via the specific α-ketoglutarate carrier. Using differential centrifugation and sedimentation techniques it has been shown that there is a 1:1 exchange for the pairs of metabolites citrate and malate (Palmieri and Quagliariello, 1968), malate and α-ketoglutarate (Papa et al., 1969b) and malate and inorganic phosphate (Papa et al., 1969a).

It has recently been demonstrated (Johnson and Chappell, 1973) that the dicarboxylate carrier may function independently of the phosphate carrier. N-Ethylmaleimide was used to inhibit phosphate transport by means of the specific phosphate carrier, in order to study the operation of the dicarboxylate carrier which can exchange malate for inorganic phosphate; this carrier may be inhibited by n-butylmalonate.

It has been established (Palmieri, Quagliariello and Klingenberg, 1970) that the accumulation of anions in the mitochondria is compensated by a movement of hydrogen ions. The mechanism of this accumulation, as well as the role of the phosphate carrier, has also been investigated (McGivan and Klingenberg, 1971). It has been shown that the energy-dependent pH gradient across the mitochondrial membrane permits the entry of phosphate by means of the operation of the specific carrier; the uptake of phosphate would drive a cascade of exchange reactions to effect the movement across the membrane of dicarboxylates, tricarboxylates and α-ketoglutarate via their specific carriers. Movement of hydrogen ions would take place in the net uptake of anions such as inorganic phosphate, malate and citrate; however, in the exchanges only the carrier of citrate would be linked to the movements of hydrogen ion. The net uptake of citrate by means of the malate–inorganic phosphate shuttle might occur according to one of two possibilities: that all the anions are transported undissociated, as anions neutralized with their corresponding protons, or that only the phosphate carrier transports the phosphate undissociated (McGivan and Klingenberg, 1971). The first possibility seems to be more consistent since the citrate–malate exchange would move one proton; the second possibility would not permit electroneutral transport. It has been suggested (Klingenberg, 1970a) that the concentration of undissociated acids is very low and it is likely that the proton accompanying the anion would bind to the carrier as H^+ and then the H^+ carrier would bind the anion. Evidence has been presented in support of this (Palmieri, Quagliariello and Klingenberg, 1970) in the sense that the entry of anions into the mitochondrion takes place coupled to the movement of H^+. It has been suggested (Klingenberg, 1970a) that the transport of inorganic phosphate by means of the specific phosphate carrier occurs without any exchange and its charge could be compensated by the movement of a hydrogen ion; this view is in opposition to a former proposal (Chappell and Crofts, 1966) according to which the dissociated phosphate anion, with one negative charge, is exchanged against hydroxide ion. Both formulations give electroneutral transport.

The maximal transport rate of inorganic phosphate via the specific phosphate carrier has recently been determined for rat liver mitochondria as 600 μmol min^{-1} per gram of protein at 25 °C (Klingenberg, Durand and Guerin, 1974). The kinetics of the dicarboxylate carrier have also been studied, in the presence of N-ethylmaleimide to inhibit the phosphate carrier, with the inhibitor-stop method using substrate analogues such as 2-phenylsuccinate, 2-n-butylmalonate or 2-benzylmalonate (Palmieri et al., 1971). Maximal rates for the dicarboxylates have been found to be approximately 70 μmol min^{-1} per gram of protein at 9 °C both for malate and succinate, with K_m values of 0.23 and 1.17 mM respectively. It has been shown that malate and succinate behave competitively in the kinetics of uptake, whereas the inhibition of malate uptake by inorganic phosphate was found to be predominantly non-competitive. From these observations it has been concluded that the carrier has two specific binding sites, one for the transport of inorganic phosphate and another for the dicarboxylates. Similar studies (Palmieri et al., 1972) on the kinetics of the tricarboxylate carrier have shown that the maximal transport rate of citrate is 23 μmol min^{-1} per gram of protein and the K_m is 0.12 mM. Competitive inhibition by phosphoenolpyruvate and by dicarboxylates has been observed. This carrier seems to have only a single binding site for tricarboxylates, dicarboxylates and phosphoenolpyruvate.

Recent studies (Sluse, Goffart and Liébecq, 1973) have also been carried out on the kinetics of the α-ketoglutarate carrier in rat heart mitochondria. It has been found that in the exchange reactions between internal malate and external α-ketoglutarate, in the presence of external malate to effect external product-inhibition, the K_m for the internal malate is unaffected by the presence of the external product whereas the K_m for external α-ketoglutarate increases with the concentration of external malate. The inhibition of the exchange by the external product is competitive when the external substrate is varied and non-competitive when the internal substrate is varied. These results give further support to a previously proposed mechanism of rapid-equilibrium random *bi bi* (Sluse, Ransom and Liébecq, 1972).

Klingenberg (1970a) has speculated on several possible models for the different carriers. More recently (Klingenberg, Durand and Guerin, 1974), the reactivity of SH reagents with the phosphate carrier has been thoroughly investigated in an effort to understand the mechanism of phosphate transport. On the basis of the data obtained, a model for the carrier has been proposed. According to this proposal the carrier in the free form is mobile and can interact with H^+; the carrier then becomes immobile as carrier–H^+. The protonated carrier can react with the mono-anionic phosphate to give the mobile phosphate–carrier–H^+ complex. The distribution of the mobile forms is controlled by the concentration of protons and phosphate ions on both sides of the membrane. The free carrier can move towards the membrane surface exposed to higher concentrations of protons and phosphate ions, whereas the phosphate–carrier–H^+ complex moves in the opposite direction, to become free carrier after reaching the membrane surface and discharging the H^+ and the phosphate. Phosphate transport would thus be an electroneutral process driven by the pH gradient.

9.5   CONCLUSIONS

The availability of oxidizable substrates within the cell permits the operation of oxidative phosphorylation, provided that ADP and inorganic phosphate are present in the mitochondria. The synthesized ATP has to be exported to the extramitochondrial part of the cell and used in energy-consuming reactions; its transport across the membrane is mediated by the translocase in a 1:1 exchange with ADP. Inorganic phosphate may enter the mitochondria by means of its specific carrier or in exchanges through the dicarboxylate carrier. The synthesis of ATP seems to be related to the exit of protons, linked to electron flow, from the mitochondrial matrix and the entry of phosphate into the mitochondria is in turn regulated by the unequal distribution of hydrogen ions across the mitochondrial membranes. When ATP is consumed a series of events will take place which tend to restore an [ATP]:[ADP] ratio that is higher in the cytoplasmic fluid than in the mitochondria.

The distribution of adenine nucleotides in the cellular compartments is of paramount importance in the control of pathways which require or utilize energy, via the modulation of key enzymes.

Future research on the control mechanisms of the specific metabolite carriers will be of great help in a better understanding of the important role of membranes in the regulation of energy flow within the cell.

REFERENCES

ATKINSON, D. E. (1966). *A. Rev. Biochem.*, **35**:85.

ATKINSON, D. E. (1968). *The Metabolic Roles of Citrate*, pp. 23–40. Ed. T. W. GOODWIN. New York; Academic Press.

ATKINSON, D. E. (1969). *A. Rev. Microbiol.*, **23**:47.

ATKINSON, D. E. (1971). *Metabolic Pathways*, 3rd edn, Vol. 5, pp. 1–21. Ed. H. J. VOGEL. New York; Academic Press.

ATKINSON, D. E. and WALTON, G. M. (1967). *J. biol. Chem.*, **242**:3239.

BRUNI, A., CONTESA, A. R. and LUCIANI, S. (1962). *Biochim. biophys. Acta*, **60**:301.

BRUNI, A., LUCIANI, S. and BORTIGNON, C. (1965). *Biochim. biophys. Acta*, **97**:434.

BRUNI, A., LUCIANI, S. and CONTESA, A. R. (1964). *Nature, Lond.*, **201**:1219.

CHAGOYA DE SÁNCHEZ, V., BRUNNER, A. and PINA, E. (1972). *Biochim. biophys. Res. Commun.*, **46**:1441.

CHAPPELL, J. B. and CROFTS, A. R. (1966). *Regulation of Metabolic Processes in Mitochondria*, pp. 293–314. Ed. J. M. TAGER, S. PAPA, E. QUAGLIARIELLO and E. C. SLATER. Amsterdam; Elsevier.

CHAPPELL, J. B. and HAARHOFF, K. N. (1967). *Biochemistry of Mitochondria*, pp. 75–91. Ed. E. C. SLATER, Z. KANIUGA and L. WOJTCZAK. London and Warsaw; Academic Press and Polish Scientific Publishers.

CHRISTIANSEN, E. N., DRAHOTA, Z., DUSZYNSKI, J. and WOJTCZAK, L. (1973). *Eur. J. Biochem.*, **34**:506.

DE HAAN, E. J. and TAGER, J. M. (1968). *Biochim. biophys. Acta*, **153**:98.

ELBERS, R., HELDT, H. W., SCHMUCKER, P. and WIESE, H. (1972). *Hoppe-Seyler's Z. physiol. Chem.*, **353**:702.

FONYO, A. (1968). *Biochem. biophys. Res. Commun.*, **32**:624.

FONYO, A. and BESSMAN, S. P. (1968). *Biochem. Med.*, **2**:145.

GUMAA, K. A., MCLEAN, P. and GREENBAUM, A. L. (1971). *Essays in Biochemistry*, Vol. 7, pp. 39–86. Ed. P. N. CAMPBELL and F. DICKENS. London; Academic Press.

HELDT, H. W., JACOBS, H. and KLINGENBERG, M. (1965). *Biochem. biophys. Res. Commun.*, **18**:174.

HELDT, H. W. and KLINGENBERG, M. (1968). *Eur. J. Biochem.*, **4**:1.

HELDT, H. W., KLINGENBERG, M. and MILOVANCEW, M. (1972). *Eur. J. Biochem.*, **30**:434.

HENDERSON, P. J. F. and LARDY, H. A. (1970). *J. biol. Chem.*, **245**:1319.

JOHNSON, R. N. and CHAPPELL, J. B. (1973). *Biochem. J.*, **134**:769.

KLINGENBERG, M. (1970a). *Essays in Biochemistry*, Vol. 6, pp. 119–159. Ed. P. N. CAMPBELL and F. DICKENS. London; Academic Press.

KLINGENBERG, M. (1970b). *FEBS Lett.*, **6**:145.

KLINGENBERG, M., DURAND, R. and GUERIN, B. (1974). *Eur. J. Biochem.*, **42**:135.

KLINGENBERG, M., GREBE, K. and FALKNER, G. (1971). *FEBS Lett.*, **16**:301.

KLINGENBERG, M., GREBE, K. and SCHERER, B. (1971). *FEBS Lett.*, **16**:253.

KLINGENBERG, M., HELDT, H. W. and PFAFF, E. (1969). *The Energy Level and Metabolic Control in Mitochondria*, pp. 236–252. Ed. S. PAPA, J. M. TAGER, E. QUAGLIARIELLO and E. C. SLATER. Bari; Adriatica Editrice.

KLINGENBERG, M. and PFAFF, E. (1965). *Regulation of Metabolic Processes in Mitochondria*, pp. 180–201. Ed. J. M. TAGER, S. PAPA, E. QUAGLIARIELLO and E. C. SLATER. Amsterdam; Elsevier.

KLINGENBERG, M. and PFAFF, E. (1968). *The Metabolic Roles of Citrate*, pp. 105–122. Ed. T. W. GOODWIN. London; Academic Press.

KLINGENBERG, M., WULF, R., HELDT, H. W. and PFAFF, E. (1969). *Mitochondria: Structure and Function*, pp. 59–77. Ed. L. ERNSTER and Z. DRAHOTA. London: Academic Press.

KLINGENBERG, M., BUCHHOLZ, M., FRIDELT, H., FALKNER, G., GREBE, K., KADNER, H., SCHERER, B., STENGEL-RUTKOWSKI, L. and WEIDEMANN, J. (1972). *Biochemistry and Biophysics of Mitochondrial Membranes*, pp. 465–486. Ed. G. F. AZZONE, E. CARAFOLI, A. L. LEHNINGER, E. QUAGLIARIELLO and N. SILIPRANDI. New York; Academic Press.

LARNER, J. (1971). *Intermediary Metabolism and its Regulation*. Englewood Cliffs, New Jersey; Prentice–Hall.

LEBLANC, P. and CLAUSER, H. (1972). *FEBS Lett.*, **23**:107.

LERNER, E., SHUG, A. L., ELSON, C. and SHRAGO, C. (1972). *J. biol. Chem.*, **247**:1513.

LOWRY, O. H., CARTER, J., WARD, J. B. and GLASER, L. (1971). *J. biol. Chem.*, **246**:6511.
MCGIVAN, J. D. and KLINGENBERG, M. (1971). *Eur. J. Biochem.*, **20**:392.
MEISNER, H. (1971). *Biochemistry*, **10**:3485.
MITCHELL, P. (1961). *Biochem. J.*, **81**:24.
MITCHELL, P. and MOYLE, J. (1967). *Biochemistry of Mitochondria*, pp. 75–91. Ed. E. C. SLATER, Z. KANIUGA and L. WOJTCZAK. New York; Academic Press.
PALMIERI, F. and QUAGLIARIELLO, E. (1968). *Mitochondria: Structure and Function.* Abstr. No. 532, Abstracts 5th FEBS Meeting. Prague; Czechoslovak Biochemical Society.
PALMIERI, F., QUAGLIARIELLO, E. and KLINGENBERG, M. (1970). *J. Biochem., Tokyo*, **17**:230.
PALMIERI, F., PREZIOSO, G., QUAGLIARIELLO, E. and KLINGENBERG, M. (1971). *Eur. J. Biochem.*, **22**:66.
PALMIERI, F., STIPANI, I., QUAGLIARIELLO, E. and KLINGENBERG, M. (1972). *Eur. J. Biochem.*, **26**:586.
PAPA, S., LOFRUMENTO, N. E., LOGLISCI, M. and QUAGLIARIELLO, E. (1969a). *Biochim. biophys. Acta*, **189**:311.
PAPA, S., D'ALOYA, R., MEIJER, A. J., TAGER, J. M. and QUAGLIARIELLO, E. (1969b). *The Energy Level and Metabolic Control in Mitochondria*, pp. 159–169. Ed. S. PAPA, J. M. TAGER, E. QUAGLIARIELLO and E. C. SLATER. Bari; Adriatica Editrice.
PFAFF, E., HELDT, H. and KLINGENBERG, M. (1969). *Eur. J. Biochem.*, **10**:484.
PFAFF, E. and KLINGENBERG, M. (1968). *Eur. J. Biochem.*, **6**:66.
PFAFF, E., KLINGENBERG, M. and HELDT, H. W. (1965). *Biochim. biophys. Acta*, **104**:312.
PURICH, D. L. and FROMM, H. J. (1972). *J. biol. Chem.*, **247**:249.
PURICH, D. L. and FROMM, H. J. (1973). *J. biol. Chem.*, **248**:461.
SCHERER, B., GREBE, K., RICCIO, P. and KLINGENBERG, M. (1973). *FEBS Lett.*, **31**:15.
SHEN, L. C., FALL, L., WALTON, G. H. and ATKINSON, D. E. (1968). *Biochemistry*, **7**:4041.
SHRAGO, E., SHUG, A., ELSON, C. and LERNER, E. (1972). *The Role of Membranes in Metabolic Regulation*, pp. 165–182. Ed. M. A. MEHLMAN and R. W. HANSON. New York; Academic Press.
SLUSE, F. E., GOFFART, G. and LIÉBECQ, C. (1973). *Eur. J. Biochem.*, **32**:283.
SLUSE, F. E., RANSON, M. and LIÉBECQ, C. (1972). *Eur. J. Biochem.*, **25**:207.
SPENCER, T. and BYGRAVE, F. L. (1972a). *Biochem. J.*, **129**:355.
SPENCER, T. and BYGRAVE, F. L. (1972b). *FEBS Lett.*, **26**:225.
STANISLAS, E. and VIGNAIS, P. M. (1964). *C. r. hebd. Séanc. Acad. Sci., Paris*, **259**:4871.
TYLER, D. D. (1969). *Biochem. J.*, **111**:665.
VAARTJES, W. J., KEMP, A., JR., SOUVERIJN, J. H. M. and VAN DEN BERGH, S. G. (1972). *FEBS Lett.*, **23**:303.
VIGNAIS, P. V., VIGNAIS, P. M. and DEFAYE, G. (1973). *Biochemistry*, **12**:1508.
VIGNAIS, P. V., VIGNAIS, P. M. and STANISLAS, E. (1962). *Biochim. biophys. Acta*, **60**:284.
VIGNAIS, P. V., VIGNAIS, P. M., DEFAYE, G., CHABERT, J., DOUSSIERE, J. and BRANDOLIN, J. (1972). *Biochemistry and Biophysics of Mitochondrial Membranes*, pp. 447–456. Ed. G. F. AZZONE, E. CARAFOLI, A. L. LEHNINGER, E. QUAGLIARIELLO and N. SILIPRANDI. New York; Academic Press.
WEBER, N. E. and BLAIR, P. V. (1970). *Biochem. biophys. Res. Commun.*, **41**:821.
WEIDEMANN, M. J., ERDELT, H. and KLINGENBERG, M. (1970a). *Eur. J. Biochem.*, **16**:313.
WEIDEMANN, M. J., ERDELT, H. and KLINGENBERG, M. (1970b). *Biochem. biophys. Res. Commun.*, **39**:363.
WOJTCZAK, L. and ZALUSKA, H. (1967). *Biochem. biophys. Res. Commun.*, **28**:76.

10

The brush border of the renal proximal tubule and the intestinal mucosa

Bertram Sacktor
Laboratory of Molecular Aging, Gerontology Research Center, National Institute on Aging, Baltimore City Hospitals, Baltimore

10.1 INTRODUCTION

The renal proximal tubule and the small intestine, which carry out transepithelial transport of solutes and fluid, are characterized by cells with demonstrable polarity. This polarity is evident functionally by the observed differences in the mechanisms by which solutes enter and leave the cell and ultrastructurally by the differentiation of the plasma membrane into two distinct entities, the brush border and basal-lateral membranes. To illustrate this, *Figure 10.1* shows the epithelium of the renal proximal tubule. It consists of a single layer of cells surrounding the tubular lumen. At the apical, or luminal, pole of the cell the plasma membrane is modified into a brush border containing a large number of microvilli. At the contraluminal, or peritubular, end of the cell the basal-lateral membrane abuts the interstitium, or plasma, side. Infoldings of the contraluminal membrane may be extensive here.

Physiological studies *in vivo* and with a variety of relatively intact intestinal and renal preparations have demonstrated the net movement of sugars, amino acids and electrolytes across a membranous barrier against the existing electrochemical gradient. These findings suggest that the brush border membrane is the site of the active transport processes. This prompted, first with the intestine and later with the kidney, the development of techniques for the isolation of the brush border membrane and the description of its chemical and enzymatic properties. Recently, the brush border membrane has been used as a model system to examine the mechanisms by which these solutes are transported. In this chapter, selected findings from these active fields of investigation will be discussed.

10.2   ISOLATION AND BIOCHEMICAL PROPERTIES OF BRUSH BORDER MEMBRANES

The brush border of the proximal tubule cell (*Figure 10.1*), or small intestine, consists of numerous finger-like processes, the microvilli, that project freely into the lumen from the apical surface of the cell. There are approximately 60 microvilli per μm^2 of luminal surface area in the rat proximal tubule (Maunsbach, 1973). The length of the microvilli varies along the proximal tubule, being 2.5 μm in the first segment and 1.5 and 3.0 μm in the second

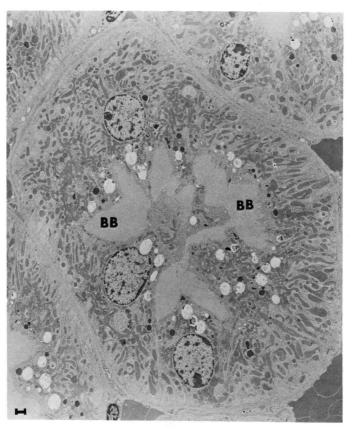

Figure 10.1   Electron micrograph of the rabbit renal proximal tubule. Brush border microvilli (BB) project from the apical pole of the epithelial cell into the lumen. The bar represents 1 μm

and third segments, respectively. The microvilli in the intestine are considerably shorter, averaging 1 μm in length with a range of 0.75–1.5 μm (Trier, 1968). The width of the microvilli is *ca.* 0.08–0.09 μm in the kidney and 0.1 μm in the intestine. It has been estimated that the microvilli increase the luminal cell surface approximately 40-fold (kidney) and 14- to 39-fold (intestine) compared with that which would be presented if the cell had a flat apical plasma membrane.

The microvillus has distinct ultrastructural features. External to the plasma membrane proper is a sialic-acid-containing, glycoproteinaceous (Groniowski, Biczyskowa and Walski, 1969), fibrillar material known as the *fuzzy coat* or *glycocalyx*, which is approximately 10 nm in width (Ito, 1965). When the microvillus is viewed in the electron microscope as a negatively stained preparation, it appears to be studded with knobs about 6 nm in diameter (Johnson, 1967). The plasma membrane of the microvillus is a triple-layered structure comprising two electron-dense layers separated by an electron-translucent space (Maunsbach, 1973). The total thickness of the trilamellate membrane is approximately 9 nm, which is thicker than the basal-lateral segment of the plasma membrane. In cross section thin filaments are seen within the microvillus. These are approximately 6 nm in diameter and run parallel to the long axis of the microvillus. In the intestinal brush border, bundles of these filaments interdigitate with and contribute to the complex meshwork of fine filaments comprising the terminal web (Trier, 1968). Upon disruption and fractionation of intestinal brush border preparations the filaments appear to co-sediment with the 'core' fraction (Eichholz and Crane, 1965). Thuneberg and Rostgaard (1969) have shown that the filament resembles actin and Rostgaard, Kristensen and Nielson (1972) have proposed that the presence of this contractile mechanism with movement of the microvilli may suggest a role for the filaments in transport.

10.2.1   Isolation of the brush border

Miller and Crane (1961), using the hamster, were the first to isolate brush border membranes of the intestinal mucosa as a morphologically distinct entity, and brush border membranes from intestines of rabbit (Porteous and Clark, 1965), guinea-pig and cat (Hubscher, West and Brindley, 1965), and pigeon (Boyd, Parsons and Thomas, 1968) have been prepared by similar methods. However, membranes isolated by these procedures are severely contaminated by other subcellular components, particularly DNA and RNA. The isolation technique has been significantly improved by Forstner, Sabesin and Isselbacher (1968) and Porteous (1968), and intestinal brush borders essentially free from nucleic acids can now be isolated. In a rat preparation (Forstner, Sabesin and Isselbacher, 1968) the sucrase : DNA ratio is more than 100 times as great as was found in the less pure earlier preparations. Currently, the procedure as developed by Forstner, Sabesin and Isselbacher (1968) is generally considered to yield the most satisfactory animal intestinal brush border membranes. A preparation of membranes which starts with isolated epithelial cells rather than mucosal scrapings, uses a single zonal centrifugation, and avoids hypotonic media is described by Connock, Elkin and Pover (1971). Schmitz *et al.* (1973) have reported a method for preparing human intestinal brush border membranes that may be applied to fresh or frozen intestine, to surgical specimens or peroral jejunal biopsies.

   Preparations of morphologically identifiable subfractions, derived from the isolated intact intestinal brush borders, have been described. Eichholz and Crane (1965) used hypertonic Tris buffer to disrupt brush borders, then separated the fragments on a glycerol gradient, and isolated particulates enriched in either the microvillus membrane or 'core' filaments. Intestinal microvillus membranes have also been prepared using hypotonic EDTA shock treatment of isolated brush borders (Forstner, Sabesin and Isselbacher, 1968). Approximately two-thirds and one-third of the total brush border membrane protein is found in the microvillus membrane and fibrillar residue, respectively. Hopfer *et al.* (1973) have substantially modified the procedure originally described by Forstner, Sabesin and Isselbacher (1968). The technique of free-flow electrophoresis of membranes (Heidrich *et al.*, 1972) has

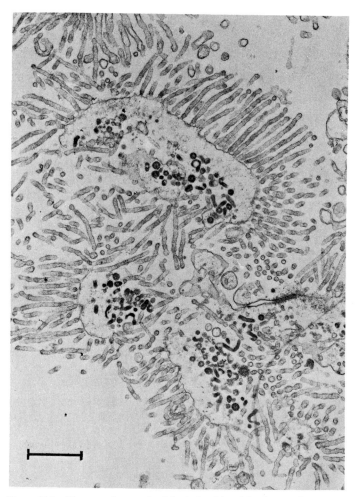

Figure 10.2   Electron micrograph of the isolated brush border membrane from rabbit renal cortex. The bar represents 1 μm

recently been adopted to separate the microvillus membrane from the basal-lateral plasma membrane of rat intestinal epithelial cells (Murer *et al.*, 1974). Douglas, Kerley and Isselbacher (1972) have described a preparation of basal-lateral segments of the enterocyte plasma membrane relatively free of brush border segments.

Various methods for the isolation of renal brush borders have been reported. Most rely on the original observation of Thuneberg and Rostgaard

(1968) that the preliminary disruption of the kidney cortex must be gentle in order to minimize the shearing of microvilli from the rest of the membrane. A detailed procedure for the preparation of rabbit renal brush borders employing sucrose density and multiple differential centrifugations has been described (Berger and Sacktor, 1970; Aronson and Sacktor, 1975). An electron micrograph of the isolated brush borders is shown in *Figure 10.2*. Brush border membranes from rat kidney have been prepared (Kinne and Kinne-Saffran, 1969; Wilfong and Neville, 1970), and a particulate fraction rich in brush border membranes from human kidney has been isolated (Scherberich *et al.*, 1974). Purified preparations of rabbit brush border membranes have also been obtained by zonal centrifugation followed by differential centrifugations (Quirk and Robinson, 1972; George and Kenny, 1973). Several investigators (Hillman and Rosenberg, 1970; Stevenson, 1972a) have used collagenase to disperse the renal cortex cells initially. Glossmann and Gips (1974) have developed a two-phase polymer system (dextran–polyethylene glycol) to isolate membranes enriched with brush borders. A method for preparing plasma membrane vesicles containing both brush border and basal-lateral membranes has been reported (Busse and Steinmaier, 1974). Free flow electrophoresis (Heidrich *et al.*, 1972) has been used to separate the two membranous entities.

10.2.2   Chemical composition of brush border membranes

Analyses of the general chemical composition of the brush border membrane have been carried out in attempts to relate its distinctive structural organization to its specialized functions. Protein comprises about 54 percent of the dry weight of the rabbit renal brush border (Stevenson, 1973), a value similar to the 58 percent reported for a rat-liver plasma membrane preparation (Emmelot and Bos, 1972). The microvillus membrane of the rat intestine contains approximately 600 μg of lipid per milligram of protein (Forstner, Tanaka and Isselbacher, 1968). Of the total lipid, approximately 20 percent is neutral lipid, 30 percent is phospholipid and as much as 50 percent has been estimated to be glycolipids. Compared with other plasma membranes from different tissues of the rat, the brush border membrane possesses the greatest amount of glycolipid and the least phospholipid. Also characteristic of the brush border membrane is a high cholesterol:phospholipid ratio. Forstner, Tanaka and Isselbacher (1968) reported a value of 1.0 mole of cholesterol per mole of phospholipid for the intact brush border, and this ratio is increased to 1.26 in the microvillus membrane subfraction. A molar ratio of about 0.5 is found in the basal-lateral segment of the rat enterocyte plasma membrane (Douglas, Kerley and Isselbacher, 1972). This latter value is comparable with those reported for a variety of other plasma membranes (Steck and Wallach, 1970) but is very different from values (i.e. below 0.1) found in endoplasmic reticulum and mitochondria (Korn, 1969). A cholesterol:phospholipid molar ratio of 0.65 is reported in rabbit renal brush borders (Quirk and Robinson, 1972).

The neutral lipid fraction of the rat intestinal microvillus membrane totals 0.119 mg per milligram of protein (Forstner, Tanaka and Isselbacher, 1968). Of this, 70 percent is cholesterol, none of which is esterified. Of the remaining neutral lipids, fatty acids and diglycerides account respectively for 13 and 11 percent of the total. Mono- and triglycerides are found only in small amounts. The total phospholipid in the rat intestinal microvillus membrane is estimated to be 0.143 mg per milligram of protein (Forstner, Tanaka and Isselbacher, 1968), a value significantly below the 0.301 mg per milligram of protein reported in the rabbit kidney brush border (Quirk and Robinson, 1972). Of the phospholipids in the intestinal microvillus membrane, phosphatidylethanolamine, phosphatidylcholine, phosphatidylserine, sphingomyelin and lysophosphatidylcholine comprise, respectively, approximately 40, 21, 14, 7 and 2 percent of the total. The glycolipids of the rat intestinal brush border membrane have been analyzed by Forstner and Wherrett (1973). The major glycosphingolipids are ceramides (mono-, di- and trihexosylceramides) and gangliosides. These workers have also estimated the sugar moieties of each class of glycolipid. The carbohydrate composition of the brush border membrane glycoprotein and the amino acid composition of the total membrane protein have been determined (Kim and Perdomo, 1974).

10.2.3   Enzyme activities of brush border membrane preparations

The epithelial cells of the kidney and intestine possess similar systems by which solutes in the tubular and intestinal lumens are transported into the cell. It is not surprising, therefore, that the two brush border membranes have enzyme components in common. On the other hand, the intestinal brush border has, in addition, a major role in digestion; hence, it is to be expected that the intestinal membrane would have enzyme activities that are not found in the renal membrane.

The determination that an enzyme is localized in the brush border membrane is based, in part, on estimates of the increase in specific activity of that enzyme in the brush border preparation relative to that in the tissue homogenate. In the renal brush border, several enzymes show large and comparable increases in relative specific activities, about 10- to 20-fold. As shown in *Table 10.1*, these include the disaccharidases trehalase and maltase, alkaline phosphatase, 5′-nucleotidase, γ-glutamyl transpeptidase and two aminopeptidases.

The disaccharidases appear to be localized exclusively in the brush border membrane, as is evident by progressive increases in specific activities at each step in the purification of the brush border and by the very low specific activities of the enzymes in non-brush-border fractions (Berger and Sacktor, 1970). Moreover, trehalase is not found in the renal medulla and papilla (Sacktor, 1968), nor in the glomerulus and interstitial areas of the cortex (Grossman and Sacktor, 1968; Stevenson, 1972b). Trehalase has been reported in the kidneys of most mammals, including rabbit, mouse, dog, primates and man. It is virtually absent, however, from the kidneys of the rat and cat (Sacktor, 1968). In these two species the disaccharidase maltase is very active. The precise function of the disaccharidases in the renal brush border remains to be clarified. A possible role for the enzymes in glucose transport has been hypothesized (Sacktor, 1968; Sacktor and Berger, 1969).

Alkaline phosphatase, which also exhibits significant increases in specific activity in the brush border relative to that in the homogenate, has been used

Table 10.1 ENZYME ACTIVITIES LOCALIZED IN RENAL BRUSH BORDERS (Measured at 37 °C except as noted)

Enzyme	Species	Activity, μmol min^{-1} per mg of protein		Relative sp. act.	Reference
		Homogenate	Brush border		
Trehalase	Rabbit	0.040*	0.485*	12	Berger and Sacktor (1970)
		0.077	1.17	15	George and Kenny (1973)
		0.084	1.18	14	Quirk and Robinson (1972)
Maltase	Rabbit	0.028*	0.395*	14	Berger and Sacktor (1970)
		0.042	0.944	22	George and Kenny (1973)
		0.059	0.951	16	Quirk and Robinson (1972)
	Rat	0.067	1.15	17	Glossmann and Neville (1972a)
Alkaline phosphatase	Rabbit	0.184	2.75	15	George and Kenny (1973)
		0.092	1.75	19	Quirk and Robinson (1972)
	Rat	0.100	1.67	17	Glossmann and Neville (1972a)
	Man	0.020	0.138	7	Scherberich et al. (1974)
5'-Nucleotidase	Rabbit	0.013	0.220	17	George and Kenny (1973)
	Rat	0.042	0.667	17	Glossmann and Neville (1972a)
Aminopeptidase (leucine)	Rabbit	0.167	2.50	15	Quirk and Robinson (1972)
	Rat	0.057	0.234	4	Glossmann and Neville (1972a)
Aminopeptidase (alanine)	Rabbit	0.245	4.58	19	Goerge and Kenny (1973)
	Man	0.014	0.157	11	Scherberich et al. (1974)
γ-Glutamyl transpeptidase	Rabbit	0.105	1.87	18	George and Kenny (1973)
		0.31	4.4	14	C. Filburn and B. Sacktor†
	Rat	0.684	3.55	5	Glossmann and Neville (1972b)

* Measured at 21 °C.
†Unpublished data, 1975.

as a marker for the renal brush border, especially by Kinne and co-workers (e.g. Heidrich *et al.*, 1972). In some studies, however, alkaline phosphatase activity, although enriched many-fold, may be increased less than that of the disaccharidases (Berger and Sacktor, 1970). The use of alkaline phosphatase as a marker for the brush border is complicated additionally by estimates that the total activity in all fractions in the preparation of the brush border fraction may sum to a value greater than 100 percent (Berger and Sacktor, 1970), by the finding of alkaline phosphatase activity in renal membranes other than the brush border (Bonting *et al.*, 1958; Reale and Luciano, 1967), and by strain-dependent differences in activity in rat brush border preparations (O'Bryan and Lowenstein, 1974). The use of 5'-nucleotidase activity as a marker for brush border membranes suffers from the inability to discriminate unequivocally between 5'-nucleotidase and alkaline phosphatase activities (George and Kenny, 1973). Aminopeptidases, although of high relative specific activity (15- to 17-fold) in the rabbit (Quirk and Robinson, 1972; George and Kenny, 1973), are reported to be enriched only 3- to 4-fold in the rat (Glossmann and Neville, 1972a; Thomas and Kinne, 1972). An apparently similar species-dependent variance in relative specific activity is reported for the enzyme γ-glutamyl transpeptidase (*Table 10.1*).

Contrasting with the enzymes that show 10- to 20-fold increases in relative specific activities and are localized predominantly in the apical region of the plasma membrane, other enzymes that are found in preparations of renal brush borders have final enrichments of only 2- to 5-fold and, therefore, are probably associated with subcellular fractions of the tubular epithelial cell in addition to the brush borders. Most prominent of the enzymes of this latter type are the different ATPases (*Table 10.2*). Mg^{2+}-ATPase and HCO_3^--stimulated Mg^{2+}-ATPase activities are found in the brush border and are known to be present in mitochondria. The possibility, however, that the activities in the brush border membranes may be accounted for by mitochondrial contamination of the brush border preparations has been ruled out (C. T. Liang and B. Sacktor, unpublished work, 1975). A Ca^{2+}-stimulated hydrolysis of ATP, assayed in the absence of added Mg^{2+}, is reported in rabbit (Berger and Sacktor, 1970) and rat (Kinne-Saffran and Kinne, 1974a) brush border preparations. The enzyme shows a 2-fold enrichment relative to the cortical homogenate. Other enzymes which have been reported in renal brush border preparations are adenyl cyclase (Wilfong and Neville, 1970; C. Filburn, R. Balakir and B. Sacktor, unpublished work, 1975), protein kinase (Insel *et al.*, 1974; E. George, R. Balakir, C. Filburn and B. Sacktor, unpublished work, 1975) and cAMP phosphodiesterase (Filburn and Sacktor, 1975).

As shown in *Figure 10.2*, electron micrographs of the renal brush border preparations indicate the occasional presence of tight junctions attached to the isolated brush border membranes. Thus, there is a question as to whether the ATPases are associated principally with the renal microvillus membrane proper or with the tags of basal-lateral membrane remaining attached to the brush borders. In earlier studies of the distribution of Na^+,K^+-ATPase activity in rat intestinal mucosa, Quigley and Gotterer (1969) isolated a plasma membrane fraction, relatively free of brush borders, that contained 85 percent of the total Na^+,K^+-ATPase with a 25- to 35-fold increase in specific activity. A small but consistent fraction of Na^+,K^+-ATPase activity (15 percent of the total, with relatively low specific activity) is still found in

Table 10.2 ATPase ACTIVITIES ASSOCIATED WITH RENAL BRUSH BORDER PREPARATIONS

Enzyme	Species	Activity, μmol min^{-1} per mg of protein		Relative sp. act.	Reference
		Homogenate	Brush border		
Mg^{2+}-ATPase	Rabbit	0.069	0.230	3	C. Liang and B. Sacktor*
	Rat	0.115	0.429	4	Quirk and Robinson (1972)
	Rat	0.123	0.239	2	Glossmann and Gips (1974)
Ca^{2+}-ATPase	Rat	0.17	0.36	2	Kinne-Saffran and Kinne (1974a)
Na$^+$,K$^+$-ATPase	Rabbit	0.040	0.168	4	George and Kenny (1973)
	Rat	0.047	0.103	2	Glossmann and Gips (1974)
HCO$_3^-$-ATPase	Rabbit	0.028	0.120	4	C. Liang and B. Sacktor*

* Manuscript in preparation, 1975.

brush borders. Subsequent studies by these investigators (Quigley and Gotterer, 1972) indicated that the two Na^+,K^+-ATPase activities, one in the
basal-lateral region and the other in the microvillus segment of the plasma
membrane, have distinct biochemical properties. Douglas, Kerley and Isselbacher (1972) have also characterized a preparation of basal-lateral membranes from rat intestine that is rich in Na^+,K^+-ATPase and poor in sucrase.
In the kidney, separation of the two plasma membrane moieties by differential
centrifugation schemes has so far proved difficult. However, Heidrich et al.
(1972) have subjected crude brush border fragments to free-flow electrophoresis and on the basis of different electrical surface charges have partially
separated renal basal-lateral from microvillus membranes. As shown in *Table
10.3*, their preparations of crude plasma membranes are enriched about 4-
fold in both alkaline phosphatase and Na^+,K^+-ATPase when compared with
the cortex homogenate. After electrophoresis, membrane fragments rich in
alkaline phosphatase tend to separate from those rich in Na^+,K^+-ATPase.
The specific activity of alkaline phosphatase in the microvillus fraction is
enriched an additional 2 times, so that it has a final relative specific activity
7 times that of the homogenate. Correspondingly, the specific activity of alkaline phosphatase in the basal-lateral fragments decreases relative to the crude
plasma membrane, although the specific activity in the purified basal-lateral
membranes is still slightly higher than that in the homogenate. The distribution of Na^+,K^+-ATPase is markedly different. Its specific activity in the purified basal-lateral portion of the plasma membrane increases relative to that
in the crude plasma membrane, whereas its specific activity decreases by one-
third in the purified microvillus membrane. Again, however, the enzyme in
the brush border fraction has a greater specific activity than that in the cortex.
The patterns of distribution for Ca^{2+}-ATPase and HCO_3^--ATPase resemble
those for Na^+,K^+-ATPase and alkaline phosphatase, respectively. However,
the separation seems to be not as distinct. For example, the relative specific
activity of Ca^{2+}-ATPase in the basal-lateral membrane is 4, whereas that
of Na^+,K^+-ATPase is over 10. A corresponding discrepancy is evident
between the relative specific activities of HCO_3^--ATPase and alkaline phosphatase in the microvillus fraction. Nevertheless, these findings suggest a
compositional and functional polarity of the renal tubular cell plasma membrane, alkaline phosphatase and HCO_3^--ATPase being predominant at the
apical pole whereas Na^+,K^+-ATPase and Ca^{2+}-ATPase are concentrated
at antiluminal portions of the membrane. Contrasting with the situation in
the kidney, it is noted that an active, vitamin D-induced, Ca^{2+}-ATPase has
been reported in purified preparations of intestinal brush borders (Melancon
and DeLuca, 1970).

   With the possible exception of one or two enzymes for which no information is as yet available, all the enzymes that have been cited as associated
with renal brush borders have been reported in intestinal brush border membranes. In addition, other enzymes, whose principal function seems to be
directly related to digestive processes, have been localized in intestinal brush
borders. The disaccharidases sucrase, isomaltase, lactase and cellobiase,
which are not found in renal brush borders (Sacktor, 1968), and maltase and
trehalase, which are localized in renal brush borders, are present in intestinal
brush borders. Forstner, Sabesin and Isselbacher (1968) have shown that the
sucrase in purified rat intestinal brush borders has a specific activity almost

Table 10.3 DISTRIBUTION OF ATPase ACTIVITIES IN MICROVILLAR AND BASAL-LATERAL SEGMENTS OF RAT RENAL PLASMA MEMBRANES (From Kinne-Saffran and Kinne, 1974a, b)

Fraction	Alkaline phosphatase		Na^+,K^+-ATPase		Ca^{2+}-ATPase		HCO_3^--ATPase	
	Activity*	Relative sp. act.	Activity*	Relative sp. act.	Activity*	Relative sp. act.	Activity*	Relative sp. act.
Homogenate	0.24	1.00	0.04	1.00	0.17	1.00	0.06	1.00
Plasma membrane	1.07	4.45	0.16	4.00	0.36	2.11	0.13	2.3
Basal-lateral	0.32	1.32	0.42	10.50	0.68	4.00	0.06	1.0
Microvillus	1.79	7.45	0.05	1.25	0.29	1.70	0.18	3.0

* Measured in milliunits per milligram of protein.

20 times that in mucosal scrapings and this specific activity is increased an additional 1.6 times when microvillus membranes are prepared from the isolated brush borders. *Table 10.4* shows the disaccharidase activities in isolated brush borders from intestines of the hamster, rat and man. Maltase is the most active disaccharidase. In the adult, lactase is the least active. The activity of trehalase varies 10-fold, depending on species. In the rat it has an activity comparable to sucrase, in man it has one-third the activity of sucrase, whereas in the hamster its activity is less than 10 percent that of sucrase. Interestingly, of these three species the highest intestinal trehalase activity is found in the rat, the very species which has no trehalase in the kidney.

Table 10.4   DISACCHARIDASE ACTIVITIES OF INTESTINAL BRUSH BORDER PREPARATIONS

Species	Enzyme	Activity, μmol min^{-1} per mg of protein	Reference
Hamster	Maltase	4.14	Benson, Sacktor and Greenawalt (1971)
	Isomaltase	1.43	
	Sucrase	1.86	
	Trehalase	0.16	
Rat	Maltase	6.5	Forstner, Sabesin and Isselbacher (1968)
	Isomaltase	1.96	
	Sucrase	1.14	
	Trehalase	1.14	
	Lactase	0.14	
Man	Sucrase	1.46	Schmitz *et al.* (1973)
	Trehalase	0.46	

In addition to disaccharidases, several other hydrolytic enzymes have been identified in intestinal brush borders. These include: aminopeptidases and enteropeptidase (enterokinase), which hydrolyze a variety of di-, tri- and poly-peptides (Rhodes, Eichholz and Crane, 1967; Maroux, Louvard and Baratti, 1973); lipase (Senior and Isselbacher, 1963); cholesterol esterase (David, Malathi and Ganguly, 1966); and retinol hydrolase (Malathi, 1967). These enzymes are presumably involved in the terminal digestion of their respective substrates. Rosenberg *et al.* (1969) suggested that folate deconjugase, which releases folate from the polyglutamate form, is located in the brush border. Isolated intestinal brush borders are also able to acylate fatty acids, indicating the presence of a thiokinase, and mono- and diglyceride acylases (Forstner *et al.*, 1965). Van den Berg and Hulsmann (1971) have reported a membrane-bound hexokinase in rat intestinal brush borders. Previously, the presence of brush-border-bound hexokinase activity in rabbit renal preparations had been noted (Berger and Sacktor, 1970).

10.2.4   **Molecular organization, development and turnover of brush border membrane components**

It is evident that the renal and intestinal brush border membranes are complex assemblies of lipids, carbohydrates and proteins and that the individual components of this membrane mosaic undergo continual synthesis and degradation throughout the lifespan of the cell. In this connection it should be emphasized that the brush border population may not be homogeneous in a preparation isolated from a tissue such as the intestine. Indeed, there is considerable evidence to indicate that the biochemical characteristics of the membranes depend on the topography of the epithelial cell from which the membranes were derived. For example, it is well known that the absorptive and digestive potential of the intestine varies with position along the tract (Fisher and Parsons, 1949). Regional variations in disaccharidase activities are also indicated from human intestinal biopsy material (Newcomer and McGill, 1966). Moreover, proximal to distal variation in the activities of different enzyme components of the brush border need not be the same for all enzymes. In the rat, sucrase activity at the proximal end of the intestine is increased 2-fold, reaching a peak at the mid-jejunum, then decreases rapidly to zero at the distal end (Harrison and Webster, 1971). Leucine aminopeptidase is relatively constant in activity up to the mid-jejunum, then it slowly decreases distally. In contrast, alkaline phosphatase shows maximal activity at the most proximal point, falling very sharply at the near-proximal area and being virtually absent from the distal end. The activity of the brush border enzyme sucrase also varies, depending on the position of the enterocyte in the villus (Doell, Rosen and Kretchmer, 1965). Maltase, leucine aminopeptidase and alkaline phosphatase activities show considerable daily rhythmic changes in rat intestine (Saito, 1972). This is associated with the feeding schedule rather than with the light–dark pattern. No such daily rhythm, however, is found with the same enzymes localized in the kidney.

Marked changes in intestinal and renal transport systems are found during development (Deren, 1968; Segal and Thier, 1973). Concomitant with these developmental patterns are large increases in total surface area principally as a result of the formation of villi and microvilli and the appearance and disappearance of specific enzymes. The increase in intestinal alkaline phosphatase generally parallels the progressive differentiation of the enterocyte (Moog, 1951). However, the activity of the enzyme may also change per unit of surface area: in the 11- to 18-day-old mouse the mucosal surface area per unit of serosal surface area increases 3-fold, whereas the increase in alkaline phosphatase activity relative to mucosal surface is 20-fold. Numerous studies indicate that the activities of the intestinal disaccharidases vary with development. In general, the intestine of the neonate contains a considerably higher specific activity of lactase than that of the adult of the same species. Conversely, the intestine of the newborn has relatively low specific activities of sucrase, maltase and trehalase as compared with the intestine of the adult (Bailey, Kitts and Wood, 1956; Dalqvist, 1961; Doell and Kretchmer, 1962). This general pattern does not hold for the human intestine, however, the α-glucosidases and β-galactosidases both being fully active at birth (Auricchio, Rubino and Murset, 1965).

Trehalase activity is a good criterion for appraising developmental changes in the renal proximal tubule brush border. As shown in *Table 10.5*, the specific activity of the enzyme in membranes isolated from the rabbit increases 4-fold from the fetal to the new-born rabbit and increases an additional 10-fold during the first 3 to 4 weeks of development (Chesney and Sacktor, 1974). There is little further enhancement in trehalase activity with maturation.

Electron micrographs show that at about 4 weeks of age the brush border membranes in the proximal tubule are essentially fully developed. Parallel studies indicate that trehalase activity in cortical homogenates of the fetus and adult rabbit are 4.5 and 49 nmol min^{-1} per milligram of protein, respectively. Perhaps significantly, the urine glucose concentration falls from 0.67 mM in the 3- to 5-day old rabbit to 0.09 mM in the month-old rabbit, whereas, concomitantly, the rate of uptake of D-glucose by isolated brush border vesicles increases in early development. Disaccharidase activity may also be a marker for changes in brush border membranes during the aging process. It has been reported that the specific activities of maltase and alkaline phosphatase in renal brush border preparations are decreased by 50 and 35 percent, respectively, in the 24-month-old rat as compared with the 3-month-old animal (O'Bryan and Lowenstein, 1974). This decrement in maltase activity is not due to a change in the K_m of the enzyme, but results from a decrease in V_{max}.

The factors regulating the enzymatic changes that occur during development are not fully understood. Neither the increase in intestinal lactase activity during the prenatal period nor the decrease following weaning seems to be related to the presence of substrate (Doell and Kretchmer, 1962).

Table 10.5   TREHALASE ACTIVITY IN RABBIT
RENAL BRUSH BORDERS DURING DEVELOPMENT
(From Chesney and Sacktor, 1974)

Age	Activity*
Fetal (-2 days)	16.7 ± 0.4
3–5 days	61.2 ± 18.9
3–4 weeks	653.3 ± 74.6
Adult	733.5 ± 55.6

* Measured in nmol min^{-1} per milligram of protein.

Further, the feeding of sucrose and isomaltose to new-born rats does not induce the precocious enhancement of sucrase and isomaltase activities (Rubino, Zimbalatti and Auricchio, 1964). Steroid administration to neonatal rats, which induces morphological and functional changes in the ileum resembling those observed at weaning, does not precipitously diminish lactase activity (Doell and Kretchmer, 1964). Hydrocortisone and ACTH, however, do stimulate the appearance of sucrase activity.

Other intestinal brush border enzymes also appear to be inducible. Administration of vitamin D to vitamin-D-deficient rats (Martin, Melancon and DeLuca, 1969) and chicks (Melancon and DeLuca, 1970) stimulates brush border Ca^{2+}-ATPase and alkaline phosphatase (Holdsworth, 1970). Activities of the disaccharidases in these animals, on the other hand, are not altered (Norman et al., 1970). There is a lag period between administration of vitamin D and the development of maximal activities of Ca^{2+}-ATPase and alkaline phosphatase. Also, the induction can be blocked by cycloheximide and actinomycin D (Norman et al., 1970). These observations suggest that vitamin D induces the syntheses of Ca^{2+}-ATPase and alkaline phosphatase de novo. The possibility that vitamin D_3 may alter the lipid composition of the brush border membrane, thus increasing the permeability of the membrane to Ca^{2+}, has been considered (Goodman, Haussler and Rasmussen, 1972).

The proteins in the brush border membranes are apparently genetically determined. Preiser *et al.* (1974) have reported that in two patients with a hereditary disorder, sucrose–isomaltose intolerance, peroral jejunal biopsies indicate the absence of sucrase and only traces of isomaltase activities. Maltase activity in the patients is also low. In contrast, lactase and trehalase activities are elevated 2-fold, whereas alkaline phosphatase, leucine aminopeptidase and γ-glutamyl transpeptidase activities are within normal ranges. Protein patterns obtained by gel electrophoresis are illustrated in *Figure 10.3*

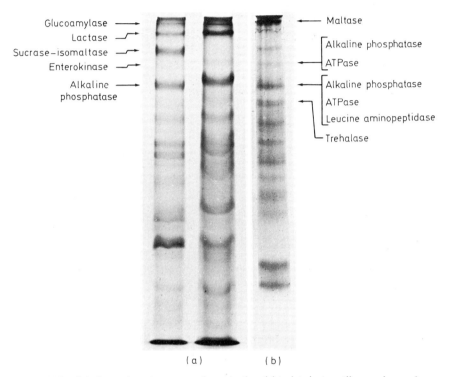

(a)   (b)

Figure 10.3   Gel-electrophoresis patterns of proteins from (a) isolated microvillus membranes from human peroral jejunal biopsies: left, normal subject; right, patient with sucrase–isomaltase deficiency; and (b) isolated rabbit renal brush border membranes. (The intestinal patterns are reproduced with the kind permission of Dr Robert K. Crane from Preiser et al., 1974, courtesy of Associated Scientific Publishers)

and confirm the absence of a band associated with sucrase–isomaltase activity in the patients. Also shown in *Figure 10.3* are gel-electrophoresis patterns of a normal human intestinal preparation (Preiser *et al.*, 1974) and the rabbit renal brush border membrane (B. Sacktor and R. Balakir, unpublished work, 1975). Similarities in the protein gel patterns are evident.

Relatively little information is available concerning the turnover of brush border membranes. When labeled leucine is injected into rats, intestinal brush border proteins attain maximum specific activity 10 hours after injection; a 2-fold increase is found between 2 and 10 hours (James *et al.*, 1971). Initially, most of the leucine is incorporated into the microsomal fraction. James *et al.* suggested that the lag in reaching peak incorporation of radioactivity into the brush borders is related to the time required for transfer of label from the microsomes to the brush border. This agrees with the findings of Forstner (1969) showing a similar delayed rise in incorporation of [^{14}C]glucosamine into intestinal brush borders, and with the observations of Quirk, Byrne and Robinson (1973) on the incorporation of lysine and glucosamine into rabbit renal brush borders and whole kidney cortex. In contrast, McNamara, Koldovsky and Segal (1974) have reported that in the rat kidney cortex, the increase in specific activity in the brush border precedes that in the homogenate. They proposed that either the precursor proteins are preferentially transferred to the membrane or protein synthesis takes place either in the kidney brush border or in close proximity to it. Obviously, this hypothesis requires further study. After maximal labeling, the half-life of proteins in the intestinal brush border is shorter (18 h) than for those in the homogenate (31 h). The separated and purified disaccharidases have an even faster turnover rate (11.5 h) than whole brush borders (James *et al.*, 1971), indicating that not all brush border proteins turn over at a uniform rate. The turnover rates of maltase, sucrase and lactase, however, are essentially the same. This suggests that the decrease in lactase activity in the adult is not due to a specific alteration in the turnover rate of this enzyme compared with sucrase and maltase. A heterogeneity in turnover rates of intestinal brush border proteins is also indicated from the observations of Alpers (1972), who found a correlation between relative degradative rates and molecular weight of the brush border protein. In general, the largest proteins (molecular weights 170000–270000), such as disaccharidases, turn over the fastest; intermediate-size proteins (53000–140000), for example alkaline phosphatase, have intermediate rates of degradation; and proteins corresponding to the filamentous 'core' of the microvilli (19000–45000) turn over the slowest. In contrast to these findings for intestinal brush borders, Quirk, Byrne and Robinson (1973) interpreted their results on the kinetics of incorporation of lysine and glucosamine into rabbit renal brush borders as indicating that the membrane components turn over in unison. The kidney and the intestine also reportedly differ in that the half-life of proteins in the rat renal cortex has been estimated to be 85 h and that in the brush border fraction to be 63 h (McNamara, Koldovsky and Segal, 1974). These authors proposed that the longer turnovers in the kidney relative to the intestine may reflect differences in cell turnover rate between the jejunum (cell life 2 days) and kidney (very little, if any, cell turnover).

Preliminary studies have attempted to determine the factors instrumental in integrating the different protein and nonprotein constituents into the mosaic comprising the brush border membrane. The technique of selective disorganization of the membrane followed by reconstitution, an approach used successfully in investigating structural–functional relationships in other membranes, has been initiated, in part, with the brush border membrane. Proteolytic enzymes, organic solvents, neutral salts, and nonionic, anionic and cationic detergents have been used to solubilize or disaggregate different protein components.

Incubation of hamster (Johnson, 1967) and rabbit (Nishi, Yoshida and Takesue, 1968) intestinal brush borders with papain releases sucrase and maltase activities. This treatment also removes the 6 nm knobs from the periphery of the microvillus membrane. It was concluded that the knobs and disaccharidases are contained in the glycocalyx, external to the plasma membrane of the microvillus. Benson, Sacktor and Greenawalt (1971), however, have questioned this conclusion and have carried out a correlated kinetic study of the treatment of hamster intestinal brush borders with papain, monitoring simultaneously the removal of disaccharidase activities and the 6-nm particles. The results clearly show that it is not possible to correlate the solubilization of disaccharidases with the release of knobs. Moreover, it is found that each of the disaccharidases (i.e. sucrase, isomaltase, maltase and trehalase) has a distinct pattern of release with incubation time; trehalase is not solubilized at all. This finding implies that each disaccharidase has a different immediate chemical environment which may play a role in integrating the enzyme in the microvillus membrane. Stevenson (1972a) has also reported that trehalase is resistant to release from rabbit renal brush borders by papain and he has confirmed that maltase is solubilized by this treatment. Trypsin fails to effect the release of either disaccharidase from renal membranes (Stevenson, 1972a). This contrasts with the results of Dahlqvist (1960) from hog intestinal brush borders, where trehalase is reported to be solubilized after prolonged exposure to trypsin. Trehalase and maltase can be released from renal membranes by butanol without loss of activity (Sacktor, 1968).

The aminopeptidases of the brush border are readily solubilized by a variety of agents, including papain, lithium chloride, hexadecyltrimethylammonium bromide, Triton X-100, deoxycholate and sodium dodecyl sulfate (Thomas and Kinne, 1972). Enteropeptidase is also preferentially released, relative to disaccharidases and alkaline phosphatase, from intestinal brush borders by papain, trypsin, chymotrypsin, carboxypeptidase, hyaluronidase and bile salts (Nordstrom, 1972). Part of the γ-glutamyl transpeptidase activity in renal brush borders is solubilized by papain (George and Kenny, 1973). Alkaline phosphatase, on the other hand, is like trehalase in being extremely difficult to solubilize with proteolytic enzymes or detergents (Scherberich et al., 1974) and in being released from the membrane by butanol (Wachsmuth and Hiwada, 1974). It is suggested that the enzymes that are solubilized readily are localized on the surface of the microvillus membrane, whereas those that are more strongly bound are localized in the matrix of the microvillus membrane (Thomas and Kinne, 1972). In early studies on disaccharide transport with intestinal loops, a similar hypothesis as to the location of trehalase relative to other disaccharidases has been proposed (Sacktor and Wu, 1971).

10.3   TRANSPORT STUDIES WITH ISOLATED BRUSH
BORDER MEMBRANES

The transport of solutes and fluid in the kidney and intestine is one of the more active fields of biological research. The preparations used in such studies have ranged from the intact animal to perfused organs and tubules, tissue slices and segments of the isolated tubule and gut. The results of these investigations, although contributing significantly to an understanding of the overall transport systems, do not provide direct information on the mechanisms of membrane transport. However, with the development of techniques for the isolation and characterization of the renal and intestinal brush borders, plasma membrane preparations have become available as model systems to examine the biochemical mechanisms of transport into the epithelial cells.

10.3.1   Transport of D-glucose and other sugars by brush border membranes

In general, two approaches have been taken to describe the molecular events involved in transport of D-glucose. In one, the binding to the membrane of phlorizin, a presumably nontransportable inhibitor of D-glucose transport, is investigated to provide insight into the properties of the proposed D-glucose carrier (Bode et al., 1970; Glossmann and Neville, 1972b; Chesney, Sacktor and Kleinzeller, 1974). However, discrepancies between the kinetics of phlorizin binding and glucose transport, for example high- and low-affinity binding sites of varying specificities, apparent differences between K_D and K_i, and Na$^+$ requirements, raise questions as to the precise relationship of the two processes (Aronson and Sacktor, 1974). In the other approach, D-glucose uptake is measured directly (Hopfer et al., 1973; Busse, Elsas and Rosenberg, 1972; Aronson and Sacktor, 1974, 1975). Such studies are complicated, too, by metabolism of the sugar (Busse, Elsas and Rosenberg, 1972), proposals of multiple saturable uptake or binding sites (Faust, Wu and Faggard, 1967; Eichholz, Howell and Crane, 1969; Chesney, Sacktor and Rowen, 1973; Chertok and Lake, 1974), and the presence of contaminating bacteria that may mask the kinetics of D-glucose uptake by brush border preparations (Garcia-Castineiras, Torres-Pinedo and Alvarado, 1973; Mitchell, Aronson and Sacktor, 1974).

It was important to resolve the question of whether the uptake of D-glucose by the brush border represents membrane binding or sugar transport across the membrane into an intravesicular space. Three lines of evidence indicate that the latter possibility is correct. First, Aronson and Sacktor (1974) have reported that phlorizin not only inhibits the uptake of D-glucose by renal brush border membranes but that the glycoside essentially completely inhibits the efflux of D-glucose from the membranes. If the interaction of D-glucose and brush border were binding only, then it would be expected that phlorizin would displace the sugar, accelerating its release, providing phlorizin binding is competitive and the number of sites are limiting. The fact that phlorizin inhibits the release of D-glucose indicates that phlorizin reacts with the membrane at a site external to the site of D-glucose accumulation. Secondly, preloading with D-glucose markedly stimulates the initial rate of D-glucose uptake (Aronson and Sacktor, 1974). The effect is stereospecific, as the rate of D-glucose uptake is not augmented by preincubation with L-glucose. Accelerated exchange diffusion cannot be explained by the binding hypothesis, because preincubation with D-glucose would fill binding sites and, if anything, would inhibit the uptake of additional D-glucose. On the other hand, models to explain accelerative exchange diffusion all involve facilitated transport across a membrane. Lastly, studies have been carried out to examine the effect of the intravesicular space on D-glucose uptake (Beck and Sacktor, 1975). In the experiment summarized by *Figure 10.4*, the intravesicular space is varied by altering the osmolarity of the medium with sucrose, an impermeable solute. The amount of D-glucose taken up at equilibrium should be dependent on the intravesicular space and this volume should be inversely proportional to the osmolarity of the medium. As shown, D-glucose uptake is proportional to the inverse osmolarity and, thus, to the intravesicular space. Moreover, extrapolation to infinite medium osmolarity (zero space) results in little, if any, uptake. Therefore, D-glucose uptake is accounted for essentially completely by transport across the renal brush border membrane into the intravesicular volume. Analogous osmolarity experiments by Hopfer and Sigrist-Nelson (1974) with intestinal microvillus membrane preparations indicate additionally that transport and binding of sugar, as previously reported, are unrelated.

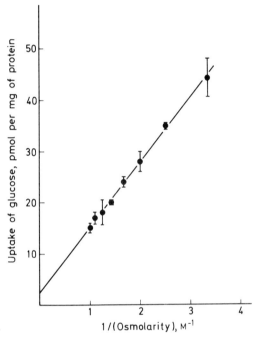

Figure 10.4   The effect of medium osmolarity on D-glucose uptake by rabbit renal brush border membrane vesicles. (From Beck and Sacktor, 1975, courtesy of the American Society of Biological Chemists)

Earlier proposals for the binding and/or 'high affinity' uptake of D-glucose by brush border membrane preparations are also refuted by subsequent studies with bacterium-free membranes (Mitchell, Aronson and Sacktor, 1974). The observation that the D-glucose taken up by the membrane vesicles is not metabolized (Aronson and Sacktor, 1974) further supports the view that the sugar is transported and accumulated intravesicularly rather than being enzymatically degraded.

It is only very recently that unequivocal studies on the mechanisms of D-glucose transport using brush border membrane preparations have become available. The initial rate of uptake of D-glucose by intestinal (Hopfer et al., 1973) and renal (Aronson and Sacktor, 1974) brush border membrane vesicles is enhanced by Na^+. This effect is specific for Na^+ (other cations are ineffective) and is stereospecific for D-glucose, since uptake of L-glucose is not stimulated. In the absence of Na$^+$, the rates of uptake of D- and L-glucose are, in fact, similar. These studies also demonstrate that it is the presence of extravesicular Na$^+$ that is important for the uptake of D-glucose; uptake of the sugar into membranes preloaded with Na$^+$ may, indeed, be inhibited.

Recent experiments point out additional significant characteristics of the Na$^+$ gradient D-glucose uptake system. *Figure 10.5* illustrates an experiment by Aronson and Sacktor (1975) in which the uptake of 50 μM D-glucose by

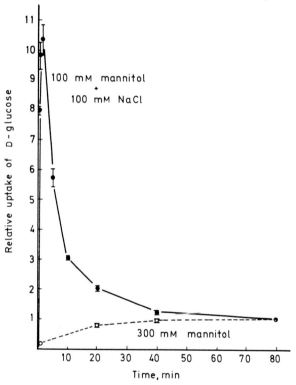

Figure 10.5   *The time course of the uptake of* D-*glucose by rabbit renal brush border membranes with and without an electrochemical gradient of* Na$^+$. (*From Aronson and Sacktor, 1975, courtesy of the American Society of Biological Chemists*)

renal brush border membrane vesicles is measured during incubations for different lengths of time, either in 300 mM buffered mannitol medium or in a medium in which mannitol is replaced isoosmotically by 100 mM NaCl at the initiation of incubation. In the absence of the Na$^+$ gradient, steady-state levels are reached in about 80 minutes. The presence of a Na$^+$ gradient between the external incubation medium and the intravesicular medium induces a marked stimulation of D-glucose transport. The initial (30 seconds) rate of D-glucose uptake with the Na$^+$ gradient is frequently greater than 40 times the initial rate in the absence of the gradient. Accumulation of the sugar in the membranes is maximal at about 2 minutes. Afterwards the amount of D-glucose in the vesicles decreases, indicating efflux of the sugar.

The final level of uptake of the sugar in the presence and absence of the Na^+ gradient is identical, however, suggesting that equilibrium has been established. Moreover, the experiment shows that at the peak of the 'overshoot', at an incubation period of 2 minutes, the uptake of D-glucose is more than 10 times the final equilibrium value. These results suggest that the imposition of a large extravesicular to intravesicular gradient of Na^+ effects the transient movement of D-glucose into renal brush border membranes against its concentration gradient. A similar observation has been reported by Murer and Hopfer (1974) with intestinal microvillar membranes.

Increasing the concentration of Na^+ increases the rate of D-glucose uptake (Aronson and Sacktor, 1975). At 100 mM NaCl, the rate of D-glucose uptake is about 40 times the rate in the absence of Na^+. Even at this concentration, only partial saturability with respect to Na^+ is evident. Studies on the effect of Na^+ on the kinetics of D-glucose uptake indicate that increasing Na^+ in the external medium lowers the apparent K_m of D-glucose for transport. The calculated apparent K_m values are 500, 220 and 80 μM D-glucose at 10, 20 and 50 mM NaCl, respectively. The value of V_{max} is unaffected. Aronson and Sacktor (1975) have also found that the effect of the Na^+ gradient on D-glucose transport, measured as influx or efflux, can be dissected into a stimulatory effect of Na^+ on transport when sugar and Na^+ are on the same side of the membrane (*cis* stimulation) and an inhibitory effect of Na^+ on transport when sugar and Na^+ are on opposite sides of the membrane (*trans* inhibition). These findings also suggest symmetry of the D-glucose 'carrier' at both faces of the luminal membrane.

With renal brush border membranes, the uptake of D-glucose, at a given concentration of sugar, reflects the sum of contributions from a Na^+-dependent transport system and a Na^+-independent system (Aronson and Sacktor, 1975). The relative stimulation of D-glucose uptake by Na^+ decreases as the sugar concentration increases. It is suggested, however, that at physiological concentrations of D-glucose the asymmetry of Na^+ across the brush border membrane fully accounts for uphill D-glucose transport.

These experimental findings on the specific effect of the Na^+ electrochemical gradient on the mediated transport of D-glucose across the luminal membrane are consistent with the Na^+ gradient hypothesis formulated by Crane (1962) and Schultz and Curran (1970) for intestinal sugar transport. This proposal suggests that the electrochemical gradient of Na^+ across the plasma membrane drives the uptake of D-glucose, the translocation of the sugar being coupled in some manner to the flux of Na^+. An important aspect of the driving force is the question of whether Na^+-dependent D-glucose transport is an electroneutral or an electrogenic process. If Na^+-dependent D-glucose transport across the membrane is electroneutral, then the positive charge associated with Na^+ flux is compensated by the co-transport of an anion or the countermovement of a cation via the same carrier. If, however, the transport process is electrogenic, then charge compensation is not made via the glucose carrier but at a different site in the membrane. In the latter case, Na^+-dependent D-glucose transport should be influenced by an electrochemical potential across the membrane. This question has been examined recently in both intestinal (Murer and Hopfer, 1974) and renal (Beck and Sacktor, 1975) brush border membrane vesicles.

Two experimental approaches have been used to regulate the membrane potential across the brush border membrane. These are first, the use of anions of different modes of permeability, and secondly, the utilization of specific ionophores and proton conductors. With renal luminal membrane preparations, Beck and Sacktor (1975) have found that the imposition of a salt gradient of either Na_2SO_4 or sodium isethionate, in contrast to one of NaCl, does not result in the accumulation of D-glucose above equilibrium. With either salt, however, the initial rate of D-glucose uptake is stimulated by the presence of Na^+, when compared with the rate in the absence of Na^+, but it is significantly less than that with NaCl. Both sulfate and isethionate anions are relatively impermeable to the luminal membrane of the proximal tubule. Therefore, little development of electrochemical potential is to be expected for driving an electrogenic Na^+-dependent D-glucose uptake. Since the same Na^+ chemical gradient is present when Na_2SO_4 or sodium isethionate is used as when NaCl is used, it is evident that the electrochemical potential generated in part by the anion is of considerable significance in the control of D-glucose transport against its concentration gradient. This view is supported additionally when salt gradients of the lipophilic anions NO_3^- and SCN^- are used. With these salts, the transient 'overshoot' of D-glucose uptake is greater and/or faster than that with NaCl. The anions NO_3^- and SCN^- penetrate biological membranes in the charged form at pH 7.5, and both are known to stimulate Na^+ transport in the toad urinary bladder to a greater extent than Cl^- (Singer and Civan, 1971). In the experiments with brush border membranes, if the Na^+-stimulated D-glucose uptake is electrogenic, diffusion of the anions into the vesicles will influence D-glucose uptake by producing an electrochemical membrane potential (negative inside). Since in the proximal tubule of the rabbit, Cl^- is three times as permeable as Na^+ (Schafer, Troutman and Andreoli, 1974), presumably Cl^- enters the intravesicular space more rapidly than Na^+ and permits development of an electrochemical potential (interior negative). Further, SCN^- and NO_3^-, which are probably more permeable than Cl^-, will facilitate the more rapid or greater development of an electrochemical potential. Murer and Hopfer (1974) have also reported that D-glucose uptake into intestinal microvillus membranes is enhanced by NaSCN relative to NaCl.

Beck and Sacktor (1975) have also shown with renal brush border membranes that for Na^+ salts whose mode of membrane translocation is electroneutral (e.g. acetate and bicarbonate), or one which on entering the vesicle dissociates to yield a proton (e.g. phosphate), there is no accumulation of D-glucose above the equilibrium value. These findings suggest that only anions which penetrate the brush border membrane and generate an electrochemical potential, negative on the inside, permit the uphill Na^+-dependent transport of D-glucose.

This suggestion is supported by determining how alterations in the electrochemical potential of the membrane induced by specific ionophores and a protein conductor affect the uptake of the sugar (Beck and Sacktor, 1975). Valinomycin, an ionophore that mediates electrogenic K^+ movements, supports the Na^+-dependent accumulation of D-glucose, provided a K^+ gradient (vesicle > medium) is present. In contrast, nigericin, which mediates an electroneutral exchange of Na^+ for K^+, does not. Sodium-dependent D-glucose uptake is diminished by ionophores that allow Na^+ to pass through the membrane via another channel, either electrogenically, as with gramicidin, or electroneutrally, as with nigericin. An electrogenic proton conductor, carbonyl cyanide *p*-fluoromethoxyphenylhydrazone, enhances D-glucose uptake in the presence of a protein gradient (vesicle > medium) in renal (Beck and Sacktor, 1975) and intestinal (Murer and Hopfer, 1974) brush border preparations. These findings demonstrate that changing the electrochemical potential across the vesicular membrane, that is, by making the interior negative, stimulates the Na^+-dependent transport of D-glucose.

These results indicate that the Na^+-dependent transport of D-glucose into renal and intestinal brush border membrane vesicles is an electrogenic process. It is suggested that in the intact system the asymmetric distribution of Na^+ across the epithelial cell and the electrochemical potential across the luminal membrane provide the energy required to transport D-glucose against its concentration gradient.

The Na^+-dependent D-glucose transport system in the isolated renal brush border membranes possesses the sugar specificities characteristic of the more physiological intact system. Of the D-glucose analogues tested only D-galactose and α-methyl-D-glucoside inhibit the Na^+-dependent transport of D-glucose (Aronson and Sacktor, 1975). None of the other sugars tested, including L-glucose, D-mannose, D-fructose, D-xylose and 2-deoxy-D-glucose, inhibits the Na^+-dependent uptake. A slight inhibition is found with 3-*O*-methyl-D-glucose. In contrast to these findings, the Na^+-independent transport system in renal brush border membranes shows a general lack of specificity with respect to the various analogues. Except for L-glucose, all the sugars inhibit D-glucose uptake. However, D-galactose and α-methyl-D-glucoside are markedly less effective as inhibitors of the Na^+-independent uptake than as inhibitors of the Na^+-dependent system. Kleinzeller (1970) has suggested that in rabbit renal cortical slices the structural requirements for the Na^+-dependent active sugar transport system are a D-pyranose or furanose ring, a hydrophilic group on C-2, a hydroxyl group on C-3 in the same configuration as that in D-glucose, a hydroxyl group on C-6, but not a hydroxyl group on C-1. The results with rabbit renal luminal membranes are generally consistent with this view with the additional specifications that the ring must be in the pyranose form and the hydroxyl group on C-2 must be in the D-glucose rather than in the D-mannose configuration. Silverman, Aganon and Chinard (1970) have also stressed the distinction *in vivo* between the D-glucose and D-mannose transport systems in the dog. The findings with isolated renal brush borders are also in general accord with elegant stop-flow microperfusion experiments in the proximal tubule of the intact rat kidney (Ullrich, Rumrich and Kloss, 1974).

The uptake of D-glucose in intestinal microvillus membranes, but not that of L-glucose, is inhibited by D-galactose (Hopfer *et al.*, 1973), and in turn D-glucose inhibits the uptake of D-galactose. As shown in *Figure 10.6*, the mutual interaction of D-galactose and D-glucose is also found in renal brush border membranes (Sacktor *et al.*, 1974). Moreover, the inhibitions are competitive in nature. Therefore, the data support the view that D-glucose and D-galactose have common or closely associated carrier sites on the brush border membranes. In contrast, brush border membranes from rat intestine have a transport system for D-fructose which is distinct from that for D-glucose (Sigrist-Nelson and Hopfer, 1974). Uptake of D-fructose is unaffected by Na^+, phlorizin, D-glucose or D-galactose.

Recent evidence tends to suggest that the sugar transport system at the luminal pole of the enterocyte is distinct from that at the basal-lateral end. Murer *et al.* (1974) have partially separated microvillus membranes from basal-lateral membranes of rat intestine. In the brush border fraction, alkaline phosphatase is increased 8-fold and Na^+,K^+-ATPase is decreased to half relative to the specific activities in the homogenate. In the basal-lateral fraction, Na^+,K^+-ATPase activity is increased 5-fold but alkaline phosphatase is increased only 2-fold over that of the homogenate. With these membrane preparations, and in the presence of 100 mM Na^+, D-glucose is taken up twice as fast as L-glucose in the brush border compared with the basal-lateral fractions; the values with microvillus membranes are 1.00 and 0.29 nmol per 30 s per milligram of protein for D- and L-glucose, respectively, and with basal-lateral membranes they are 0.90 and 0.45 for D- and L-glucose, respectively.

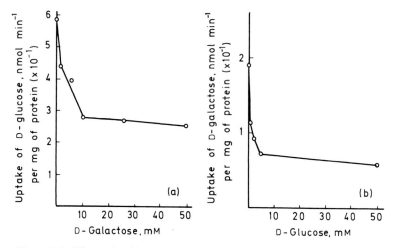

Figure 10.6   The uptake of 1 mM D-glucose or D-galactose in the presence of different concentrations of the other sugar. (a) The effect of D-galactose on D-glucose uptake; (b) the effect of D-glucose on D-galactose uptake

Also, the uptake of D-glucose by the brush border membranes is stimulated by Na^+ to a greater extent, 104 percent as compared with 22 percent, and is inhibited more by phlorizin, 71 percent as compared with 27 percent, than is the uptake by the basal-lateral membranes. In a somewhat different approach, Bihler and Cybulsky (1973) have blocked the active sugar transport system (Na^+- and phlorizin-sensitive) of isolated mouse epithelial cells with $HgCl_2$ and have shown that the uptakes of D-mannose, D-fructose and 2-deoxy-D-glucose are not inhibited by D-glucose and D-galactose. The uptakes of D-glucose and D-galactose in cells poisoned with $HgCl_2$ are sensitive to neither Na^+ nor phlorizin. These authors postulate that uptake of sugars in the presence of $HgCl_2$ is mediated via the basal-lateral portion of the enterocyte. In dog kidney, Silverman (1974a), using the multiple indicator dilution technique *in vivo*, has distinguished sugars being transported across the luminal from those being transported across the antiluminal membranes. He suggests that for the brush border membrane the specificity characteristics consist of a pyranose ring with hydroxyl groups on C-3 and C-6 arranged as in the configuration of D-glucose. For the antiluminal membrane, the specificities are a pyranose ring, hydroxyl groups on C-1 and C-2, and hydroxyl groups, if present on C-3 and C-6, oriented equatorially as in the configuration of D-glucose. If these findings suggesting polarity of the transport function are substantiated at the membrane level, it is postulated that the transcellular 'active' transport of D-glucose consists of a Na^+-dependent uphill influx at the luminal pole and Na^+-independent downhill efflux of sugar at the basal-lateral region of the epithelial cell.

Figure 10.7 illustrates diagrammatically a tentative model for the Na^+-coupled D-glucose transport in the proximal tubule or intestine. Sodium ion

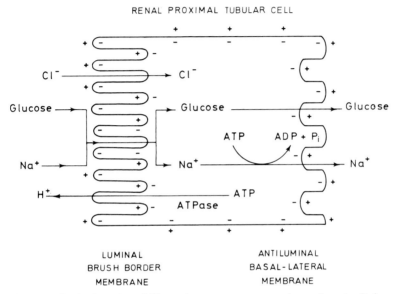

Figure 10.7   A schematic model for D-glucose transport across the renal proximal tubular cell. (From Beck and Sacktor, 1975, courtesy of the American Society of Biological Chemists)

and D-glucose are translocated from the lumen across the brush border membrane into the cell by an electrogenic process, with the transmembrane electrochemical potential (interior negative) providing the driving force. This electrochemical membrane potential may be maintained in part by an active chloride pump transporting the anion into the cell (Field, Fromm and McColl, 1971), the extrusion of H^+ from the cell across the luminal membrane by an HCO_3^--stimulated ATPase (Liang and Sacktor, 1975), and by the extrusion of Na^+ from the cell across the basal-lateral membrane by a ouabain-sensitive Na^+,K^+-ATPase localized in this membrane (Quigley and Gotterer, 1969; Heidrich et al., 1972). D-Glucose leaves the cell via the basal-lateral membrane, presumably by a downhill Na^+-independent process (Murer et al., 1974). Thus, the asymmetric distribution of Na^+ across the epithelial cell and the electrochemical potential across the brush border membrane provide the energy needed to transport D-glucose against its concentration gradient.

Transport of D-glucose by intestinal (Hopfer et al., 1973) and renal (Busse,

Elsas and Rosenberg, 1972; Chesney, Sacktor and Rowen, 1973; Aronson and Sacktor, 1974) brush border membranes is inhibited by phlorizin. However, in these studies significant inhibition of the uptake by membranes required phlorizin concentrations of 10^{-4}–10^{-3} M, whereas concentrations of 10^{-5}–10^{-7} M inhibit D-glucose reabsorption *in vivo* (Chan and Lotspeich, 1962). Aronson and Sacktor (1974) have suggested that this apparent discrepancy may be due to the possibility that only a Na^+-dependent component of the D-glucose uptake is highly sensitive to phlorizin. That this is indeed the case has been demonstrated recently by Aronson and Sacktor (1975). Phlorizin (1 mM) gives 97 percent inhibition of the Na^+-dependent uptake of 50 μM D-glucose, whereas the same concentration gives only 59 percent inhibition of the Na^+-independent uptake. Kinetics consistent with competitive inhibition are found, with a K_i for phlorizin of 7 μM. The K_i for phlorizin corresponds exceptionally well to the dissociation constant of 7–8 μM which has been reported for the Na^+-dependent high-affinity phlorizin binding site in similar preparations of rabbit renal brush border membranes (Chesney, Sacktor and Kleinzeller, 1974; Mitchell, Aronson and Sacktor, 1974), although it is somewhat higher than the dissociation constants of 0.2 (Glossmann and Neville, 1972b) to 3.4 μM (Bode *et al.*, 1970) which have been reported in rat renal brush borders.

Although phlorizin is a potent inhibitor of the uptake of D-glucose, the uptake of L-glucose into renal (Chesney, Sacktor and Rowen, 1973) and intestinal (Hopfer *et al.*, 1973) brush border membranes is not affected. Inhibition of D-glucose transport by phlorizin is competitive, as shown with isolated membranes (Chesney, Sacktor and Rowen, 1973; Aronson and Sacktor, 1975) as well as with relatively intact preparations (Alvarado and Crane, 1962; Diedrich, 1963). Additionally, inhibition occurs in the absence of appreciable levels of the glycoside in cells (Stirling, 1967; Heath and Aurbach, 1973), suggesting that the site of phlorizin action is at the membrane level. Further, it is postulated that phlorizin interacts with and binds to the proposed D-glucose carrier (Frasch *et al.*, 1970). Thus, studies on the interaction of phlorizin with brush border membranes have been prompted by expectations that phlorizin binding may serve as a marker of the D-glucose 'carrier', essential for the isolation and chemical characterization of this membrane component.

Kinetic analyses of the relationship between the concentration of phlorizin in the incubation medium and the binding of phlorizin to rabbit renal brush border membrane preparations indicate two classes of receptor sites (Chesney, Sacktor and Kleinzeller, 1974). One class, comprising high-affinity sites, reaches saturation at 20–25 μM phlorizin, has a K value for phlorizin of 8 μM, and 8×10^{-2} nmol binds to 1 mg of brush border membrane protein. The other class, comprising low-affinity sites, has a K value for phlorizin of 2.5 mM, and the number of binding sites is 1.25 nmol per milligram. Generally similar findings are reported for rat kidney preparations (Frasch *et al.*, 1970; Glossmann and Neville, 1972b; Heath and Aurbach, 1973). Sodium ion is required for the binding of phlorizin at the high-affinity sites (Frasch *et al.*, 1970; Chesney, Sacktor and Kleinzeller, 1974). It decreases the apparent K_m for phlorizin; the apparent V_{max} is not altered (Chesney, Sacktor and Kleinzeller, 1974). In contrast, Na^+ is not required for binding at low-affinity sites. Chesney, Sacktor and Kleinzeller (1974) have also shown that D-glucose and D-galactose are competitive inhibitors of the high-affinity binding of phlorizin. The apparent K_i for D-glucose is 1 mM. D-Glucose inhibits noncompetitively at the low-affinity sites, while L-glucose does not affect phlorizin binding. D-Mannose and 2-deoxy-D-glucose at concentrations of 0.4 mM have no effect on phlorizin binding, but they do inhibit at higher concentrations (Bode, Baumann and Diedrich, 1972). Glossmann and Neville (1972b) have reported that D-mannose, D-fructose and L-arabinose inhibit only the low-affinity binding. These findings suggest that the low-affinity binding is relatively nonspecific. Indeed, Silverman (1974b), from studies with the intact dog, and Ozegovic et al. (1974), working with plasma membranes from rat kidney, have suggested that the brush border membranes contain only high-affinity phlorizin sites whereas the basal-lateral membranes possess the nonspecific low-affinity phlorizin sites. Thus, the distinct localizations and specificities of the phlorizin binding sites may have their counterpart in the distribution and characteristics of the D-glucose transport systems.

Preliminary attempts have been made to isolate and characterize the phlorizin-binding protein (D-glucose carrier?). The high-affinity, Na^+-dependent binding of phlorizin is destroyed by trypsin and papain, but not by carboxypeptidase A and B or neuraminidase (Glossmann and Neville, 1973). Glossmann and Neville suggested that the phlorizin receptor is a glycoprotein but the terminal sialic acid is not important to phlorizin binding. Treatment with the proteolytic enzymes increases the low-affinity, nonspecific, phlorizin binding. It is proposed that the latter binding is localized in a deeper, more hydrophobic layer in the brush border membrane. Thomas (1972) has used papain and deoxycholate treatments, under isotonic conditions, to solubilize the D-glucose-sensitive phlorizin-binding component in rat renal brush border membranes. Gel electrophoresis of the solubilized membranes shows a region containing two protein bands which are enriched in phlorizin binding. Comparison of deoxycholate and sodium dodecyl sulfate gel patterns indicates that the phlorizin receptor may be a dimer consisting of two 30 000-dalton units (Thomas, 1973).

10.3.2   Transport of amino acids by brush border membranes

There is nearly a complete void of published information on the transport of amino acids by isolated brush border membrane preparations. Hillman and Rosenberg (1970) have described a rat kidney membrane preparation that contains some brush borders as seen by phase microscopy and is enriched in the non-brush-border enzyme Na^+,K^+-ATPase, which is reported to bind L-proline. However, their argument to distinguish between surface binding and transport is not persuasive. 'Binding' of L-proline is enhanced by Na^+. Faust and Shearin (1974) have reported the isolation of a protein from the 'core' filaments of hamster jejunum brush borders that binds L-histidine and D-glucose. They claimed that this common binding protein accounts for the competitive inhibitory effects between intestinal sugar and amino acid transports. It is also postulated that 'core' filaments act as a conduit for the Na^+-dependent active transport of all monosaccharides and all amino acids. The proposed relationship between this binding and transport of solutes has been seriously questioned, however (Mitchell, Aronson and Sacktor, 1974; Hopfer and Sigrist-Nelson, 1974).

Very recent investigations in our laboratory have focused on the transport of amino acids using the same well-characterized renal brush border membrane preparations that have been used for studies of sugar transport. *Figure 10.8* shows that the uptake of L-alanine into the membrane vesicles exhibits Na^+-dependent and -independent components (S. Fass, M. Hammerman and B. Sacktor, work submitted for publication, 1975). In the absence of Na^+, the rate of uptake is essentially linearly related to the concentration of L-alanine from 20 μM to 100 mM, perhaps suggesting passive diffusion. In the

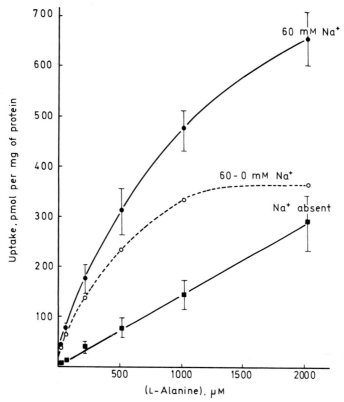

Figure 10.8   The uptake of different concentrations of L-alanine by rabbit renal brush border membrane vesicles, with and without a Na^+ gradient. (From Fass and Sacktor, 1975, unpublished work)

presence of a Na^+ gradient, L-alanine uptake is stimulated. If, at each L-alanine concentration, the Na^+-free uptake is subtracted from the uptake obtained in the presence of Na^+, a curve is described which is consistent with a hypothesis for a Na^+-dependent L-alanine transport system in renal brush border membranes that saturates at about 2 mM and has an apparent K_m of 300 μM (at 60 mM Na^+). When the uptake of 20 μM L-alanine by renal brush border membranes is measured during incubations for different lengths of time in the absence of a Na^+ gradient, uptake increases gradually and a steady state is reached in about 40 minutes. In contrast, when transport is determined in the presence of 60 mM Na^+, Fass, Hammerman and Sacktor found that the initial rate of L-alanine uptake is markedly enhanced.

Maximum accumulation is attained in 2 to 5 minutes. The level then decreases, indicating efflux, and approaches the same level as with no Na^+, suggesting that equilibrium has been established. At the peak of the 'overshoot', the uptake is several times that at equilibrium. Thus, these results demonstrate that the imposition of a large extravesicular to intravesicular gradient of Na^+ effects the transient movement of L-alanine into renal brush border membranes against its concentration gradient, in a fashion similar to that seen for D-glucose (Aronson and Sacktor, 1975).

Other experiments by Fass, Hammerman and Sacktor (submitted for publication, 1975) show that increasing Na^+ concentrations from 7.5 to 100 mM increase the initial rate of L-alanine uptake 2- to 5-fold with no evidence for saturation with respect to Na^+. Other cations, such as K^+, Li^+ or choline, do not stimulate uptake. Table 10.6 gives data demonstrating the stereospecificity of the Na^+-dependent transport of alanine by renal brush border membrane vesicles. In the absence of Na^+, L- and D-alanine are taken up at the same initial rate. In the presence of 60 mM Na^+, the rate of L-alanine uptake is increased 3.4-fold, whereas the rate of D-alanine uptake, although enhanced, is increased only 1.4-fold.

Table 10.6 STEREOSPECIFICITY OF ALANINE UPTAKE (From S. Fass, M. Hammerman and B. Sacktor, work submitted for publication, 1975)

Incubation	Uptake*
L-Alanine (20 μM)	2.5 ± 0.4
D-Alanine (20 μM)	2.5 ± 0.4
L-Alanine (20 μM) + Na^+ (60 mM)	8.4 ± 1.2
D-Alanine (20 μM) + Na^+ (60 mM)	3.4 ± 0.4

* Measured in pmol per 30 s per milligram of protein.

Mutual inhibition of transport between amino acids and sugars has been observed in intestine (Schultz and Curran, 1970) and kidney (Genel, Rea and Segal, 1971). Several mechanisms have been proposed to explain the phenomenon, including formation of toxic metabolites, competition for metabolic energy, stimulation of efflux from the cell, allosteric interactions for a common polyfunctional carrier, and competition of the sugar and amino acid transport systems for Na^+. An experiment (Fass, Hammerman and Sacktor, work submitted for publication, 1975) bearing on this question is summarized in Table 10.7. The uptake of L-alanine by renal brush border membrane vesicles is not affected by D-glucose in the absence of a Na^+ gradient. In

Table 10.7 COMPETITION BETWEEN L-ALANINE AND D-GLUCOSE (From S. Fass, M. Hammerman and B. Sacktor, work submitted for publication, 1975)

Incubation	Uptake*
50 μM L-Alanine	7.3 ± 0.7
50 μM L-Alanine + 5 mM D-glucose	9.0 ± 1.6
50 μM L-Alanine + 100 mM NaCl	44.6 ± 2.1
50 μM L-Alanine + 5 mM D-glucose + 100 mM NaCl	31.7 ± 2.6

* Measured in pmol per 30 s per milligram of protein.

contrast, in the presence of a Na^+ gradient, the sugar significantly inhibits the transport of L-alanine. Thus, it is of considerable significance that the inhibition first reported with relatively intact preparations is also seen with isolated luminal membrane vesicles. Moreover, the first three postulates cannot apply to the transport across the isolated purified brush border membrane vesicles which lack respiratory and glycolytic enzymes and into which uptake of solutes is insensitive to metabolic energy in the form of ATP (Aronson and Sacktor, 1974). The finding that D-glucose and L-alanine do not compete in the absence of Na^+ argues counter to the simple allosteric interaction of carriers. On the other hand, the inhibition observed in the presence of Na^+ is consistent with the view of competition for Na^+. This hypothesis is presently expanded by suggesting that sugars and amino acids compete for the electrochemical Na^+ gradient or membrane potential.

The mechanism of transport of the basic amino acid, L-arginine, in renal brush border membrane vesicles has recently been examined (M. Hammerman and B. Sacktor, work submitted for publication, 1975). The presence of a positively charged guanido group is a prerequisite for the recognition of arginine. Transport is stereospecific and L-arginine can be distinguished from D-arginine by the membranes. In contrast to the uptake of L-alanine, the uptake of L-arginine is inhibited by a Na^+ gradient (medium > vesicle). However, K^+ and other cations added to the external medium also inhibit. On the other hand, the presence of a positively charged gradient (vesicle > medium) stimulates the initial rate of L-arginine uptake.

Meister (1973) has proposed a cyclic mechanism for the transport of amino acids by renal and intestinal cells. The initial enzymatic step of the scheme is a γ-glutamyl transpeptidase reaction in which the γ-glutamyl moiety of glutathione is somehow transferred to the external amino acid being taken up by the cell. The γ-glutamylamino acid dipeptide and cysteinyglycine, the products of the transpeptidase reaction, are consequently to be found on the inside of the cell. The remaining sequences of reactions of the cycle are concerned with the liberation of the transported amino acid and the regeneration of glutathione. According to the hypothesis, only the transpeptidase reaction occurs at the brush border membrane. That this enzyme is indeed found in the membrane has been noted above. The other enzymes, essential to the scheme, have been described from the cytosol fraction of the cell. Although considerable evidence which tends to be consistent with the hypothesis has been gathered by Meister and his colleagues, definitive experiments to validate this interesting proposal remain to be reported.

10.3.3   Transport of ions by brush border membranes

Except for some studies on the uptake of Ca^{2+}, virtually nothing has been reported on the mechanisms of transport of ions by isolated brush border membrane preparations. The presence of a Ca^{2+}-stimulated ATPase in rabbit renal brush border membranes has been noted (Berger and Sacktor, 1970). The pattern of distribution for Ca^{2+}-ATPase in rat renal plasma membrane fractions resembles that for Na^+,K^+-ATPase, both enzymes being localized predominantly in the basal-lateral membrane (Kinne-Saffran and Kinne, 1974a). It has been proposed that the Ca^{2+}-stimulated ATPase is involved in the active trans-tubular transport of Ca^{2+} by a mechanism analogous to that indicated for Na^+,K^+-ATPase and the translocation of Na^+ from lumen to plasma. According to this hypothesis, at the brush border membrane an electrochemical potential which favors the influx of Ca^{2+} into the cell exists, because the interior of the cell is negative relative to the tubular lumen and because the concentration of intracellular Ca^{2+} is less than that in the filtrate (Borle, 1971). Thus, Ca^{2+}, by following its electrochemical potential, can cross the brush border membrane by a downhill mechanism. On the other hand, Kinne-Saffran and Kinne (1974a) have suggested that at the basal-lateral region of the tubular cell Ca^{2+} has to overcome an electrochemical potential difference between the cellular and interstitial fluids. This requires an uphill transport mechanism presumably mediated by the Ca^{2+}-ATPase. That iso-lated renal plasma membranes can, indeed, accumulate Ca^{2+} has been shown by Moore *et al.* (1974).

The localization of Ca^{2+}-ATPase and the mechanism of Ca^{2+} transport in the intestine seemingly differ from those in the kidney. Kinetic studies have led Patrick (1973) to conclude that, in the intestine, entry across the brush border membrane is rate-limiting for Ca^{2+} absorption. Moreover, Ca^{2+}-ATPase is found in highly purified brush border membrane preparations from rat (Martin, Melancon and DeLuca, 1969) and chick (Melancon and DeLuca, 1970; Holdsworth, 1970). Significantly, vitamin D elicits a marked increase in brush border Ca^{2+}-ATPase activity (Martin, Melancon and DeLuca, 1969). The time course of the appearance of Ca^{2+}-ATPase activity correlates with the increase in Ca^{2+} transport (Melancon and DeLuca, 1970). Additionally, Ca^{2+} has little effect on ATPase in vitamin-D-deficient chicks but strikingly stimulates ATPase in vitamin-D-replete animals. Norman *et al.* (1970) have reported that cholecalciferol also increases the level of alkaline phosphatase in chick (rachitic) intestinal brush border membranes. Other brush border enzymes, such as the disaccharidases, are not increased. The simultaneous time course of appearance of increased levels of brush border alkaline phosphatase and of increased rates of Ca^{2+} transport, measured *in vitro* across ileal segments, has prompted these investigators to suggest a functional involvement of alkaline phosphatase in vitamin-D-mediated Ca^{2+} transport. Ca^{2+}-ATPase activity is also found in the basal-lateral segments of the plasma membrane of rat intestine (Birge, Gilbert and Alvioli, 1972). The ATPase in the basal-lateral membrane, but not in the brush border mem-brane, is inhibited by ethacrynic acid, but not ouabain, and is activated by Na^+. Birge and co-workers claim that the basal-lateral Ca^{2+}-ATPase may be part of the translocation system for Ca^{2+}, in the intestine, and Na^+ may have a role in activating the enzyme.

10.4   GENERAL CONCLUSION

In this chapter the properties of the brush border membrane systems are de-tailed and the role of the membrane in mediating transport *in vitro* is de-scribed. Each functional feature of the isolated brush border membrane has its direct counterpart in transport systems *in vivo*. The excellent correlation between the mechanisms of transport in the brush border membrane vesicle and in the intact organism strongly supports the view that the isolated membranes serve advantageously as a model system in studies of the molecular basis of transport in the kidney and intestine.

REFERENCES

ALPERS, D. H. (1972). *J. clin. Invest.,* **51**:2621.

ALVARADO, F. and CRANE, R. K. (1962). *Biochim. biophys. Acta,* **56**:170.

ARONSON, P. S. and SACKTOR, B. (1974). *Biochim. biophys. Acta,* **356**:231.

ARONSON, P. S. and SACKTOR, B. (1975). *J. biol. Chem.,* **250**:6032.

AURICCHIO, S., RUBINO, A. and MURSET, G. (1965). *Pediatrics, Springfield,* **35**:944.

BAILEY, C. B., KITTS, W. D. and WOOD, A. J. (1956). *Can. J. agric. Sci.,* **36**:51.

BECK, J. C. and SACKTOR, B. (1975). *J. biol. Chem.,* **250**:8674.

BENSON, R. L., SACKTOR, B. and GREENAWALT, J. W. (1971). *J. Cell Biol.,* **48**:71..

BERGER, S. J. and SACKTOR, B. (1970). *J. Cell Biol.,* **47**:637.

BIHLER, I. and CYBULSKY, R. (1973). *Biochim. biophys. Acta,* **298**:429.

BIRGE, S. J., JR., GILBERT, H. R. and ALVIOLI, L. V. (1972). *Science, N.Y.,* **176**:168.

BODE, F., BAUMANN, K. and DIEDRICH, D. F. (1972). *Biochim. biophys. Acta,* **290**:134.

BODE, F., BAUMANN, K., FRASCH, W. and KINNE, R. (1970). *Pflügers Arch. ges. Physiol.,* **315**:53.

BONTING, S. L., POLLAK, V. E., MUEHRCKE, R. C. and KARK, R. M. (1958). *Science, N.Y.,* **127**:1342.

BORLE, A. B. (1971). *Cellular Mechanisms for Calcium Transfer and Homeostasis,* pp. 151 seq. Ed. G. NICHOLS, JR. and R. H. WASSERMAN. New York; Academic Press.

BOYD, C. A. R., PARSONS, D. S. and THOMAS, A. V. (1968). *Biochim. biophys. Acta,* **150**:723.

BUSSE, D., ELSAS, L. J. and ROSENBERG, L. E. (1972). *J. biol. Chem.,* **247**:1188.

BUSSE, D. and STEINMAIER, G. (1974). *Biochim. biophys. Acta,* **345**:359.

CHAN, S. S. and LOTSPEICH, W. D. (1962). *Am. J. Physiol.,* **203**:975.

CHERTOK, R. J. and LAKE, S. (1974). *Biochim. biophys. Acta,* **339**:202.

CHESNEY, R. W. and SACKTOR, B. (1974). *Pediatric Res.,* **8**:180.

CHESNEY, R., SACKTOR, B. and KLEINZELLER, A. (1974). *Biochim. biophys. Acta,* **332**:263.

CHESNEY, R. W., SACKTOR, B. and ROWEN, R. (1973). *J. biol. Chem.,* **248**:2182.

CONNOCK, M. J., ELKIN, A. and POVER, W. F. R. (1971). *Histochem. J.,* **3**:11.

CRANE, R. K. (1962). *Fedn Proc. Fedn Am. Socs exp. Biol.,* **21**:891.

DAHLQVIST, A. (1960). *Acta chem. scand.,* **14**:9.

DAHLQVIST, A. (1961). *Nature, Lond.,* **190**:31.

DAVID, J. S. K., MALATHI, P. and GANGULY, J. (1966). *Biochem. J.,* **98**:662.

DEREN, J. J. (1968). *Handbook of Physiology,* Section 6, Vol. 3, pp. 1099–1123. Ed. C F. CODE. Washington, DC; American Physiological Society.

DIEDRICH, D. F. (1963). *Biochim. biophys. Acta,* **71**:688.

DOELL, R. G. and KRETCHMER, N. (1962). *Biochim. biophys. Acta,* **62**:353.

DOELL, R. G. and KRETCHMER, N. (1964). *Science, N.Y.,* **143**:42.

DOELL, R. G., ROSEN, G. and KRETCHMER, N. (1965). *Proc. natn. Acad. Sci. U.S.A.,* **54**:1268.

DOUGLAS, A. P., KERLEY, R. and ISSELBACHER, K. J. (1972). *Biochem. J.,* **128**:1329.

EICHHOLZ, A. and CRANE, R. K. (1965). *J. Cell Biol.,* **26**:687.

EICHHOLZ, A., HOWELL, K. E. and CRANE, R. K. (1969). *Biochim. biophys. Acta,* **193**:179.

EMMELOT, P. and BOS, C. J. (1972). *J. Membrane Biol.,* **9**:83.

FAUST, R. G. and SHEARIN, S. J. (1974). *Nature, Lond.,* **248**:60.

FAUST, R. G., WU, S. L. and FAGGARD, M. L. (1967). *Science, N.Y.,* **155**:1261.

FIELD, M., FROMM, D. and MCCOLL, I. (1971). *Am. J. Physiol.,* **220**:1388.

FILBURN, C. and SACKTOR, B. (1975). *Archs Biochem. Biophys.,* in press.

FISHER, R. B. and PARSONS, D. S. (1949). *J. Physiol., Lond.,* **110**:36.

FORSTNER, G. G. (1969). *Am. J. med. Sci.,* **258**:172.

FORSTNER, G. G. and WHERRETT, J. R. (1973). *Biochim. biophys. Acta,* **306**:446.

FORSTNER, G. G., SABESIN, S. M. and ISSELBACHER, K. J. (1968). *Biochem. J.,* **106**:381.

FORSTNER, G. G., TANAKA, K. and ISSELBACHER, K. J. (1968). *Biochem. J.,* **109**:51.

FORSTNER, G. G., RILEY, E. M., DANIELS, S. J. and ISSELBACHER, K. J. (1965). *Biochem. biophys. Res. Commun.,* **28**:83.

FRASCH, W., FROHNERT, P. P., BODE, F., BAUMANN, K. and KINNE, R. (1970). *Pflügers Arch. ges. Physiol.,* **320**:265.

GARCIA-CASTINEIRAS, S., TORRES-PINEDO, R. and ALVARADO, F. (1973). *FEBS Lett.,* **30**:115.

GENEL, M., REA, C. F. and SEGAL, S. (1971). *Biochim. biophys. Acta,* **241**:779.

GEORGE, S. G. and KENNY, A. J. (1973). *Biochem. J.*, **134** :43.

GLOSSMANN, H. and GIPS, H. (1974). *Naunyn-Schmiedebergs Arch. exp. Path. Pharmak.*, **282** :439.

GLOSSMANN, H. and NEVILLE, D. M., JR. (1972a). *FEBS Lett.*, **19** :340.

GLOSSMANN, H. and NEVILLE, D. M., JR. (1972b). *J. biol. Chem.*, **247** :7779.

GLOSSMANN, H. and NEVILLE, D. M., JR. (1973). *Biochim. biophys. Acta*, **323** :408.

GOODMAN, D. B., HAUSSLER, M. R. and RASMUSSEN, H. (1972). *Biochem. biophys. Res. Commun.*, **46** :80.

GRONIOWSKI, J., BICZYSKOWA, W. and WALSKI, M. (1969). *J. Cell Biol.*, **40** :585.

GROSSMAN, I. W. and SACKTOR, B. (1968). *Science, N.Y.*, **161** :571.

HARRISON, D. D. and WEBSTER, H. L. (1971). *Biochim. biophys. Acta*, **224** :432.

HEATH, D. A. and AURBACH, G. D. (1973). *J. biol. Chem.*, **248** :1577.

HEIDRICH, H. G., KINNE, R., KINNE-SAFFRAN, E. and HANNIG, K. (1972). *J. Cell Biol.*, **54** :232.

HILLMAN, R. E. and ROSENBERG, L. E. (1970). *Biochim. biophys. Acta*, **211** :318.

HOLDSWORTH, E. S. (1970). *J. Membrane Biol.*, **3** :43.

HOPFER, U. and SIGRIST-NELSON, K. (1974). *Nature, Lond.*, **252** :422.

HOPFER, U., NELSON, K., PERROTTO, J. and ISSELBACHER, K. J. (1973). *J. biol. Chem.*, **248** :25.

HUBSCHER, G., WEST, G. R. and BRINDLEY, D. N. (1965). *Biochem. J.*, **97** :629.

INSEL, P., FILBURN, C., GEORGE, E., BALAKIR, R. and SACKTOR, B. (1974). *Proceedings 2nd International Conference on Cyclic AMP, Vancouver.* TUP-19, p. 48.

ITO, S. (1965). *J. Cell Biol.*, **27** :475.

JAMES, W. P. T., ALPERS, D. H., GERBER, J. E. and ISSELBACHER, K. J. (1971). *Biochim. biophys. Acta*, **230** :194.

JOHNSON, C. F. (1967). *Science, N.Y.*, **155** :1670.

KIM, Y. S. and PERDOMO, J. M. (1974). *Biochim. biophys. Acta*, **342** :111.

KINNE, R. and KINNE-SAFFRAN, E. (1969). *Pflügers Arch. ges. Physiol.*, **308** :1.

KINNE-SAFFRAN, E. and KINNE, R. (1974a). *J. Membrane Biol.*, **17** :263.

KINNE-SAFFRAN, E. and KINNE, R. (1974b). *Proc. Soc. exp. Biol. Med.*, **146** :751.

KLEINZELLER, A. (1970). *Biochim. biophys. Acta*, **211** :264.

KORN, E. D. (1969). *Fedn Proc. Fedn Am. Socs exp. Biol.*, **28** :6.

MALATHI, P. (1967). *Gastroenterology*, **52** :1106.

MAROUX, S., LOUVARD, D. and BARATTI, J. (1973). *Biochim. biophys. Acta*, **321** :282.

MARTIN, D. L., MELANCON, M. J. and DELUCA, H. F. (1969). *Biochem. biophys. Res. Commun.*, **35** :819.

MAUNSBACH, D. B. (1973). *Handbook of Physiology*, Section 8, pp. 31–79. Ed. J. ORLOFF and R. W. BERLINER. Washington, DC; American Physiological Society.

MCNAMARA, P. D., KOLDOVSKY, O. and SEGAL, S. (1974). *FEBS Lett.*, **41** :139.

MEISTER, A. (1973). *Science, N.Y.*, **180** :33.

MELANCON, M. J. and DELUCA, H. F. (1970). *Biochemistry*, **9** :1658.

MILLER, D. and CRANE, R. K. (1961). *Biochim. biophys. Acta*, **52** :293.

MITCHELL, M. E., ARONSON, P. S. and SACKTOR, B. (1974). *J. biol. Chem.*, **249** :6971.

MOOG, F. (1951). *J. exp. Zool.*, **118** :187.

MOORE, L., FITZPATRICK, D. F., CHEN, T. S. and LANDON, E. J. (1974). *Biochim. biophys. Acta*, **345** :405.

MURER, H. and HOPFER, U. (1974). *Proc. natn. Acad. Sci. U.S.A.*, **71** :484.

MURER, H., HOPFER, U., KINNE-SAFFRAN, E. and KINNE, R. (1974). *Biochim. biophys. Acta*, **345** :170.

NEWCOMER, A. D. and MCGILL, D. B. (1966). *Gastroenterology*, **51** :481.

NISHI, Y., YOSHIDA, T. O. and TAKESUE, Y. (1968). *J. molec. Biol.*, **37** :441.

NORDSTROM, C. (1972). *Biochim. biophys. Acta*, **268** :711.

NORMAN, A. W., MIRCHEFF, A. K., ADAMS, T. H. and SPIELVOGEL, A. (1970). *Biochim. biophys. Acta*, **215** :348.

O'BRYAN, D. and LOWENSTEIN, L. M. (1974). *Biochim. biophys. Acta*, **339** :1.

OZEGOVIC, B., MCNAMARA, P. D., GOLDMANN, D. R. and SEGAL, S. (1974). *FEBS Lett.*, **43** :6.

PATRICK, G. (1973). *Nature, Lond.*, **243** :89.

PORTEOUS, J. W. (1968). *FEBS Lett.*, **1** :46.

PORTEOUS, J. W. and CLARK, B. (1965). *Biochem. J.*, **96** :159.

PREISER, H., MENARD, D., CRANE, R. K. and CERDA, J. J. (1974). *Biochim. biophys. Acta*, **363** :279.

QUIGLEY, J. P. and GOTTERER, G. S. (1969). *Biochim. biophys. Acta*, **173** :456.

QUIGLEY, J. P. and GOTTERER, G. S. (1972). *Biochim. biophys. Acta*, **255** :107.

QUIRK, S. J., BYRNE, J. and ROBINSON, G. B. (1973). *Biochem. J.*, **132** :501.

QUIRK, S. J. and ROBINSON, G. B. (1972). *Biochem. J.*, **128** :1319.

REALE, E. and LUCIANO, L. (1967). *J. Histochem. Cytochem.*, **15** :413.

RHODES, J. B., EICHHOLZ, A. and CRANE, R. K. (1967). *Biochem. biophys. Acta*, **135** :959.

ROSENBERG, I. H., STREIFF, R. R., GODWIN, H. A. and CASTLE, W. B. (1969). *New Engl. J. Med.*, **280** :985.

ROSTGAARD, J., KRISTENSEN, B. I. and NIELSEN, L. E. (1972). *J. Ultrastruct. Res.*, **38** :207.

RUBINO, A., ZIMBALATTI, F. and AURICCHIO, S. (1964). *Biochim. biophys. Acta*, **92**:305.

SACKTOR, B. (1968). *Proc. natn. Acad. Sci. U.S.A.*, **60**:1007.

SACKTOR, B. and BERGER, S. J. (1969). *Biochem. biophys. Res. Commun.*, **35**:796.

SACKTOR, B. and WU, N. C. (1971). *Archs Biochem. Biophys.*, **144**:423.

SACKTOR, B., CHESNEY, R. W., MITCHELL, M. E. and ARONSON, P. S. (1974). *Recent Advances in Renal Physiology and Pharmacology*, pp. 13–26. Ed. L. G. WESSON and G. M. FANELLI, JR. Baltimore; University Park Press.

SAITO, M. (1972). *Biochim. biophys. Acta*, **286**:212.

SCHAFER, J. A., TROUTMAN, S. L. and ANDREOLI, T. E. (1974). *J. gen. Physiol.*, **64**:582.

SCHERBERICH, J. E., FALKENBERG, F. W., MONDORF, A. W., MULLER, H. and PFLEIDERER, G. (1974). *Clinica chim. Acta*, **55**:179.

SCHMITZ, J., PREISER, H., MAESTRACCI, D., GHOSH, B. K., CERDA, J. J. and CRANE, R. K. (1973). *Biochim. biophys. Acta*, **323**:98.

SCHULTZ, S. G. and CURRAN, P. F. (1970). *Physiol. Rev.*, **50**:637.

SEGAL, S. and THIER, S. O. (1973). *Handbook of Physiology*, Section 8, pp. 653–676. Ed. J. ORLOFF and R. W. BERLINER. Washington, DC; American Physiological Society.

SENIOR, J. R. and ISSELBACHER, K. J. (1963). *J. clin. Invest.*, **42**:187.

SIGRIST-NELSON, K. and HOPFER, U. (1974). *Biochim. biophys. Acta*, **367**:247.

SILVERMAN, M. (1974a). *Biochim. biophys. Acta*, **332**:248.

SILVERMAN, M. (1974b). *Biochim. biophys. Acta*, **339**:92.

SILVERMAN, M., AGANON, M. A. and CHINARD, F. P. (1970). *Am. J. Physiol.*, **218**:743.

SINGER, I. and CIVAN, M. M. (1971). *Am. J. Physiol.*, **221**:1019.

STECK, T. L. and WALLACH, D. F. H. (1970). *Meth. Cancer Res.*, **5**:93.

STEVENSON, F. K. (1972a). *Biochim. biophys. Acta*, **266**:144.

STEVENSON, F. K. (1972b). *Biochim. biophys. Acta*, **282**:226.

STEVENSON, F. K. (1973). *Biochim. biophys. Acta*, **311**:409.

STIRLING, C. E. (1967). *J. Cell Biol.*, **35**:605.

THOMAS, L. (1972). *FEBS Lett.*, **25**:245.

THOMAS, L. (1973). *Biochim. biophys. Acta*, **291**:454.

THOMAS, L. and KINNE, R. (1972). *Biochim. biophys. Acta*, **255**:114.

THUNEBERG, L. and ROSTGAARD, J. (1968). *Expl Cell Res.*, **51**:123.

THUNEBERG, L. and ROSTGAARD, J. (1969). *J. Ultrastruct. Res.*, **29**:578.

TRIER, J. S. (1968). *Handbook of Physiology*, Section 6, Vol. 3, pp. 1125–1175. Ed. C. F. CODE. Washington, DC; American Physiological Society.

ULLRICH, K. J., RUMRICH, G. and KLOSS, S. (1974). *Pflügers Arch. ges. Physiol.*, **351**:35.

VAN DEN BERG, J. W. O. and HULSMANN, W. C. (1971). *FEBS Lett.*, **12**:173.

WACHSMUTH, E. D. and HIWADA, K. (1974). *Biochem. J.*, **114**:273.

WILFONG, R. F and NEVILLE, D. M., JR. (1970). *J. biol. Chem.*, **245**:6106.

Index